LUFTWAFFE
SECRET PROJECTS
GROUND ATTACK
& SPECIAL PURPOSE AIRCRAFT

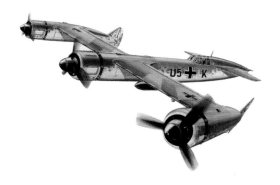

MIDLAND

An imprint of
Ian Allan Publishing

Luftwaffe Secret Projects –
Ground Attack & Special Purpose Aircraft

ISBN 1 85780 150 4

© Dieter Herwig and Heinz Rode 2002, 2003

First published 2002 in Germany by
Motorbuch Verlag, Stuttgart.

Translation from original German text
by Ted Oliver

English language edition published 2003
and this second impression 2004 by
Midland Publishing
4 Watling Drive, Hinckley, LE10 3EY, England
Tel: 01455 254 490   Fax: 01455 254 495

Worldwide distribution (except North America):
Midland Counties Publications
4 Watling Drive, Hinckley, LE10 3EY, England
Tel: 01455 233 747   Fax: 01455 233 737
E-mail: midlandbooks@compuserve.com
www.midlandcountiessuperstore.com

North America trade distribution by:
Specialty Press Publishers & Wholesalers Inc.
39966 Grand Avenue
North Branch, MN 55056, USA
Tel: 651 277 1400 Fax: 651 277 1203
Toll free telephone: 800 895 4585
www.specialtypress.com

Design concept and editorial layout
© Midland Publishing and
Stephen Thompson Associates.

Printed in Hong Kong
via World Print Limited.

LUFTWAFFE
SECRET PROJECTS
GROUND ATTACK
& SPECIAL PURPOSE AIRCRAFT

Dieter Herwig and Heinz Rode

Contents

Foreword

As a continuation of the technological and historical series of Luftwaffe Secret Projects, this volume presents the developments, prototypes and projects in the realm of ground attack, multi and special purpose aircraft covering the period 1935-1945 as well as earlier antecedents. Whilst some of these attained operational status, most did not progress further than the drawing board. At the time they were conceived, the projects had no equivalent counterparts abroad, and after 1945, became largely the basis for modern developments, chiefly in the USA and USSR.

It need hardly be mentioned that already in the 1930s, the Luftwaffe possessed Schlacht-flugzeuge (literally 'battle aircraft' – a term later superseded to mean 'close-support' or 'ground attack' aircraft) and Zerstörer (literally 'destroyer'- a generic term applied to the twin-engined heavier and more powerfully-armed bomber escort-fighter intended to clear the skies of enemy aircraft ahead, as well as acting as a home-defence interceptor).

The Schlachtflugzeuge were a special class of aircraft that had been developed from 1917 onwards from fighter and artillery-support aircraft. They were specifically employed in the ground attack or close-support role to attack enemy supply columns, field positions and tank concentrations. This, however, formed only a portion of the tasks which from 1935 led to the development of the new generation of ground attack aircraft.

In a similar manner, development of the other combat type also took place – the Kampfzerstörer (literally 'battle destroyer'), a term adopted from the Navy and which, with its combat capabilities, epitomised the 'strategic fighter' concept.

Operational use of both of these weapons carriers took place on all fronts during the Second World War and especially for the defence of Reich territory. A few months before the end of the war, the RLM or Reichs-sluftfahrtministerium (German Air Ministry) termed these aircraft as Arbeits- (literally 'work') or Heeresflugzeuge (soldier or army aircraft) which, commensurate with their multi-purpose capability, were to be employed as tactical bombers, jettisonable weapons carriers, reconnaissance and ground attack aircraft.

Although this volume documents their development, the main emphasis lies on projects that either did not or had not yet attained flight status. Additionally, a whole series of other highly-interesting developments are covered about which – with the exception of the Bachem Natter (Viper) – very little has been reported to date. These involve the piloted Sprengstoffträger (explosive-laden aircraft), Bombensegler (bomb-carrying gliders), Schleppjäger (towed fighters), manned Flak-Raketen (anti-aircraft rocket interceptors), Sonderflugzeuge (special aircraft) and Bord- and Sonderwaffen (airborne and special weapons) that were under development.

Individual types of aircraft are described largely in chronological order of evolution. The colour illustrations depict how they 'might have looked' in flight and in service use, and are based exclusively on manufacturers' drawings, mock-ups or other original documentation. In order to provide the reader with a sense of realism and to emphasise their documentary character, the illustrations are depicted in the colour schemes and markings in effect at that time.

Reasons have also been discussed as to why the RLM and leading personalities had forbidden, terminated or halted further design and construction of specific aircraft developments. Was it stupidity, lack of appreciation, or sabotage? In the RLM offices of the Generalluftzeugmeister (Chief of Air Procurement and Supply) alone, there were over 4,000 bureaucrats who were responsible for everything yet not answerable for anything at all.

On the other hand, not every drawing board study and project turned out to be suitable for practical implementation. Firms often deliberately engaged valuable staff members and highly-qualified designers with unusual projects in order to retain their indispensability and thus preserve them from call-up for military service. These, however, were exceptions. The constantly-deteriorating war situation for Germany towards the end resulted in a wealth of ideas and an enormous impulse to development that has not been matched anywhere else in history.

Dieter Herwig & Heinz Rode
September 2002

Acknowledgements

As already mentioned in *Luftwaffe Secret Projects: Strategic Bombers 1935-1945*, the photographs, outline drawings and data in this volume originate exclusively from the archives of the Deutsche Studienbüro für Luftfahrt (German Aviation Study Centre) in Frankfurt/Main.

The professional support received by the author from friends and colleagues of the Deutsche Gesellschaft für Luftfahrtdokumentation (German Society for Aviation Documentation) enables an overview to be presented of the Schlachtflugzeuge and Kampfzerstörer projects in Germany up until 1945.

My special thanks are expressed to Günter Sengfelder, Theodor Mohr, Karlheinz Kens and Manfred Griehl who assisted me with their advice on difficult questions. Thanks and recognition are equally due to my artist and co-author Heinz Rode for his colour illustrations that were often generated under difficult conditions, using heavily-damaged works documents in order to present the reader with a feeling of reality of project developments. The unit codes, colour schemes and tactical symbols are pure fiction, and mainly indicate how they 'might have looked' in flight and in service use, using markings in effect at that time.

The authors welcome any criticism, suggestions and questions which should be directed to the Deutsche Studienbüro für Luftfahrt, Postfach 551037, 60400 Frankfurt am Main, Germany.

Dieter Herwig
Frankfurt, September 2002

Publisher's Note

During the course of the translation of the German original of this book into this English version, our translator Ted Oliver carefully studied the material as supplied and published in the German, and with our approval, has made some significant changes in sequence and content which we believe enhance the readability, accuracy and completeness of the finished English article.

In the original German edition, the Schlachtflugzeuge were all listed in a single chapter, but in largely no chronological order of manufacturer or date of origin. To reflect the interim design periods, these have been rearranged into two chapters (2 and 3) covering the periods 1942-1943 and 1944-1945.

The later part of the book was a rather jumbled array of projects, now regrouped into three chapters on aircraft (7, 8 and 9) and a final chapter (10) on airborne weapons and special devices. As in earlier chapters, these are related in chronological order of project date or alphabetical order of manufacturer, except for the weapons in last chapter, grouped under related headings.

Throughout the book, Ted Oliver has taken the opportunity to eliminate some obvious contradictions in the German original, and to add to or improve information by using sources (principally captured German document sources in the UK and the USA) which were apparently not available to the author.

In some instances the extent of the additional material available to the translator has necessitated more extensive additional text. Numbered footnotes are almost all translator's remarks and necessary explanations. These also provide commentary on the author's erroneous text repeated from sources known to be incorrect or speculative.

Examples are the Blohm & Voss P 214 and the DFS 'Rammjäger Eber' where the author has contradicted published information from authoritative sources.

Asterisked footnotes* are usually the author's or translator's reference to another published work, to which readers may like to refer for further information.

The Glossary and Notes pages in this volume were compiled by the Translator and we hope that they will prove to be as valuable to readers as the similar entries in the first two volumes of this series. In particular, we believe that the information presented here on the 9-1000 and 9-10000 series of designations has not been published by anyone before as it was taken from post-war CIOS Reports; likewise the comprehensive jet-propulsion code-letter list is also believed to be fresh.

Glossary & Notes

Enterprise name followed by:
Flugzeug aircraft, airplane (USA).
bau construction, manufacturing.
fabrik factory, works, plant.
werke plant, works, enterprise.
AG Aktien Gesellschaft
– joint-stock company.
IG Interessengemeinschaft
– community of interests.
KG Kommandit Gesellschaft
– limited partnership company.
GmbH Gesellschaft mit beschränkter Haftung
– limited liability company.
eV eingetragener Verein
– registered company.

Principal types of aircraft or aerial vehicle
(stand-alone terms)
Aufklärer reconnaissance.
Bomber bomber.
Erkunder observation, scout.
Flugboot flying boat.
Flugkörper unmanned missile.
Gleiter, Segler glider, sailplane.
Heimatschützer home-protector (fighter).
Hubschrauber helicopter.
Jäger fighter, pursuit (USA).
Jagdbomber (Jabo) fighter-bomber.
Lastensegler load-, troop-carrying glider.
Luftschiff airship, dirigible.
Pulkzerstörer bomber-formation destroyer.
Tragschrauber autogyro, rotorcraft.
Waffenträger weapons carrier.
Zerstörer twin-engined heavy fighter.

Functional terms preceding the word flugzeug –
aircraft. Some are also applied before: bomber/jäger/
munition/torpedo/waffe
Arbeits- army co-operation, medium bomber.
Artillerie- artillery support.
Aufklärungs- reconnaissance.
Begleit- accompanying, escort.
Bord- air-launched, -lifted, airborne.
Düsen- jet-propelled.
Erkundungs- observation, scout.
Erdkampf- ground attack, close-support.
Fern- long-range, remote.
Gleit- glider.
Groß; Großraum large; large-capacity.
Heeres- army or soldier aircraft.
Infanterie- infantry support.
Jagd- fighter, pursuit (USA).
Kampf- combat, bomber (50).
Klein- small.
Kleinst- miniature, smallest.
Langstrecken- long-range.
Marine- naval.
Mehrzweck- multi-purpose.

Nachtjagd- night-fighter.
Nahaufklärungs- short-range reconnaissance.
Nahkampf- close-combat.
Panzer- armoured, anti-tank.
Raketen- rocket-propelled.
Reise- touring, communications.
Segel- sailplane, glider.
Schlacht- ground attack, close-support
(formerly: battle aircraft).
Schnell- fast, high-speed.
Schnellst- super-fast.
See- water-based, floatplane.
Seenot- air-sea rescue.
Sonder- special purpose.
Strahl- jet-propelled.
Schlepp- tow-craft, towing aircraft.
  Deichselschlepp pole-tow.
  Mistelschlepp (mistletoe) composite, carry-tow.
  Starrschlepp rigid-tow.
  Tragschlepp carry-tow.
Schul- basic trainer.
Selbstopferungs- self-sacrifice (suicide).
Selbstvernichtungs- self-destruction (suicide).
Träger- carrier-based.
Tiefangriffs- low-level attack.
Verbindungs- liaison, communications.
Verkehrs- civil, commercial.
Versuchs- experimental, test.
Volks- people's (pre-war: popular).
Zwillings- twin-fuselage.

More than 100 types of flugzeug designations were used
before 1939. For obvious reasons, prefixes such as Land-,
Post-, Sport-, Transport-, and Zivil- have been omitted.

Other terms:
Abteilung Department.
APZ Automatischer Peilzusatz – automatic
supplementary direction finding equipment.
Attrappe mock-up.
AWG Auswertegerät – plotting device.
Behälter container.
BK Bordkanone – fixed aircraft cannon.
Dipl-Ing Diplomingenieur – literally diploma-ed
engineer, equivalent to Diploma of Engineering.
Doppelreiter Literally 'double-rider', wing fuel fairings.
E Entwurf – project.
EF Entwicklungsflugzeug – development aircraft.
EiV Eigenverständigungsanlage – crew intercom.
ESK Entwicklungssonderkommission
– Special Development Commission.
ETC Elektrische Trägervorrichtung für
Cylinderbomben – electrically-operated
carrier device for cylindrical bombs.
EZ Einheitszielvorrichtung
– standard sighting device.
FDL Ferngerichtete Drehringlafette
– remotely controlled barbette.

FHL   Ferngerichtete Hecklafette
       – remotely controlled tail barbette.

Flak (Fliegerabwehrkanone)   anti-aircraft cannon.

Fl-E   Flugzeugentwicklung – Aircraft Development
       Department within the TLR.

Flitzer   Literally Dasher or Whizzer, single-seater
       jet fighter.

Forschungsanstalt   research institute.

FuBl   Funk-Blindlandeanlage
       – radio blind-landing equipment.

FuG   Funkgerät – radio or radar set.

FZG   Fernzielgerät – remote aiming
       device/bombsight.

General der Jagdflieger   Luftwaffe Rank –
       Air Officer Commanding fighters.

Generalleutnant   Luftwaffe rank –
       equivalent to Air Vice Marshal (RAF)
       or Major General (USAAF).

Generalluftzeugmeister   Luftwaffe Rank –
       Chief of Air Procurement and Supply.

Generalmajor   Luftwaffe rank –
       equivalent to Air Commodore (RAF)
       or one-star General (USAAF).

Generalstab   General Staff.

Gerät   device, apparatus, equipment.

Geschoß   projectile.

GM-1   Nitrous oxide.

Gruppe   Luftwaffe equivalent to Wing (RAF)
       or Group (USAAF).

IFF   Identification, friend or foe.

Jägerstab   Fighter Staff.

Kanone   cannon (20mm calibre and above).

Lotfe (Lotfernrohr)   telescopic bombsight.

Maschinen-Gewehr   machine-gun (below 20mm calibre).

MG   Maschinengewehr – machine gun;
       later also cannon.

MK   Maschinenkanone – machine cannon.

Munition   ammunition.

MW 50   Methanol-water mixture.

NJG   Nachtjagdgeschwader – night fighter group.

Obergruppenführer   SS rank –
       equivalent to Lieutenant General.

Oberleutnant   Luftwaffe rank –
       equivalent to Flying Officer (RAF)
       or 1st Lieutenant (USAAF).

Oberstleutnant   Luftwaffe rank –
       equivalent to Wing Commander (RAF)
       or Lieutenant Colonel (USAAF).

OMW   Otto Marder Works.

'Otto-Jäger'   Piston-engined fighter.

P   Projekt – project.

PeilG   Peilgerät – direction finding set.

Pulk   Luftwaffe term for USAF bomber box.

Rb   Reihenbildkamera – automatic aerial camera.

RfRuK   Reichsministerium für Rüstung und
       Kriegsproduktion – Reich's Ministry of
       Armament and War Production.

Revi (Reflexvisier)   reflection, reflex gunsight.

Rüstsatz   Field conversion set.

SC   Splitterbombe – fragmentation bomb.

Schräge Musik   Luftwaffe term for oblique upward-firing
       armament (literally oblique or jazz music).

SD   Splitterbombe, Dickwand
       – fragmentation bomb, thick-walled.

Technisches Amt   Technical Office (of the RLM).

TL   Turbinenluftstrahl-Triebwerk – turbojet engine.

UKW   Ultrakurzwelle – VHF.

Waffe   weapon.

W/nr   Werk nummer
       – construction (or airframe serial) number.

Wurf   jettison, drop, release
       (used with behälter/gerät/geschoß/munition).

ZVG   Zielflugvorsatzgerät – homer attachment device.

Firms, Organisations and Research Institutes:

AEG   Allgemeine Electrizitäts-Gesellschaft mbH, Berlin.

AVA   Aerodynamische Versuchsanstalt, Göttingen.

BBC   Brown-Boveri Company, Mannheim.

BMW   Bayerische Motoren-Werke GmbH, München.

DFS   Deutsche Forschungsinstitut für Segelflug
       'Ernst Udet' e.V.

DVL   Deutsche Versuchsanstalt für Luftfahrt, Berlin.

DWM   Deutsche Waffen- und Munitionswerke,
       Lübeck-Schlutup.

EHK   Entwicklungshauptkommission
       – Development Main Committee.

FGZ   Forschungsinstitut Graf Zeppelin, Stuttgart-Ruit.

GWF   Gothaer Waggonfabrik, Gotha.

HFW   Henschel Flugzeugwerke AG, Berlin.

HWA   Heereswaffenamt – Army Ordnance Office.

HWK   Hellmuth Walter KG, Kiel.

JFM   Junkers Flugzeug- & Motorenwerke, Dessau.

KHD   Klöckner-Humboldt-Deutz AG, Köln-Deutz.

LFA   Luftfahrtforschungsanstalt 'Hermann Göring',
       Braunschweig.

LFM   Luftfahrtforschungsanstalt München-Ottobrunn.

LFW   Luftfahrtforschungsanstalt Wien (Vienna).

OKH   Oberkommando des Heeres.

OKL   Oberkommando der Luftwaffe.

OKM   Oberkommando der Marine.

OKW   Oberkommando der Wehrmacht.

RLM   Reichsluftfahrtministerium – German Air Ministry.

Rh.-B.   Rheinmetall.-Borsig AG., Düsseldorf.

RPF   Reichspost Forschungsanstalt, Berlin.

RVM   Reichsverkehrsministerium (predecessor of RLM).

RWM   Reichswehrministerium – Armed Forces Ministry.

SAM   Siemens Apparate- & Maschinenbau GmbH, Berlin.

TLR   Technische Luftrüstung
       – Technical Air Armaments Board.

WFG   Weser Flugzeugbau GmbH, Weser.

WNF   Wiener-Neustädter Flugzeugwerke GmbH, Vienna.
       (also Wn but WNF used to avoid confusion with w/nr).

ZMe   Zeppelin/Messerschmitt AG
       (aircraft joint co-operation).

ZSO   Zeppelin / SNCASO (France)
       (aircraft joint co-operation) (142).

RLM Designation Systems

Aircraft, aero-engine and component manufacturers from their inception had adopted internal company designation systems for their products. Even before the establishment of the RLM on 27th April 1933, a standard designation system for aircraft had apparently been introduced under its predecessor the RVM. Aircraft contracted for bore the fixed prefix 8 followed by two numerals, the system beginning with 8-10 onwards. Glider aircraft bore the fixed prefix 108, eg DFS 108-30. The manufacturers were identified by two letters (or more) of the company name. During the period of the Third Reich, the following aircraft and engine identification letters had been applied:

Al	Albatros Flugzeugwerke GmbH.
Ao	AGO Flugzeugwerke GmbH.
Ar	Arado Flugzeugwerke GmbH.
As	Argus Motorenwerke GmbH.
Ba	Bachem Werke GmbH.
Bf	Bayerische Flugzeugwerke AG (later Me).
Bü	Bücker Flugzeugbau GmbH.
BV	Blohm & Voss Schiffswerft, Abteilung Flugzeugbau.
DB	Daimler-Benz AG.
DFS	Deutsche Forschungsanstalt für Segelflug.
Do	Dornier Werke GmbH.
Dz	Klöckner-Humbold-Deutz AG.
Fa	Focke, Achgelis & Co KG.
Fh	Flugzeugwerke Halle.
Fi	Gerhard Fieseler Werke GmbH.
Fk	Flugzeugbau Kiel GmbH.
Fl	Anton Flettner GmbH.
Fw	Focke-Wulf Flugzeugbau GmbH.
Go	Gothaer Waggonfabrik AG.
Ha	Hamburger Flugzeugbau GmbH (later BV).
He	Ernst Heinkel AG.
HM	Hirth Motoren GmbH (later He-Hirth).
Ho	Horten Flugzeugbau.
Hs	Henschel Flugzeugwerke AG.
Hü	Dipl-Ing Ulrich Hütter GmbH.
Ju	Junkers Flugzeug (& Motorenwerke) AG.
Ka	Dipl-Ing Albert Kalkert.
Kl	Hans Klemm Flugzeugbau GmbH.
LZ	Luftschiffbau Zeppelin GmbH.
Me	Messerschmitt AG (formerly Bf).
NR	Bruno Nagler & Hans Rolz.
Sh	Siemens & Halske AG (later SAM).
Si	Siebel Flugzeugwerke KG.
Sk	Skoda-Kauba Flugzeugbau.
So	Dipl-Ing Heinz Sombold.
Ta	Prof Dipl-Ing Kurt Tank (at Fw).
Wn	Wiener-Neustädter-Flugzeugwerke GmbH.
Z	Zündapp.
BMW	Bayerische Motorenwerke.
Bramo	Brandenburgische Motorenwerke GmbH.
Jumo	Junkers Motorenwerke GmbH.
WASAG	Westphälisch-Anhaltische Sprengstoff AG.

Aero-engines bore the fixed prefix 9, followed by three digits, except that each manufacturer was accorded a first digit identifier. The series ran as follows:
9-090 series: (small firms, eg Breuer Werke GmbH 9-091; Zundapp Werke GmbH 9-092)
9-100 series: BMW (later 800-series)
9-200 series: Junkers-Jumo
9-300 series: Bramo (absorbed by BMW June 1939)
9-400 series: Argus
9-500 series: Hirth
9-600 series: Daimler-Benz
9-700 series: Deutz (KHD)
9-800 series: BMW

Aero-engine accessories such as gas-turbine starters bore the prefix 19 followed by three digits, the best known being the Hirth 19-518 and 19-526. The prefix 9 followed by four digits were applied to regular accessories, for example:
9-1000 series: exhaust units, valve seatings, turning jigs
9-2000 series: fuel and lubricant feeds
9-4000 series: magnetos, radiators, spark plugs
9-5000 series: manifolds, air ducts
9-7000 series: starting gear (Bosch and Hirth types)
9-9000 series: airscrew motivation gear

Propellers and spare parts for same, including variable-pitch mechanisms, gearings, etc, bore designations in the 9-10000 series, several firms having a specific range, such as Junkers 9-30000 and 9-36000 wooden-blade v.p. propellers; VDM 9-41000; Argus and Messerschmitt 9-60000 and 9-70000 etc.

To distinguish jet-propulsion engines from orthodox airscrew engines, the RLM in the Second World War introduced the temporary prefix 109 followed by three digits in the series 109-001 to 999. At first, the numbers were applied in consecutive order of contract, thus:
109-001 Heinkel (HeS 8)
109-002 BMW (P.3304)
109-003 BMW (P.3302)
109-004 Junkers-Jumo (T1)
109-005 Porsche
109-006 Heinkel (HeS 30)
109-007 Daimler-Benz (ZTL 5000)
109-009 Heinkel (HeS 9)
109-010 Heinkel (HeS 10)
109-011 Heinkel-Hirth (HeS 11)

After the last number had been assigned, the system was changed whereby each manufacturer was allocated a fixed end digit as identifier, corresponding to the reciprocating engine series. Under the revised nomenclature, the end digits were allocated as follows:
1 Heinkel-Hirth Examples:109-011, 109-021,109-051
2 Junkers-Jumo Examples:109-012, 109-022
4 Argus Examples:109-014, 109-024, 109-044
5 Porsche (reserved)
6 Daimler-Benz Examples: 109-016
8 BMW Examples: 109-018, 109-028

The other unused end digits would probably have been assigned to teams that were engaged on turbojet and turboprop projects at the AEG Berlin, BBC Mannheim, FKFS Stuttgart, Oesermaschinen GmbH and elsewhere.
Code letters were also introduced for the different types of reaction-propulsion -Antriebe (drives) or -Triebwerke (engines), the following list not claiming to be complete:
L   Luftstrahl (jet) or Lorin (ramjet). Latter in honour of its inventor Réné Lorin (1913).
R   Rakete (rocket).
PR   Pulver-Rakete (solid-fuel);
FR   Flüssig-Rakete (liquid-fuel).
LR   Lorin-Rakete (ramjet-rocket).
          HWK development.
RL   Rakete-Lorin (rocket-ramjet).
          HWK development.

ERL   Einstoff RL (single-fuel RL).
          HWK development.
I L   Intermittierendes Luftstrahl
          (intermittent combustion unit pulsejet).
ML   Motor-Luftstrahl
          (engine-driven jet unit without airscrew).
PML   Propeller-ML (ML unit with airscrew drive)
TL   Turbinen-Luftstrahl (turbojet).
PTL   Propeller-TL (turboprop)
ZTL   Zweikreis-TL (two-circuit)
TL   ducted-fan or bypass turbojet eg 109-007, 010.
TLR   TL + Rakete (turbojet-rocket combination).
          eg 109-003R, 011R, 018R.
GTW   Gas-Turbine mit Wärmetäuscher (turboprop with heat-exchanger). AEG & BBC-type

When R-Geräte (Rauch = smoke-trail devices, a code name for the still secret development of Raketen = rocket motors) began to be developed by the HWA Kummersdorf and the HWK Kiel for aircraft use in 1935, these were divided into two categories, prefixed by the identifiers RI for short-duration rocket motors, and RII for longer-duration sustainer motors. Individual units were assigned a three-digit number, beginning in the 100-series. Initially, each manufacturer was assigned a leading digit as identifier, as follows:
RI-100 series: HWA. Examples: RI-101; RII-101a
RI-200 series: HWK. Eg: RI-201, 203, 260; RII-203, 211, 213
RI-300 series: BMW. Eg: RI-301, 302, 305; RII-301, 302, 303
RI-500 series: Rh.-B. Examples: RI-501, 502, 505

This designation system was still in use in December 1942, after which time it was replaced by numbers in the 109-500 to 109-999 series. Like the turbojet designations, these were initially applied sequentially as follows:
109-500 to -503 HWK; 109-505 Rheinmetall-Borsig; 109-506 WASAG; 109-507 HWK; 109-509 HWK; 109-510 and -511 BMW. Thereafter, a fixed end digit was assigned to each manufacturer:
2. WASAG. Examples: 109-512, 522, 532
3. Schmidding. Examples: 109-513, 533, 603
5. Rh.-Borsig. Examples: 109-515, 525
8. BMW. Examples: 109-548, 558, 708, 718
9. HWK. Examples: 109-559, 709, 729

The examples cited here by no means comprise the whole list, as many more rocket motors were under development by the firms concerned. In addition, rocket motors were being worked upon by Dr-Ing Eugen Sänger at the Raketen-technisches Forschungsinstitut, Trauen (until 1942), as well as at the LFA Braunschweig; Elektromechanische-Werke, Karlshagen; the Oberbayerische Forschungsanstalt Dr Konrad, Oberammergau, and of course, the HWA at Peenemünde.
Fuels and oxidants for rocket motors were also assigned code designations, those mentioned in this volume being C-Stoff (30% hydrazine hydrate, 57% methanol and 13% water); T-Stoff (80% hydrogen peroxide), and the catalyst Z-Stoff C (calcium permanganate) or Z-Stoff N (sodium permanganate). Others had code names such as Visol. Salbei, Tonka, and so on.
Armament manufacturers were also identified by the first numeral of a three-digit designation, namely:
1 Rheinmetall-Borsig, 2 Mauser, 3 & 4 Krieghoff, beginning in the 100-series.
Where not stated in the text, airborne weapons developed had the following calibres:

7.9mm MG15	20mm MG FF	30mm MG 213/C	45mm SG113
7.9mm MG17	30mm MK101	37mm BK 3.7	30mm SG116
7.9mm MG 81	30mm MK103	50mm BK 5	30mm SG117
13mm MG131	30mm MK108	75mm BK7.5	30mm SG118
15mm MG151/15	55mm MK112	50mm MK 214	30mm SG119
20mm MG151/20	55mm MK114	30mm MK 303	50mm SG 500
20mm MG 204	55mm MK115	55mm MK 412	88mm Düka 88

Messerschmitt Aircraft Designations
This brings us to the inevitable 'chestnut' of 'Bf' or 'Me' for Messerschmitt's earlier designs. This work is very much a study of official documentation and RLM nomenclature has been adhered to.
For the RLM the transition from 'Bf' to 'Me' occurs between the unsuccessful Bf162 Jaguar (whose number was subsequently allocated to the He162 Volksjäger) and the Me163 Komet. The 'Me'155 avoids the issue by having been transferred at a very early stage to Blohm und Voss.
Then there is the 'grey area' of projected developments of 'Bf' types (eg the Bf109G) initiated after the change to 'Me'. This is tempting (eg me109H) but since later operational variants of the '109 retained the 'Bf' prefix (Bf109K circa 1944-45) this work has standardised on all Messerschmitt types below the RLM number 162 being prefixed 'Bf' and all those from 163 and upwards being prefixed 'Me'.

Measurements
All German aircraft measurements etc are given in decimal (or SI – Système International d'Unités, established in 1960) units, with an Imperial (of British FPSR – foot, pound, second, Rankine) figure second. This is reversed where UK or US-built aircraft are quoted. The following may help:

aspect ratio   wingspan and chord – expressed as a ratio. Low aspect ratio, short, stubby wing; high aspect ratio, long, narrow wing.

ft   feet – length, multiply by 0.305 to get metres (m). For height measurements involving service ceilings and cruise heights, the figure has been 'rounded'.

ft²   square feet – area, multiply by 0.093 to get square metres (m²).

fuel   measured in both litres/gallons and kilograms/pounds. The specific gravity (sg) of German fuel varied considerably during the war and conversions from volume to weight and vice versa are impossible without knowing the specific gravity of the fuel at the time.

gallon   Imperial (or UK) gallon, multiply by 4.546 to get litres. (500 Imperial gallons equal 600 US gallons.)

hp   horse power – power, measurement of power for piston and turboprop engines. Multiply by 0.746 to get kilowatts.

kg   kilogram – weight, multiply by 2.205 to get pounds (lb). Fuel load frequently given in kilograms.

kg/m²   kilograms per square metre – force, measurement of wing loading, multiply by 0.204 to get pounds per square foot.

km/h   kilometres per hour – velocity, multiply by 0.621 to get miles per hour (mph).

kP   kilopascal – force, for measuring thrust, effectively a kilogram of static thrust. Present day preferences are for the kilo-newton (kN), one kN equalling 224.8lb or 101.96kg.

kW   kilowatt – power, measurement of power for piston and turboprop engines. Multiply by 1.341 to get horse power.

lb   pound – weight, multiply by 0.454 to get kilograms (kg). Also used for the force measurement of turbojet engines, with the same conversion factor, as pounds of static thrust.

lb/ft²   pounds per square foot – force, measurement of wing loading, multiply by 4.882 to get kilograms per square metre.

litre   volume, multiply by 0.219 to get Imperial (or UK) gallons.

m   metre – length, multiply by 3.28 to get feet (ft).

m²   square metre – area, multiply by 10.764 to get square feet (ft²)

mm   millimetre – length, the bore of guns is traditionally a decimal measure (eg 30mm) and no Imperial conversion is given.

mph   miles per hour – velocity, multiply by 1.609 to get kilometres per hour (km/h).

Chapter One

Ground Attack Aircraft – Development of a New Weapon 1917-1941

The birth of the first Schlachtflugzeuge actually dates from 18th November 1916, when the Army Supreme Command through its Flying Corps Inspectorate (Idflieg) ordered on a trial basis from the Junkers firm, a heavily-armoured infantry aircraft capable of dropping lightweight hand-held bombs and cluster-type weapons onto enemy positions. A thick armour-plated shell protected the powerplant and crew compartment from machine-gun fire aimed at the aircraft from below.

Junkers only reluctantly accepted the commission for the armoured two-seater since in their view, the armoured ground attack aircraft required by the Idflieg to be of biplane configuration, appeared to be a retrograde step. Conceived jointly by Dr-Ing Otto Mader, Dipl-Ing Hans Steudel and Dipl-Ing Ernst Brandenburg, the aircraft in its final form was a one-and-a-half-wing biplane which made use of the newly-developed duralumin material, employed for the first time in the Junkers J 3 on which construction work was stopped before it was completed. Under the company designation J 4 (Army designation Junk J1), this first all-metal infantry aircraft was powered by a 200hp Benz Bz IV motor. The three-part wing featured a fixed centre section with removable outer sections made up of a duralumin tubular frame covered with a corrugated skin. Only the upper wings had ailerons, the forward fuselage consisting of a 5mm (0¼in) thick armoured pannier or 'bathtub' outer layer which enclosed the engine and crew compartments. The rear fuselage portion consisted of fabric-covered duralumin tubing, although an experimental prototype with sheet metal covered fuselage was also built.

Initial ground 'hops' made with the J 4 prototype on 17th January 1917 followed by its maiden flight on 28th January, were undertaken by Junkers pilot Arved von Schmidt in Döberitz. Because the forward armoured skinning had not been installed, the aircraft exhibited tail-heaviness. On 17th May 1917, the J 4 acquired its military type-test certificate in Berlin-Adlershof. In frontline service, it

Title page of a special issue of the magazine *Der Adler* (The Eagle) of July 1939, depicting an aircraft in a low-level dive attacking tanks with its machine-guns – as envisaged by the artist of how a close-support mission was undertaken in the First World War. He was not so far off the mark, had the ground attack flyers been able to prove themselves against enemy frontline and artillery positions, troops and combat-ready emplacements. In 1917, they became the precursors of a new breed of fighter and artillery-support aviators.

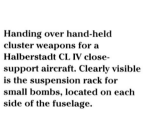

Handing over hand-held cluster weapons for a Halberstadt CL IV close-support aircraft. Clearly visible is the suspension rack for small bombs, located on each side of the fuselage.

soon gained a reputation of being indestructible, and despite its long take-off run and poor forward field of vision, was very much liked by its crews. Up to the Armistice of November 1918, a total of 227 examples had been manufactured.

The Junkers firm, a pioneer of the cantilever all-metal monoplane built of duralumin, brought out the J10 first flown on 4th May 1918, and under its military designation Junk CL 1, it also entered frontline service, albeit in the form of only a few experimental prototypes. Capable of higher speeds and possessing good flying characteristics, the lightly-armoured J10 with its weapons and bombs, proved to be the ideal ground attack aircraft for infantry close-support duties.

As a rule, the CL aircraft were equipped with two fixed machine-guns synchronised to fire through the propeller disc and for the rear observer/gunner, a traversible Parabellum-

MG on a dorsal turret mounting. In relation to its total weight, the J10 carried an appreciable amount of expendable weapon loads. With it, a special type of ground attack tactic was soon developed: approaching at a height of 300m (984ft), the target was attacked with batteries of machine-guns, bombs and hand grenades from a height of 50m (165ft). When the requirement for ground attack missions became ever louder at the front, the Army leadership at the beginning of 1918 began to form Schlachtfliegerstaffeln (ground attack squadrons) that within a short time were expanded or consolidated into Schlachtfliegergeschwadern (groups). A Geschwader consisted of four Staffeln each of six aircraft and formed a tactical unit.

With the introduction of ground attack operations, the aircraft had become an independent airborne weapon and with it began the 'space war' at a time when combat aces

such as Manfred von Richthofen, Oswald Boelcke and Max Immelmann achieved their greatest successes in man-to-man aerial combat.

It was only after 1933, when the build-up of the new still-secret Luftwaffe (German Air Force) soon become an independent arm of the armed forces, that recourse was again made to this proven weapon. As early as 1928, at a time when the manufacture and possession of this type of aircraft was still strictly forbidden in Germany, the Junkers firm in its AB Flygindustri subsidiary factory in Sweden had performed sterling preparatory work. The firm had evolved the Junkers K 47 (civil designation A 48) two-seat fighter which could also function as a Sturzbomber (dive-bomber) and Schlachtflugzeug (ground attack) aircraft. A group of engineers led by Dipl-Ing Hermann Pohlmann were foremost in its development whose end result could be clearly seen: a high-performance aircraft intended for the ground attack role that in terms of flight performance, was able to compete on equal terms with the best fighters of the day.

The Junkers K 47 consisted of a cable-braced low-wing monoplane of sheet-metal construction, powered by a British 600hp Bristol Jupiter VII supercharged radial engine. Besides the internally and externally balanced ailerons, the novel fixed undercarriage with its attachment points for the underslung bombs were the most notable features of the design. The Junkers firm's intention to manufacture the K 47 in series, however, was not fulfilled. The Reichswehrministerium (State Ministry of Defence) could not make up its mind to purchase the K 47 since it had exerted no influence on K 47 development and doubted its suitability as a ground attack aircraft.

The confirmation letter dated 19th February 1917 from the Aircraft Engineering Centre in Charlottenburg for the urgent initial manufacture of 100 Panzer-Flugzeuge (armoured aircraft) which 'for purely formal reasons, could not previously be officially announced. The Inspectorate has a pressing interest that the firm of Junkers & Co should not wait for the formal construction contract, but should immediately commence with all its man-power and resources to make arrangements for mass-production in its own and other suitable centres.'

Junkers J 4

1917 to 1918

Junkers J 4 – data

Powerplant

1 x 200hp Benz Bz IV motor driving a two-bladed wooden propeller.
Flat, rectangular radiator at the wing centre-section; fuel in fuselage.

Dimensions

Span	16.00m	52ft 6in
Length	9.10m	29ft 10¼in
Height	3.40m	11ft 2in
Wing area	49.00m²	527.42ft²

Weights

Empty weight	1,766kg	3,893 lb
Loaded weight	2,176kg	4,797 lb

Performance

Max speed, at sea level	155km/h	96mph
Range	210km	130 miles
Service ceiling	3,000m	9,840ft

Armament

2 x 7.9mm MG 08/15* machine-guns mounted beside the engine and firing through the propeller disc, plus a movable 7.9mm Parabellum-MG with 500 rounds, mounted on dorsal fuselage turret (B-Stand).

* This First World War machine-gun is the origin of the common German expression 'null-acht-fünfzehn' when referring to an obsolete 'Heath Robinson' contraption.

Use: Infantry-support and ground attack aircraft.

Construction: Cantilever sesquiplane of all-metal construction with a fuselage of octagonal cross-section. Powerplant and crew of two were enclosed in a welded 5mm-thick armour-plated shell of chrome-nickel steel.

Junkers J 4 with 220hp Daimler-Mercedes D IV a motor and canted exhaust stacks.

Left: **Junkers J 4 with 200hp Benz Bz IV motor and vertical exhaust stacks. Photo dated 3rd May 1918.**

Below left: **Junkers J 4 with Daimler-Mercedes D IV motor.**

Below: **Junkers J 4 armoured trough of 5mm-thick chrome-nickel steel. It weighed 470kg (1,036 lb).**

Below: **Junkers J 4 with Benz Bz IV motor.**

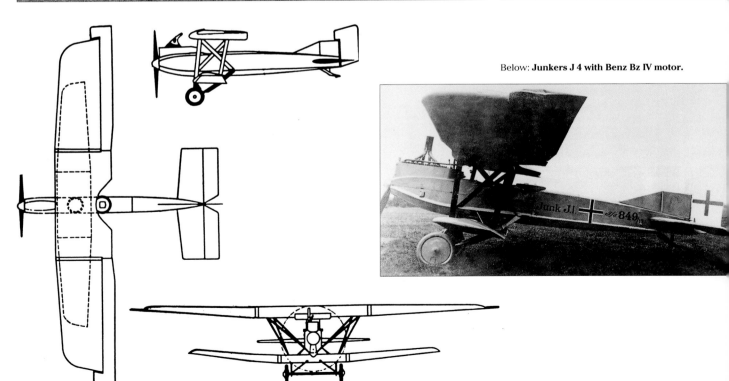

Junkers J 10

1918

Military designation: Junk CL 1

Use: Low-level attack, protection, and ground attack aircraft seating a crew of two.

Construction: Cantilever low-wing monoplane of all-metal construction with rectangular cross-section fuselage. Constant-chord wings of duralumin corrugated sheeting attached to the centre section built integral with the fuselage. Quadratic tail surfaces with one-piece vertical fin and one-piece rudder without aerodynamic balance.

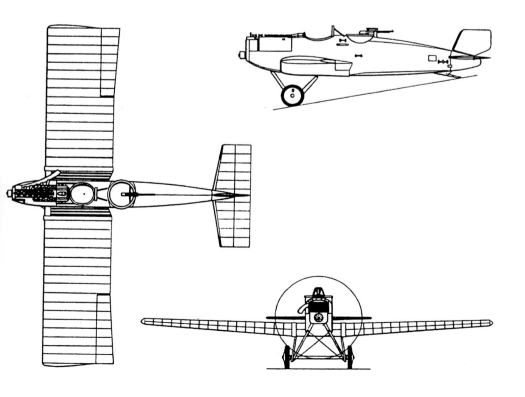

Above: **The Junkers J10, photographed here on 19th June 1918, was not only an extraordinary ground attack aircraft but was also the world's first all-metal combat aircraft.**

Left: **A rare photograph of the Junkers J10 in flight.**

Junkers J10 – data

Powerplant

1 x 180hp Daimler D III aü or 1 x 185hp BMW IIIa; two-bladed wooden propeller with adjustable front radiator behind it. Fuel tanks in fuselage.

Dimensions

Span	12.15m	39ft 10¼in
Length	7.90m	25ft 10.0in
Height	3.10m	10ft 2.0in
Wing area	23.70m²	255.10ft²

Weights

Empty weight	735kg	1,620 lb
Loaded weight	1,155kg	2,546 lb

Performance

Max speed, at sea level	190km/h	118mph
Range	700km	435 miles
Service ceiling	5,200m	17,060ft

Armament   2 x 7.9mm MG 08/15 machine-guns above the engine firing through the propeller disc, plus a movable 7.9mm Parabellum-MG on dorsal turret (B-Stand).

Junkers K 47

1926 to 1928

Use: Two-seat fighter, dive-bomber and ground attack aircraft.

Construction: Cantilever mid-wing monoplane of all-metal construction with thin rectangular wings of constant thickness/chord ratio. Its braced undercarriage differed from the usual Junkers type of construction. Fixed supporting struts formed the load-bearing element between the fuselage and wheels. The fuselage was built of steel sheet, with wings and tail surfaces of corrugated skinning. It was the first aircraft to feature underwing dive-brakes.

Junkers K 47 – data

Powerplant	1 x 600hp Bristol Jupiter VII.
	Several other engines up to 600hp were also tested

Dimensions		
Span	12.40m	40ft 8¼in
Length	8.55m	28ft 0½in
Height	2.80m	9ft 2¼in
Wing area	22.80m²	45.41ft²

Weights		
Empty weight	1,035kg	2,282 lb
Loaded weight	1,650kg	3,638 lb

Performance		
Max speed	290km/h at 3,500m	180mph at 11,485ft
Range	480km at 5,000m	298 miles at 16,400ft
Service ceiling	8,500m	27,890ft

Armament

Intended were two fixed forward-firing machine-guns with 200 rounds/gun plus one movable dorsal MG (B-Stand) and 2 x 250kg (551 lb) bombs.

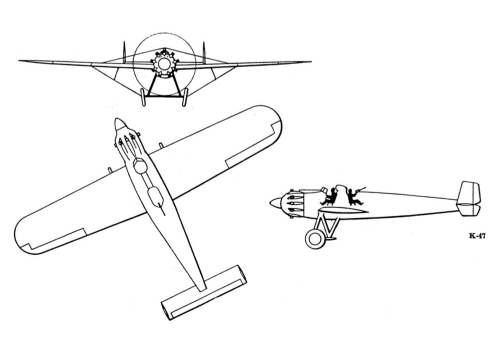

K-47

Left and centre left: **The Junkers K 47 in Reichswehr service with civil registration D-2012.**

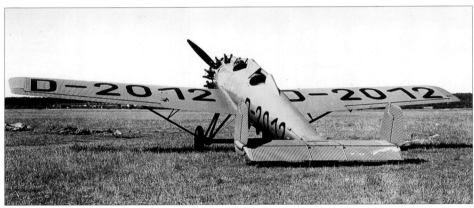

Left: **The Junkers K 47 as a low-level attack training aircraft. It was used until after 1939.**

Below left: **Bomb attachment gear between undercarriage and wing of the Junkers K 47.**

Below right: **Suspended bomb on the ETC racks of the Junkers K 47.**

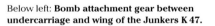

Henschel Hs 123 –
From the Dive-Bomber to the
Ground Attack Aircraft

In October 1932, the Reichswehr compiled a catalogue which encompassed the technical requirements for the new equipment period between 1933 and 1937. In it was a requirement for a single-seat fighter capable of being used for dive-bombing and low-level attack that envisaged an aircraft of robust construction, armed with fixed MGs and a 250kg (551 lb) bomb and which could attain a speed of up to 350km/h (217mph).

Upon the establishment of the RLM (German Air Ministry) on 27th April 1933, the Technisches Amt (Technical Office) Development Department LC II began to work out details of a two-stage plan which envisaged the development of this aircraft capable of dive-bombing and low-level attack. From the resulting RLM specification issued on 11th February 1934, Henschel commenced design work on an unbraced sesquiplane bearing the designation Hs 123, making the utmost use of modern technical manufacturing possibilities. The competing firm, Fieseler, built a two-seat biplane designated Fi 98. For both designs, a safety load factor of 12 in diving flight and pullout was specified, as also the installation of an air-cooled BMW 132A nine-cylinder radial engine.

The mock-up of the Hs 123 had already been approved by a technical committee in June 1934 as also that of the Fi 98. The latter accomplished its maiden flight at the end of April 1935, the Hs 123 being made flight-ready a few days later, on 8th May 1935. On that day, Ernst Udet publicly revealed the Hs 123 on the occasion of an airshow in Berlin-Johannisthal. Although the Fi 98 met the predetermined requirements, it dropped out of the running at the end of May 1935: the day of the wire-braced biplane was over. Henschel had concentrated among other things on the specification requirement for the aircraft's use in the low-level and ground attack role. After flight-testing began with the Hs 123 V1 in April 1935, the V2 prototype later attained a speed of 367km/h (228mph), powered by the American Wright SGR-1820 Cyclone nine-cylinder radial producing 770hp at 2,150rpm. With the exception of the new smooth engine cowling, the V2 was in other respects similar to its V1 predecessor that was powered by a BMW 132A radial of 725hp at 2,050rpm.

The results of flight-testing motivated the RLM to place an order with Henschel for a single-seat fighter variant of the Hs 123 having a weight of over 2,000kg (4,409 lb). Powered by

Henschel Hs 123 with four underwing bombs.

Henschel Hs 123 with four underwing bombs.

Henschel Hs 123 – data

Powerplant	1 x 660hp BMW 135 nine-cylinder radial engine	
Dimensions		
Span, upper wing	10.50m	34ft 5½in
lower wing	8.00m	26ft 3in
Length	8.66m	28ft 3in
Height, tail down	3.65m	11ft 11¾in
Wing area	24.85m²	267.48ft²
Weights		
Empty weight	1,498kg	3,302 lb
Loaded weight	2,020kg	4,453 lb
Performance		
Max speed	332km/h at 4,000m	206mph at 13,120ft
Range	860km	534 miles
Service ceiling	9,100m	29,855ft
Armament	2 x 7.9mm MG17 and 4 x SC 50 bombs or 2 containers with 92 x SC 2 anti-personnel bombs, or 2 x MG FF in underwing mounts.	

a nine-cylinder BMW 135 (licence-built Pratt & Whitney Hornet) radial of 660hp, it reached a speed of 332km/h (206mph). The trend towards the cantilever monoplane, however, induced Henschel to lay aside manufacture in accordance with an RLM decision. The experienced flyer Ernst Udet, who recognised the multifarious combat applications of the Hs 123 on the other hand, exerted his influence for the continual manufacture of a limited quantity of the aircraft.

At the beginning of the war with Poland on 1st September 1939, the Luftwaffe possessed only 40 operational examples of the Hs 123. Although it represented the first combat dive-bomber developed for the Luftwaffe, it was never employed in this role at any phase of the war; its possibilities as an infantry-support

and ground attack aircraft remained unrecognised for much too long.

The concluding chapter of the Hs 123 can be speedily told: In January 1942, out of the II. Schlachtflugzeug-Gruppe of Lehrgeschwader 2 (Instructional Group 2) or II. Schlacht LG 2, the I. Gruppe of Schlachtgeschwader 1 (I./SG 1) as well as a 2nd Gruppe (II./SG 2) were equipped for the first time with the Hs 129 – the successor of the Hs 123. But there still remained a Verband (formation) which continued to operate the Hs 123 in combat operations – the II. Gruppe of Schlacht-geschwader 'Immelmann 2' (II. SG Immelmann 2) which successfully operated in the southern sector of the Eastern Front. In mid-1944, however, II./SG 2 also had to dispense with its robust Hs 123.

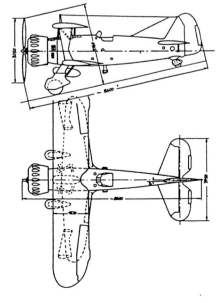

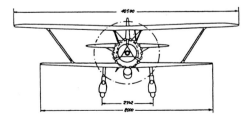

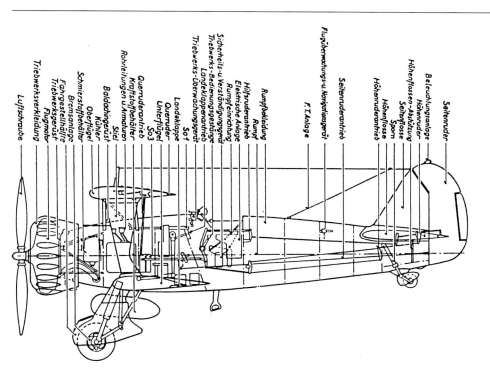

Luftschraube — Triebwerksverkleidung — Flugmotor — Triebwerksgerüst — Fahrgestellhälfte — Bremsanlage — Schmierstoffbehälter — Oberflügel — Kühler — Baldachingerüst — Stiel — Rohrleitungen u. Armaturen — Kraftstoffbehälter — Querruderantrieb — Unterflügel — Soₒ3 — Querruder — Landeklappe — Soₒ1 — Triebwerks-Überwachungsgerät — Landeklappenantrieb — Landeklappen-Bedienungsgestänge — Triebwerks-Bedienungsgestänge — Rumpfeinrichtung — Elektrische Anlage — Hilfsruderantrieb — Rumpf — Rumpfbekleidung — F.T.Anlage — Flugüberwachungs-u.Navigationsgerät — Seitenruderantrieb — Höhenruderantrieb — Höhenflosse — Sporn — Seitenflosse — Höhenflossen-Abstützung — Höhenruder — Höhenleitwerk — Beleuchtungsanlage — Seitenruder

Henschel Hs 123 annotated sectional side-view works drawing.

The Henschel Hs 132 V1 (D-ILUA)

A formation flight of three Hs 123A-1s in 1937.

Fieseler Fi 98

Second of the contestants in response to the RLM specification of 11th February 1934 for a dive-bomber and low-level attack aircraft was the Fieseler Fi 98 whose construction was to exhibit high structural strength. As with the Hs 123, the requirement called for a robust biplane. A development group under the leadership of Dipl-Ing Reinhold Mewes designed the Fi 98 as a two-seat biplane of metal construction conforming to the requirements laid down. The fuselage, of oval cross-section, was built in the form of an all-metal shell, the wing and tail surfaces con-

sisting of a metal framework with a fabric covering. A noteworthy feature of the design was the dual horizontal tail surfaces, the smaller fixed surface mounted at the apex of the fin. Despite its good flying characteristics, the RLM displayed no interest in the design, as in the meantime, the monoplane configuration of the Junkers Ju 87 had prevailed. Following completion of the sole prototype first flown in early 1935, Fieseler in 1936 terminated work on the remaining two prototypes of the three contracted for.

The Fieseler Fi 98 contestant to the RLM specification of 11th February 1934.

Fieseler Fi 98 – data

Powerplant	1 x 650hp BMW 132A-2 nine-cylinder radial engine	
Dimensions		
Span	11.50m	37ft 8¾in
Length	7.40m	24ft 3½in
Height	3.00m	9ft 10in
Wing area	24.50m²	263.71ft²
Weights		
Empty weight	1,450kg	3,197 lb
Loaded weight	2,160kg	4,762 lb
Performance		
Max speed	295km/h at 2,000m	183mph at 6,560ft
Range	470km	292 miles
Service ceiling	9,000m	29,530ft
Armament (proposed)	2 x 7.9mm MG 17 fixed forward-firing, plus 4 x 50kg (110 lb) bombs.	

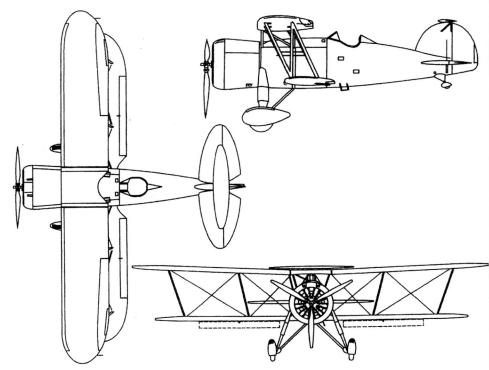

Hamburger Flugzeugbau Ha137

The HFB or Hamburger Flugzeugbau, established in June 1933 as a subsidiary of the ship-building firm of Blohm und Voss, began work on aircraft design in mid-October that year. In summer 1934, the HFB took part in the RLM competition for a dive-bomber and ground attack aircraft with an all-metal low-winged monoplane design by Dr-Ing Richard Vogt.

Bearing the designation Ha137, the deep gull-winged aircraft used flush riveting for the first time. The oval cross-section fuselage was of all-metal construction with sheet skinning, the trousered main undercarriage units being

hydraulically operated and fitted with brakes. It was first flown on 13th May 1935. During the course of extensive flight tests conducted at the Luftwaffe E-Stelle Rechlin in June 1936 where the Hs123 was also undergoing its trials at the same time, the latter aircraft showed up far more favourably in comparative trials. Despite this, Blohm und Voss received a construction contract for an improved version of the Ha137 intended to serve as a fighter. However, neither the Daimler-Benz DB600 nor the DB601 engines proposed for installation were available, and the

Junkers-Jumo Da-1 of only 610hp was not powerful enough to provide sufficient performance for a fighter, so that following the RLM decision in favour of the Junkers Ju87, the prospects of series production of the Ha137 disappeared. In addition, the RLM considered the maximum bombload of 200kg (441 lb) insufficient. It was only its proposed armament of two 20mm MG FF cannon and the two fixed forward-firing MG17s above the engine that offered the prospect of a construction contract as a ground attack aircraft, but was unsuccessful against the Ju87. Of the six prototypes built, the Ha137 V4 from May 1935 was engaged over a long period at the E-Stelle Tarnewitz on armament trials with the Rheinmetall-Borsig RZ65 rocket projectiles.

HFB Ha137 V4 with Junkers-Jumo 210A in-line engine.

Ha137 V4 – data

Powerplant
1 x 690hp Junkers-Jumo 210 Da-1 12-cylinder liquid-cooled engine

Dimensions

Span	11.15m	36ft 7.0in
Length	9.46m	31ft 0½in
Height	4.00m	13ft 1½in
Wing area	23.50m²	252.92ft²

Weights

Empty weight	1,815kg	4,001 lb
Loaded weight	2,485kg	5,478 lb

Performance

Max speed	330km/h at 2,000m	205mph at 6,560ft
Range	580km	360 miles
Service ceiling	7,000m	22,965ft

Armament (proposed)
2 x MG FF cannon in undercarriage fairings, plus 2 x MG17 fixed in fuselage, plus 4 x SC50 underwing bombs.

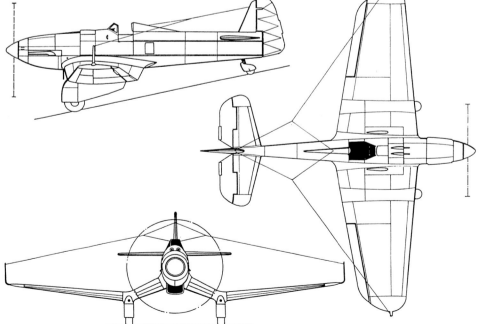

HFB Ha 137 V2 with BMW 132 radial engine.

HFB Ha 137 V1 with 640hp Rolls-Royce Kestrel V in-line engine.

Henschel Hs 129
1937 to 1944

Henschel Hs 129B-3/Wa with BK 7.5 cannon.

In April 1937, the RLM Technisches Amt issued a requirement for a light but strongly-armoured ground attack aircraft that would replace the then current single-engined attack aircraft. Following proposals submitted by Focke-Wulf and Henschel on 1st October 1937, the RLM decided in favour of the Hs 129. The only point of disagreement by the RLM with Henschel was on the company's proposal to use the weaker 465hp Argus As 410 engine. Due to lack of availability of other powerplants, it was in fact installed, if only temporarily.

The first prototype of the single-seat twin-engined Hs 129 V1 made its first flight at the beginning of 1939. Following a series of alterations made as a result of flight tests, improvements were made to the armoured cockpit as well as to engine power. A small test series of the Hs 129A-0 model, delivered to service units, were rejected by the pilots. Whereas the armour protection was recognised as being useful, the field of vision turned out to be completely inadequate. After a short period of operation, the Hs 129A-0 series of aircraft were withdrawn and following overhaul, were sold to the Rumanian Air Force.

In the Henschel project office there appeared a new, enlarged Hs129 variant with larger vision panels and a more powerful motor under the designation P76. The Technisches Amt, however, rejected the P76 because of the development proposals that were connected with project realisation and decided in favour of retention of the most important construction elements of the Hs129A-series.

After the conclusion of the Western Campaign in 1940, the Luftwaffe took over the large quantity of ready-to-use Gnôme-Rhone GR14M 24-cylinder twin-row radials available in France. Tests performed on engine testbeds had shown that the GR14M engine was suitable for installation in the Hs129, as a result of which the RLM ordered immediate changes and installation of this unit in the Hs129A.

The new model, designated Hs129B, received a new armoured crew compartment, larger vision panels and two GR14M engines each delivering 700hp at take-off.

The cockpit was manufactured as an enclosed armour-plated trough. Behind the cockpit, the fuselage main fuel tank was armour-protected, additional protection being provided for the other tanks, ammunition containers, carburettors, oil coolers, and the engines themselves. In addition, the previously installed two 20mm MG FF cannon were replaced by two 20mm MG151/20s. The necessary alterations to the Hs129B were completed in autumn 1941. Subsequent typetests with the new model were concluded successfully, so that series production of the Hs129B-1 was able to commence in December 1941. At the beginning of May 1942, the improved Hs129B-2 arrived for operational use on the Eastern Front. Its offensive armament was the same as that of the Hs129B-1 but with the addition of a 30mm MK103 cannon in a Rüstsatz (field conversion pack) in a ventral pannier. The MK103, developed by Rheinmetall-Borsig, had proved itself admirably suitable for combating armoured ground targets. Further weapon Rüstsätze that could be carried were the 3.7cm BK3.7 in the Hs129B-2/Wa or the 7.5cm BK7.5 Bordkanone in the Hs129B-3/Wa.

Production of the Hs129 was terminated in September 1944 in favour of the so-called Jägernotprogramm (Fighter Emergency Programme). In all, the Henschel plants delivered a total of 867 Hs129 aircraft in its B-1, B-2 and B-3 models to the Luftwaffe.

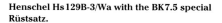

Henschel Hs129B-3/Wa with the BK7.5 special Rüstsatz.

Henschel Hs129B-2 – data

Powerplant	2 x 750hp Gnôme-Rhone twin-row radials with opposite-rotating propellers	
Dimensions		
Span	14.20m	46ft 7in
Length	9.75m	32ft 0in
Height	3.25m	10ft 8in
Wing area	29.00m²	312.15ft²
Weights		
Empty weight	4,020kg	8,162 lb
Loaded weight*	5,110kg	11,266 lb
Performance		
Max speed*	408km/h at 3,830m	253mph at 12,565ft
Combat speed*	320km/h at 3,000m	199mph at 9,840ft
Range*	560km	348 miles
Service ceiling	9,000m	29,530ft

* Loaded weight is without Rüstsatz; maximum speed is without external loads; combat speed is with Rüstsatz 4 (1 x BK7.5), and range with 2 x MK103 cannon.

Henschel Hs129B-2 with US captured Foreign Equipment number FE-4600.

A Henschel Hs129B-2 in flight. Clearly recognisable is the trapezoidal fuselage of this cantilever low-wing aircraft.

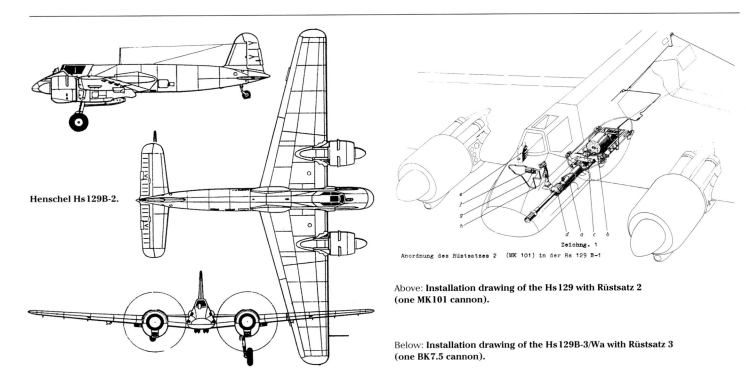

Henschel Hs 129B-2.

Anordnung des Rüstsatzes 2 (MK 101) in der Hs 129 B-1

Zeichng. 1

Above: **Installation drawing of the Hs 129 with Rüstsatz 2 (one MK 101 cannon).**

Below: **Installation drawing of the Hs 129B-3/Wa with Rüstsatz 3 (one BK 7.5 cannon).**

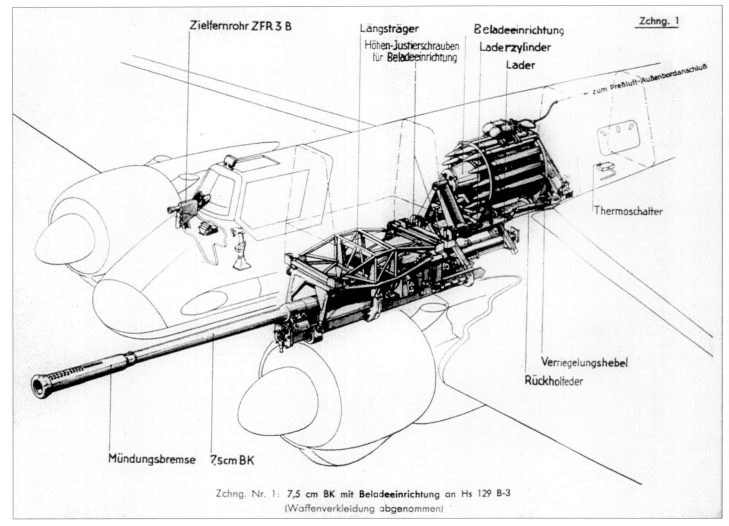

Zielfernrohr ZFR 3 B

Längsträger
Höhen-Justierschrauben
für Beladeeinrichtung

Beladeeinrichtung
Laderzylinder
Lader

Zchng. 1

zum Preßluft-Außenbordanschluß

Thermoschalter

Verriegelungshebel
Rückholfeder

Mündungsbremse   7,5 cm BK

Zchng. Nr. 1: **7,5 cm BK mit Beladeeinrichtung an Hs 129 B-3**
(Waffenverkleidung abgenommen)

Focke-Wulf Fw 189 V1b

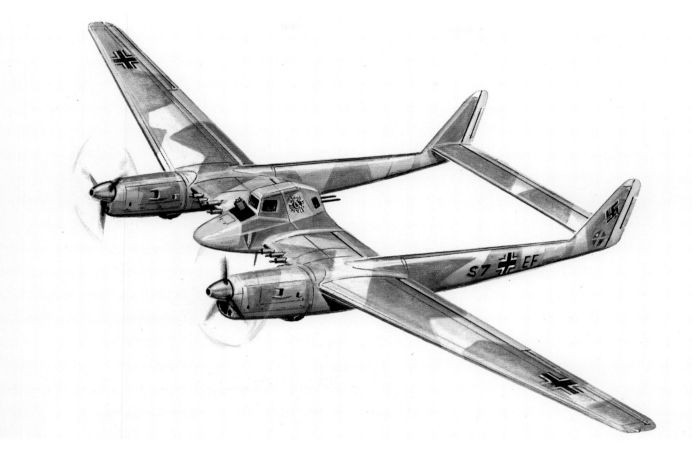

Focke-Wulf Fw 189 V6.

At the end of 1938, Focke-Wulf withdrew the Fw 189 V1 reconnaissance prototype from flight trials in order to enter it in the competition for a new ground attack aircraft. Redesignated as the Fw 189 V1b, it featured a new central fuselage and armoured crew compartment for the pilot and air gunner. Flight tests conducted in 1939 soon revealed that its flying characteristics as a ground attack aircraft were unsatisfactory; the Fw 189 V1b was simply too heavy and crew visibility was insufficient. Following modifications to the armoured crew compartment and the fitting of larger armour-glass vision panels, the Fw 189 V1b resumed its flight trials. Despite an improved opinion of it by the E-Stelle Rechlin, the RLM still preferred the Henschel Hs 129 and decided against series production.

On 1st April 1940, Focke-Wulf proposed an aerodynamically improved Fw 189 which, as the V6 prototype, was to serve as model for the Fw 189C-0 ground attack variant. Although test flights with its Bordwaffen had been successfully conducted at the E-Stellen Rechlin and Tarnewitz, the RLM decided not to place the Fw 189C in production in addition to the Henschel Hs 129.

Focke-Wulf Fw 189A-2 – data

Powerplant	Two Argus As 410A-1s of 465hp at 3,100rpm	
Dimensions		
Span	18.40m	60ft 4½in
Length	11.90m	39ft 0½in
Height	3.10m	10ft 2in
Wing area	38.00m²	409.02ft²
Empty weight	2,830kg	6,239 lb
Loaded weight	3,950kg	8,708 lb
Weights		
Loaded weight, max	4,170kg	9,193 lb
Performance		
Max speed	350km/h at 2,400m	217mph at 7,875ft
Range, normal	670km	416 miles
maximum	940km	584 miles
Service ceiling	7,300m	23,950ft
Armament	2 x MG 81Z in flexible dorsal mount	
	2 x MG FF fixed, forward-firing	
	4 x MG 17 forward-firing, 2 in tail	
	4 x SC 50 bombs	

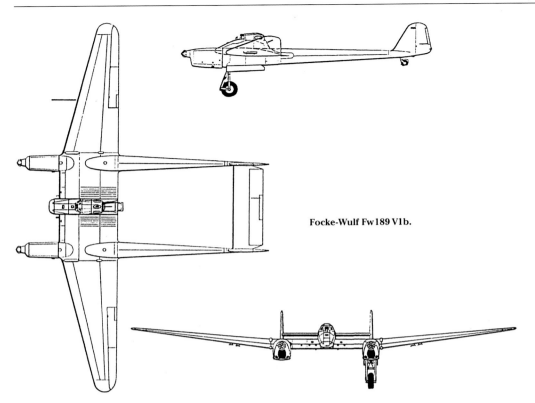

Focke-Wulf Fw 189 V1b.

Above left: **Focke-Wulf Fw 189 V1b with two-seat armoured crew compartment.**

Above right: **Focke-Wulf Fw 189 V6 with three-seat armoured crew compartment.**

Left: Front view of the armoured crew cabin and inner wing sections. The weapon ports can be clearly seen on the **Focke-Wulf Fw 189 V6.**

Above: **Another view of the three-seat Focke-Wulf Fw 189 V6 with opened cockpit hood.**

Henschel Hs P 87

1941

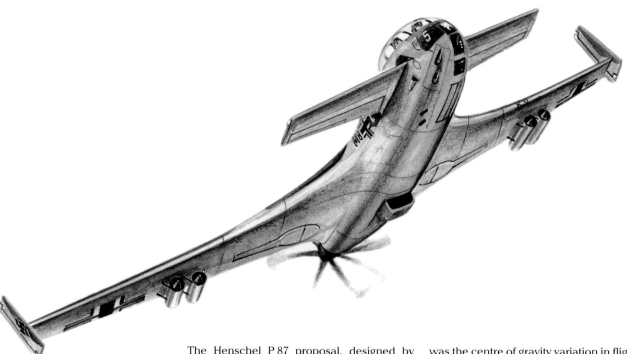

The Henschel P 87 proposal, designed by Chief Designer Dipl-Ing Friedrich Nicolaus, like the earlier P 75, featured the controversial tail-first layout that held promise of reaching higher speeds. This type of configuration offered the possibility of housing the proposed powerplant, the 24-cylinder Daimler-Benz DB 610 coupled engine in the fuselage rear. Developing 2,200hp, it was to have had eight-bladed contraprops which, as pushers, promised an improvement in flight performance compared to conventional aircraft. This engine arrangement also offered the possibility of accommodating a heavy weapon installation in the forward fuselage free of hindrances. One problem, however,

was the centre of gravity variation in flight. By careful attention paid during mock-up investigations, the problem was solved as to how the pilot and air gunner were provided with good visibility in a cockpit showing the least head drag. Emergency exit for the crew was by explosive ejection of the entire crew compartment so as to prevent collision with the eight-bladed contraprops. Although design work and mock-up construction had reached an advanced stage with the agreement of the Technisches Amt, it was rejected from the highest authorities with the lame excuse that 'pilots would not become accustomed to having the propellers at the rear and the tailplane in front'.

The Henschel Hs P 87 was designed as a two-seat cantilever low-wing monoplane of all-metal construction, the wing, with an aspect ratio of 6.2 and swept 30° on the leading edge, having endplate fins and rudders. All three undercarriage members – the nose and mainwheels, were to retract hydraulically into the fuselage and wing roots respectively.

Henschel Hs P 87 – data

Powerplant	1 x Daimler-Benz DB 610 (2 coupled DB 605s)	
Dimensions		
Span	14.00m	45ft 11¼in
Length	12.15m	39ft 10¼in
Height	2.80m	9ft 2¼in
Wing area	31.70m²	341.21ft²
Weights		
Loaded weight	9,000kg	19,841 lb
Performance		
Max speed	750km/h at 7,000m	466mph at 22,965ft
Armament	4 x MK 108 in fuselage nose	

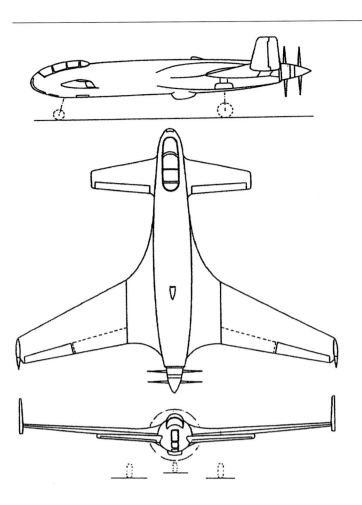

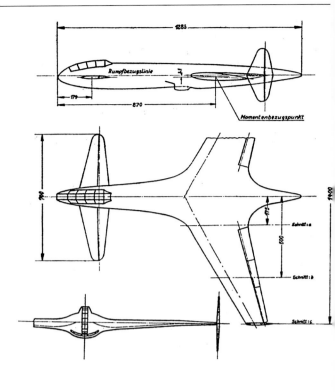

Left: **Henschel Hs P 87.**

Above: **Henschel Hs P 87 drawing of a 1/10th-scale wind-tunnel model. Dimensions visible include a span of 14m (45ft 11⅛in), an overall length of 12.85m (42ft 2in) and canard tailspan of 7m (22ft 11½in).**

Hütter Hü 136

Hütter Hü 136 (Stubo 1) long-range dive-bomber with one 500kg (1,102 lb) underfuselage bomb.

At this point, the previously largely unknown dive-bomber and ground attack projects of the brothers Wolfgang and Ulrich Hütter are published for the first time anywhere. Between 1938 and 1944, the two glider aircraft designers participated with military developments to meet RLM specifications. These projects, which ran partly under the code name Ostmark, are so novel that they should not be withheld from the reader.

Within the framework of the Luftwaffe's new equipment programme which defined dive-bombing and ground attack requirements for surface and point targets, the RLM Technisches Amt accordingly issued specifi-

attack and dive-bomber role, capable of carrying a 500kg (1,102 lb) bombload and armed with machine-guns. Its performance was expected to be similar to the then current fighter aircraft. The second type – the Stubo 2, envisaged a single-engined two-seater with the same capabilities, but carrying a 1,000kg (2,205 lb) bombload.

The projects conceived by the Hütter brothers initially did not quite achieve all the RLM specification requirements for both types in relation to range and bombload. In order to combat more distant targets and be able to carry a larger quantity of fuel, they dispensed with a conventional undercarriage and as in

Hütter Hü 136 (Stubo 1).

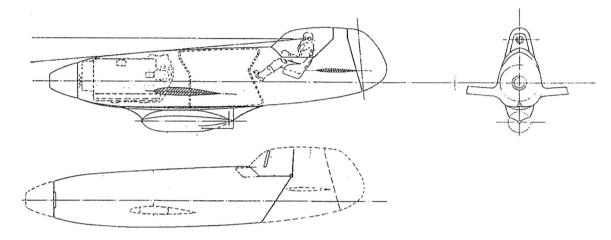

cations for such developments. The requirements for flight performance and airframe structural strength were particularly highly emphasised, for which reason the RLM instituted a strength regulation in connection with further manufacturing guidelines, the latter being notified to the aircraft industry together with the development contracts.

The specifications envisaged two types of aircraft. For the first – the Stubo 1 (an abbreviation for Sturzbomber = dive-bomber), the RLM required a heavily-armoured single-engined single-seat aircraft for the ground

sailplane practice, adopted skids instead. Take-off was to have been accomplished with the aid of a jettisonable take-off dolly, with landing on an extensible skid fitted with newly-developed surface brakes. Prior to landing, the propeller was to be blown off and descend on a parachute over the airfield.

Wolfgang Hütter, who between 1935 and 1944 had been engaged at the Wolf Hirth GmbH, also participated in other tenders to RLM requirements, among them, for the further development of the Heinkel He 219 as a high-performance night fighter. The two dive-bomber and ground attack projects did not reach the construction stage as the RLM had decided in favour of the Henschel Hs 129.

Hütter Hü 136 (Stubo 1) – data

Powerplant	1 x 1,200hp Daimler-Benz DB 601 in-line engine	
Dimensions		
Span	6.50m	21ft 4in
Length	7.20m	23ft 7½in
Weights		
Loaded weight	3,700kg	8,157 lb
Performance		
Max speed	560km/h	348mph
Range	2,000km	1,242 miles
Service ceiling	9,500m	31,170ft

Hütter Hü 136 (Stubo 2) armoured dive-bomber.

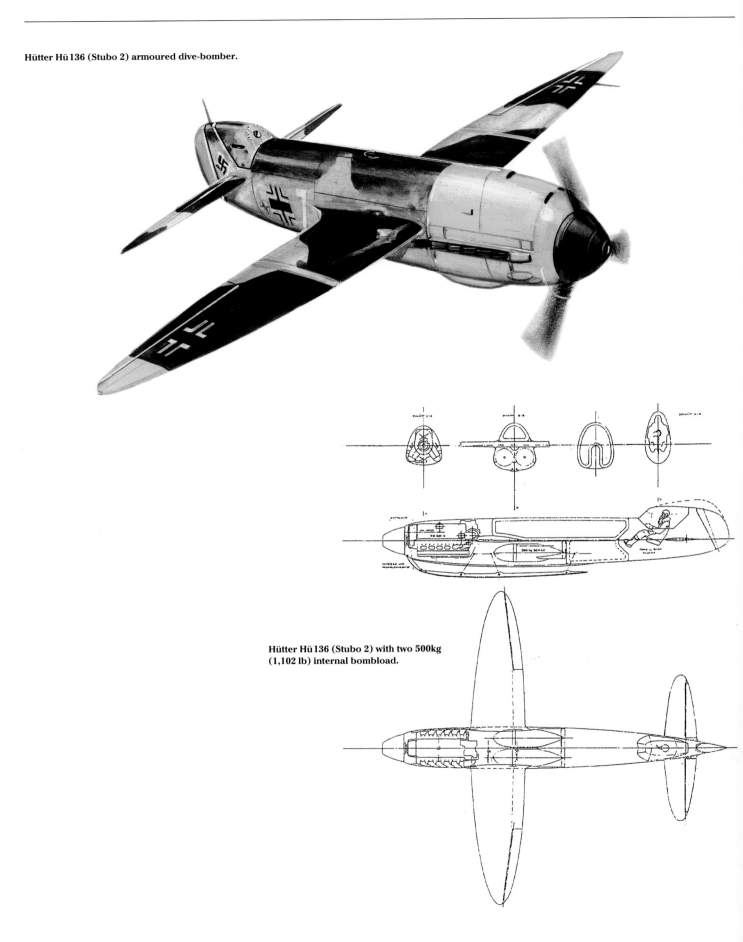

Hütter Hü 136 (Stubo 2) with two 500kg (1,102 lb) internal bombload.

Junkers EF 82
1939 to 1940

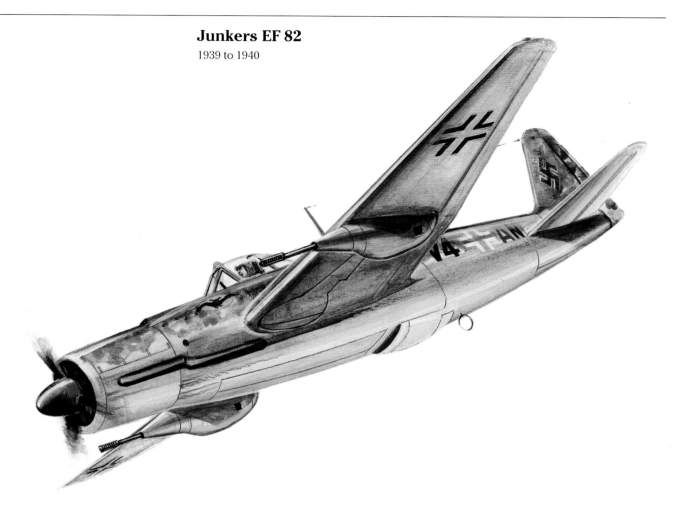

Under the Junkers Entwicklungsflugzeug (Development Aircraft) project designation EF 82 appeared, the 1939/40 design for a mid-wing single-seat aircraft intended for the ground attack and dive-bomber role. As opposed to the well-known Ju 87 Stuka dive-bomber, the proposed bombload of 1,000kg (2,205 lb) was to be housed internally in the fuselage. A wide-track main undercarriage retracting into the wing roots together with a cable-snatching tail hook was to enable its use aboard aircraft carriers. A special feature of the design was that it was to be powered by a Junkers-Jumo 214 engine – a secret development of the Junkers aero-engine division whose further advancement was not supported by the RLM, and this at a time when the aircraft industry was crying out for high-performance powerplants.

The Jumo 214 featured a front annular radiator and was housed beneath along cylindrical engine cowling. Noteworthy were the two long exhaust ducts which were not a normal feature of in-line engines. Details of dimensions and performance of the EF 82 are just as little known as the RLM judgement of the project.

Junkers EF 82 scale model.

Junkers Ju 187

1941

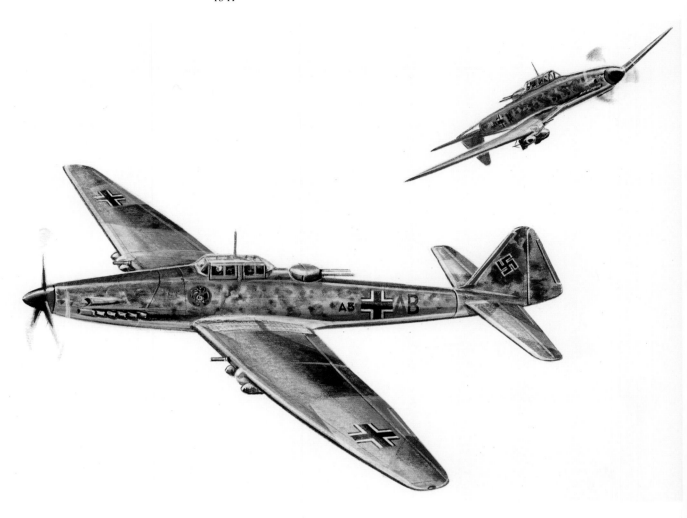

A pair of Junkers Ju 187s. Note that the tail surfaces of the rear aircraft have been rotated so as to provide a free rearward field of fire for the dorsal turret.

Junkers Ju 187 – data

Powerplant
1 x 1,750hp Junkers-Jumo 213A driving a three-bladed propeller

Dimensions

Span	14.20m	46ft 7in
Length	11.90m	39ft 1in
Height	3.95m	12ft 11½in

Armament
2 x MG 131 in wing roots, plus 2 x MG 151/20 in dorsal turret, plus
1 x 1,000kg (2,205 lb) bomb, plus 4 x 250kg (551 lb) external bombs

As a further development of the Junkers Ju 87, the Ju 187 project, by means of aerodynamic improvements, was to furnish greatly-improved flight performance to overcome the ever-increasing enemy air and ground defensive measures. In overall appearance it resembled the Ju 87 but differed from it in having a rearward retractable mainwheel undercarriage and a straight-tapered wing.[1] In the two-man crew compartment, the rear gunner operated the dorsally-mounted (B-Stand) weapon mount housing one or two MG 151/20s with the aid of an automatic alignment system. A special feature of the design was the ability to rotate the entire tail assembly through 180° so that the rearward-firing weapon turret had an unrestricted field of fire. Exactly how this alteration would have affected the aircraft's behaviour in flight is not known. With this aircraft, the maximum cruising speed could be increased from 300km/h (186mph) to 400km/h (242mph) and bombload to 2,000kg (4,409 lb). Without

giving any reason, the RLM ordered all work on the project to be terminated in autumn 1941.

[1] The Ju 187 model photos seen here show a variant with compound taper on the wing leading and trailing edges. This, and the straight-tapered variant in the three-view drawing both have provision for a large external underfuselage bomb and a pair of smaller bombs on either side of the mainwheel underwing nacelles.

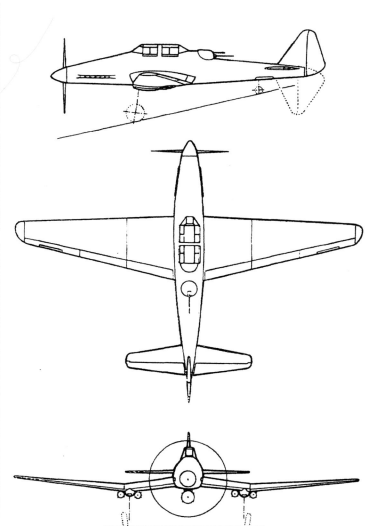

Top left: **Junkers Ju 187 model showing normal tail position.**

Centre left: **Junkers Ju 187 model showing rotated tail position.**

Above left: **View of the Junkers Ju 187 mock-up cockpit area.**

Top right: **Full-scale Junkers Ju 187 mock-up.**

Right: **Junkers Ju 187.**

Junkers Nameless Ground Attack Project

1941

Junkers Nameless Ground Attack Project – data

Powerplant	Two Daimler-Benz DB 007 turbojets	
Dimensions		
Span	14.60m	47ft 10¾in
Length	11.85m	38ft 10⅜in
Height	3.85m	12ft 7½in
Armament	4 x 20mm MG151/20, plus 4 x 30mm MK103	

In mid-1941, the Development Department of the Junkerswerke in Dessau commenced work on a project for a low-level and ground attack aircraft as a replacement for the Henschel Hs129. The project study involved a rather plump-looking mid-wing aircraft with two wingroot-mounted turbojets. According to works documentation, the turbojets were to have been two Daimler-Benz 109-007 ZTL units which allowed a considerable increase in performance at a reduced fuel consumption. Designed by Prof Dr-Ing Karl Leist, head of the Abteilung Sondertriebwerk (Special Engines Department) at the Daimler-Benz AG, the two-circuit or bypass turbojets had a larger air intake and overall diameter than the single-circuit BMW 003 and Jumo 004 turbojets.

Besides this new type of turbojet, strong armour plating was to have been provided for the fuselage and powerplants. As a ground attack aircraft, it was to have been equipped with four 30mm MK103 and four 20mm MG151/20 cannon. The undercarriage mainwheels were to retract forwards into the fuselage sides as shown in the three-view drawing. As little experience had been gath-ered with nosewheels which for a long time had been rejected by the RLM as too 'American', a retractable pneumatically-sprung skid replaced the nosewheel.

The long gestation period of turbojet development at Daimler-Benz that resulted in the first turbojet test-bed runs only in March 1943, led to termination of the project. Several decades later, this project served as the fore-runner for the US Fairchild A-10A Thunderbolt (also known as the Warthog)* ground attack and low-level combat aircraft which cannot deny its resemblance to the nameless Junkers ground attack aircraft.

* For a more comprehensive history of its development, see Mike Spick: *A-10 Thunderbolt II, Modern Combat Aircraft 28*, Ian Allan Ltd, London, 1987

Above: **A Fairchild A-10A Thunderbolt prototype. Its similarity to the Junkers design scheme is unmistakable. The propulsion units, mounted in lateral fuselage nacelles were two General Electric TF 34-GE 100 bypass turbojets. The main weapon on the A-10A consisted of a seven-barrel revolver cannon for attacking ground targets, and fired, among others, uranium-filled armour-piercing shells. Various types of bombs and rockets could be mounted on eight underwing stations. During the Gulf War against Iraq, the USAF A-10As were used with considerable success, the ground- attack aircraft accounting for innumerable columns of retreating Iraqi formations on the 'death road to Basra'.**

Below: **Schematic arrangement of a two-circuit or bypass turbojet. This type of turbojet, where the hot exhaust gases were surrounded by a jacket of cold air, enabled a considerable increase in thrust to be attained.**

Bottom: **The Daimler-Benz 109-007 bypass turbojet suspended on its engine test-bed.**

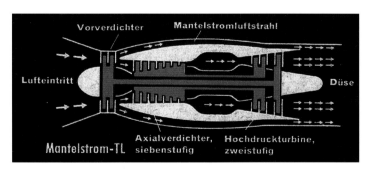

Lippisch P 09 (TL)
1941 to 1942

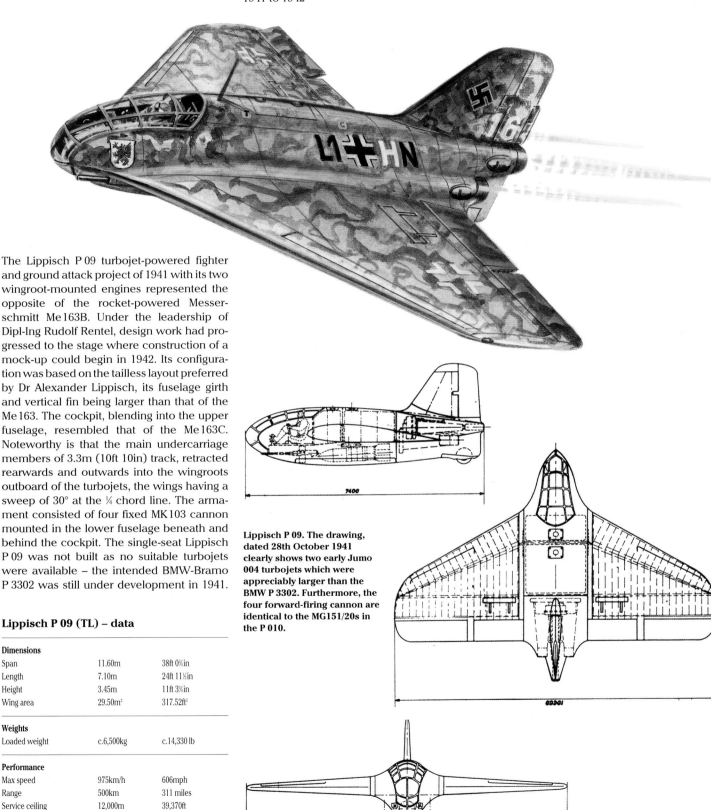

The Lippisch P 09 turbojet-powered fighter and ground attack project of 1941 with its two wingroot-mounted engines represented the opposite of the rocket-powered Messerschmitt Me 163B. Under the leadership of Dipl-Ing Rudolf Rentel, design work had progressed to the stage where construction of a mock-up could begin in 1942. Its configuration was based on the tailless layout preferred by Dr Alexander Lippisch, its fuselage girth and vertical fin being larger than that of the Me 163. The cockpit, blending into the upper fuselage, resembled that of the Me 163C. Noteworthy is that the main undercarriage members of 3.3m (10ft 10in) track, retracted rearwards and outwards into the wingroots outboard of the turbojets, the wings having a sweep of 30° at the ¼ chord line. The armament consisted of four fixed MK 103 cannon mounted in the lower fuselage beneath and behind the cockpit. The single-seat Lippisch P 09 was not built as no suitable turbojets were available – the intended BMW-Bramo P 3302 was still under development in 1941.

Lippisch P 09. The drawing, dated 28th October 1941 clearly shows two early Jumo 004 turbojets which were appreciably larger than the BMW P 3302. Furthermore, the four forward-firing cannon are identical to the MG151/20s in the P 010.

Lippisch P 09 (TL) – data

Dimensions

Span	11.60m	38ft 0¾in
Length	7.10m	24ft 11½in
Height	3.45m	11ft 3¾in
Wing area	29.50m²	317.52ft²

Weights

Loaded weight	c.6,500kg	c.14,330 lb

Performance

Max speed	975km/h	606mph
Range	500km	311 miles
Service ceiling	12,000m	39,370ft

Armament — 4 x MG151/20 in lower fuselage

Lippisch P 010 (TL)

1941 to 1942

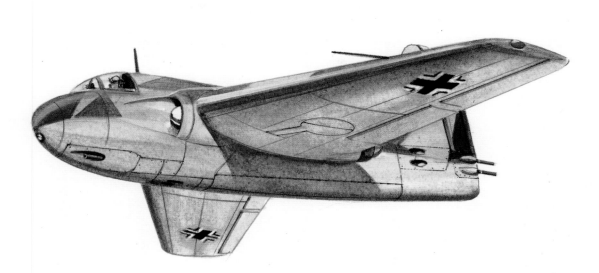

This Kampfzerstörer and Schnellbomber project drawn up in November 1941 resembled the Me 163 in overall appearance but was dimensionally larger. The project was laid out by Dipl.Ing Rudolf Rentel as a cantilever mid-wing aircraft housing two turbojets buried in the wing roots. Aft of the cockpit, the fuselage held the two large fuel tanks, beneath which space was reserved for the internal 1,000kg (2,205 lb) bombload. The fixed tail machine-gun barbette was remote controlled by the pilot via a rear-view periscope. As with the preceding P 09 project, the mainwheels retracted into the wings outboard of the turbojets. Originally intended to carry a crew of two, this was later reduced to one.

Lippisch P 010 – data

Powerplant	Two early Junkers-Jumo 004 turbojets	
Dimensions		
Span	13.40m	43ft 11½in
Length	8.15m	26ft 9in
Height	3.80m	12ft 5½in
Wing area	c.39.40m²	c.424.09ft²
Performance		
Max speed	850km/h	528mph
Armament	2 x MG 151/20 in nose, plus 2 x MG 151/20 in tail plus 1 x 1,000kg (2,205 lb) bomb	

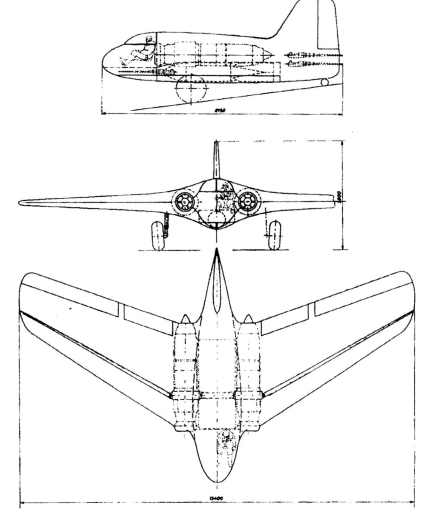

Lippisch P 010 – drawing dated 26th November 1941.

Messerschmitt Me 328

1940 to 1945

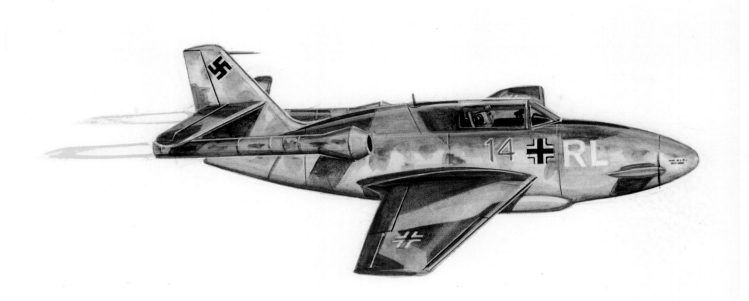

Under the project number P1079/16, the Messerschmitt AG initiated a series of studies for a cheap, high-performance fighter that could also be used on low-level attack missions. The first variant of the P1079/16, which had in the meantime been accorded the RLM designation Me 328, according to the RLM decision of 16th March 1941, was handed over to the DFS* which in co-operation with the Jacobs-Schweyer Flugzeugbau of Darmstadt, took over further development and construction of the Me 328A-series.

The first test flights as an unpowered glider took place on 5th November 1942 piloted by works pilots Flugkapitän Karl Baur (head of flight-testing at Messerschmitt) and Dipl-Ing Rudolf 'Gretchen' Ziegler of the RLM. According to Messerschmitt documents, however, the maiden flight of the unpowered Me 328 V1 prototype had already taken place on 23rd July 1941 at Wels near Linz by Rudolf Ziegler. The Me 328A-1 entered the flight-test phase in spring 1944. As foreseen by the specification, the glider was to be towed aloft in

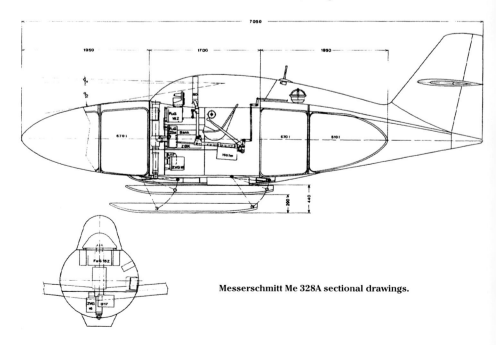

Messerschmitt Me 328A sectional drawings.

Mistelschlepp and brought to the vicinity of enemy formations, but changes to its proposed use as also the possibility of employing it as a Selbstopferjäger (suicide fighter) packed with a quantity of explosive in the fuselage, led to modifications.

The Me 328B version was powered by two Argus Versuchsschubrohr (experimental thrust duct) VSR 7 pulsejets – a precursor of the later As 014, enabling it to carry a 500kg (1,102 lb) bombload. In flight trials with the Argus units, considerable damage was sustained by the entire airframe caused by vibrations from the intermittent combustion method of operation. Upon reaching the critical frequency, these vibrations led to the crash of the first Me 328B prototype. The Me 328C version, powered by a Porsche 109-005 turbojet above the rear fuselage, was expected to commence flight-testing at the beginning of 1945.[2]

* For a detailed account of the DFS and its projects see Horst Lommel: *Geheimprojekte der DFS 1935-1945. Vom Höhenaufklärer bis zum Raumgleiter (DFS Secret Projects. From the high-altitude reconnaissance aircraft to the space glider)*, Motorbuch Verlag, Stuttgart, 2000.

[2] Published illustrations of the so-called Me 328C show it with a Junkers-Jumo 004B turbojet mounted in a nacelle beneath the fuselage. The expendable Porsche 109-005 turbojet had not reached the flight-test stage at the end of the war. It had a design static thrust of 500kg (1,102 lb) at 14,500rpm, dry weight 180kg (397 lb) and specific fuel consumption (sfc) 1.38kg/kg thrust/hr. External diameter was 65cm (25½in) and overall length 2.85m (9ft 4¼in). The turbojet outline in the drawing is actually that of the Porsche Type 300, which had a design static thrust of 400kg (882 lb) at 12,500rpm, weight without suspension 325kg (716 lb) and sfc 2.0kg/kg thrust/hr. External diameter was 57cm (22½in) and overall length 2.33m (7ft 7⅝in). In October 1944, the Chef TLR reported that the T 300 turbojet was expected, even with increased personnel, to require at least eight months to become a reality.

For known details of the Porsche 109-005, see Anthony L Kay: *German Jet Engine & Gas Turbine Development*, Airlife Publishing Ltd, Shrewsbury, 2002, pp.153-155 and for the Porsche T 300 turbojet, see Wilhelm Hellmold: *Die V1, Eine Dokumentation*, Bechtermünz Verlag, Esslingen & München, 1999, p.198 and unnumbered Attachments (two pages).

Me 328A – data as of 15.12.1942

Powerplant	2 x 300kg (661 lb) thrust Argus As 014 pulsejets	
Dimensions		
Span	6.90m	22ft 7⅜in
Length	7.17m	23ft 6¼in
Height, skid up	2.43m	7ft 11¾in
skid down	2.87m	9ft 5in
Wing area	8.50m²	91.49ft²
Weights		
Empty weight	1,510kg	3,329 lb
Loaded weight, clean	3,240kg	7,143 lb

Me 328B – data as of 15.12.1942

Powerplant	2 x 300kg (661 lb) thrust Argus As 014 pulsejets	
Dimensions		
Span	8.60m	28ft 2½in
Length	7.05m	23ft 1½in
Height, skid down	2.87m	9ft 5in
Wing area	9.40m²	101.18ft²
Weights		
Empty weight	1,510kg	3,329 lb
Loaded weights	3,740kg (8,245 lb) with 1 x 500kg bomb	
	4,250kg (9,370 lb) with 1 x 1000kg bomb	
Performance		
Max speeds at sea level	700km/h (435mph) without bomb	
	600km/h (373mph) with 1 x SC 500	
	530km/h (329mph) with 1 x SC 1000	
Range at sea level, clean	630km	391 miles
Service ceiling,	4,000m	13,120ft
after bomb release	6,800m	22,310ft

With 2 x 400kg (882 lb) thrust pulsejets, max sea level speeds were raised to 810km/h (503mph), 700km/h (435mph) and 630km/h (391mph) with the same loads respectively. Service ceiling was raised to 5,500m (18,045ft) and 7,100m (23,290ft) after bomb release.

Messerschmitt Me 328 V2 with two Argus VSR 7 pulsejets.

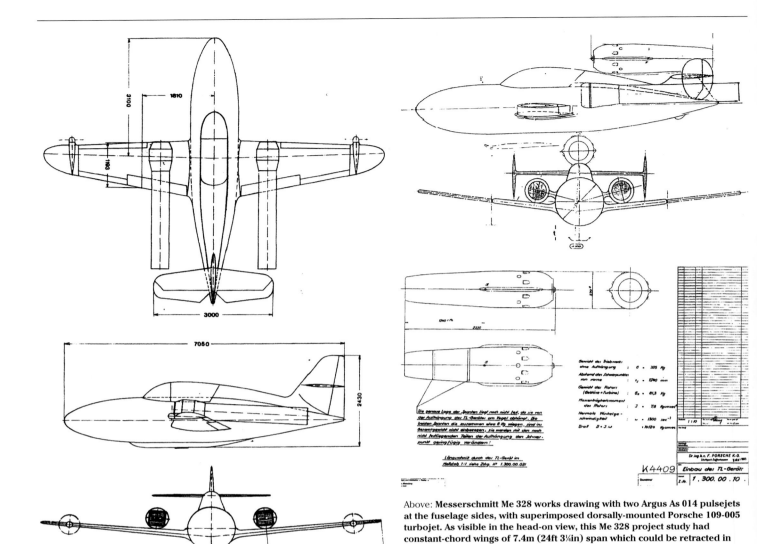

Above: Messerschmitt Me 328 works drawing with two Argus As 014 pulsejets at the fuselage sides, with superimposed dorsally-mounted Porsche 109-005 turbojet. As visible in the head-on view, this Me 328 project study had constant-chord wings of 7.4m (24ft 3¼in) span which could be retracted in flight to only 4m (13ft 1½in) to increase its maximum speed. Overall length was 7.25m (23ft 9½in) and tailplane span 3.2m (10ft 6in), the horizontal tailplane having a leading-edge sweep of 38°.

Above: **Messerschmitt Me 328B** – drawing dated 15th December 1942.

Far left: **Dipl-Ing Rudolf 'Gretchen' Ziegler (1944)** was the first RLM test pilot who flew the Messerschmitt Me 328 and the piloted V-1.

Left: **Dipl-Ing Felix Kracht, DFS design engineer,** gained appreciable recognition for the Mistelschlepp towing method and for the further development and flight-testing of the Me 328.

Chapter Two

Ground Attack Aircraft 1942-1943

Junkers Ju 87D-3 and D-5 Nachtschlachtflugzeug

Much has already been published concerning the Junkers Ju 87 – the well-known Stuka dive-bomber and ground attack aircraft, including three books by the authors Günther Just, Georg Brütting and Manfred Griehl,* so that only the Ju 87D-3 and Ju 87D-7 nocturnal ground attack variants are mentioned here.

In the summer of 1942, the Junkerswerke in Bernburg conducted initial trials with what was unusual armament for an aircraft. The modified Ju 87D-3 was equipped with an offensive armament of two underwing-mounted 3.7cm Flak 18 cannon to combat strongly-armoured point targets. These newly-converted point target and anti-tank aircraft designated Ju 87D-5 left the assembly lines in the summer of 1943 under the series designation Ju 87G-2. Operational use of the 'Flying Flak' became especially known through the successes of Oberst Hans-Ulrich

Rudel who destroyed over 500 enemy tanks with his weapons and bombs. From the experience gained with the Ju 87D-5 in night missions, there appeared the Ju 87D-7 Nachtschlachtflugzeug (nocturnal ground attack aircraft) equipped with the more powerful Junkers-Jumo 211P motor and two fixed forward-firing 20mm MG151/20 cannon in place of the previously installed 7.9mm MG17 machine-guns. The dive brakes were also removed. With constant improvements and subsequent equipment changes, series production terminated with the Ju 87D-8 in autumn 1944.

*Günther Just: *Stuka-Oberst Hans-Ulrich Rudel*, 12th Ed, Motorbuch Verlag, Stuttgart, 1983.
Georg Brütting: *Das waren die deutschen Stuka-Asse (Those were the German Stuka Aces) 1939-1945*, 8th Edition, Motorbuch Verlag, Stuttgart, 1995.
Manfred Griehl: *Junkers Ju 87 'Stuka'. Sturzkampf-bomber, Schlachtflugzeug, Panzerjäger (Dive-bomber, Ground Attack Aircraft, Anti-Tank Fighter)*, Motorbuch Verlag, Stuttgart, 1995.

Top right and centre right: **Aerial tank-buster: The Junkers Ju 87D-3 with two 3.7cm Bordkanone.** As reloading in flight was not possible on the Ju 87, each weapon had 6-round clips making a total of 12 rounds on the underwing weapon mounts. The special projectile with its tungsten core only exploded after it had penetrated the tank's armour plating. This weapon was also fitted to the Hs 129B-2 and Bf 110G-2 in an underfuselage pannier, each housing one BK 3.7, whereas the Ju 88P-3 underfuselage pannier housed two.

'The Father of the Stuka' Dipl-Ing Hermann Pohlmann (born 1894) was considerably involved in the design of the Junkers K 47, Ju 160 and Ju 87. In 1923, he was deputy head of the Junkers design office; in 1928, head of the 'New Designs Department' (Ju 88, Ju 90), and in 1940, left to become Deputy Chief Designer at Blohm & Voss, Hamburg.

The Combat Service Clasp for Ground Attack Pilots

On 30th January 1941, the Luftwaffe Commander-in-Chief introduced the Combat Service Clasp citation for operational missions. It was initially awarded in three versions – for Jäger (fighter), Kampfflieger (bomber) and Sturzkampfflieger (dive-bomber), and Aufklärer (reconnaissance) crews. With the continuation of the war, a separate version was awarded from 13th May 1942 to the Zerstörer crews and finally, on 12th April 1944 to Schlachtflieger (ground attack) crews. Each airman received the clasp in bronze after 20, the silver after

60, and the gold after 100 combat missions. Combat missions were defined as those where enemy defensive fire was encountered or on which the flight penetrated more than 30km (18 miles) behind enemy lines.

On 29th April 1944, the Commander-in-Chief instituted the Pendant to the Golden Service Clasp for more than 200 combat missions. The rectangular gold-coloured Pendant, encircled by laurel leaves, was inscribed with the number of missions flown.

Golden Service Clasp for Zerstörer crews.

Silver Service Clasp for Schlachtflieger crews.

Tanks from the Air
or: 100,000 Roubles on the head of the 'Adler der Ostfront'

The operational story of Oberst Hans-Ulrich Rudel is not only unique in the history of the Second World War, but in the history of warfare itself. In 2,350 operational sorties, the Stuka pilot and 'tank-buster' destroyed from the air no less than 519 Soviet tanks, 800 vehicles, 150 field howitzers, 4 armoured columns, 70 landing craft, one naval destroyer, one cruiser, one battleship, and numerous bridges and supply columns. He was also credited with 9 confirmed aerial victories – 7 fighters and 2 ground attack aircraft. Rudel was shot down by flak 30 times and was 5 times wounded. In spite of this, none of the Soviet flak crews were able to 'earn' the 100,000 Roubles award which Stalin had placed on the head of the Adler der Ostfront (Eastern Front Eagle). Even after his attempt on 20th March 1944 to rescue Stuka crews that had been shot down behind enemy lines and thus came into Soviet captivity, together with his gunner Oberfeldwebel Erwin Hentschel, he still managed to escape despite a shoulder injury. His companion Hentschel, however, lost his life when he drowned whilst swimming in the icy waters of the Dnjestr river. Hounded by tracker dogs and patrols on horseback and wounded by a bullet in the shoulder, he forced his way through 50km (31 miles) of enemy territory to reach his own lines; the last 15-20km (9-12 miles) traversed on foot. His motto 'The loser is one who he himself gives up' provided the self-example for this Silesian-born pilot.

Hans-Ulrich Rudel was an extraordinary personality – as an individual, sportsman, and airman. In 1945, he was the sole individual to be awarded the highest German decoration – das Goldene Eichenlaub zum Eichenlaub mit Schwertern und Brillanten des Ritterkreuzes (The Golden Oakleaves to the Oakleaves with Swords and Diamonds of the Knight's Cross). Also, the Goldene Frontflugspange mit Brillanten und Anhänger (Golden Combat Mission Clasp with Diamonds and Pendant) for 2,000 operational sorties was awarded only once – to Hans-Ulrich Rudel.

His life history
Born on 2nd July 1916 as the son of a protestant vicar in Konradswaldau/Niederschlesien (Lower Silesia). Attended the Volkshochschule of the humanistic Gymnasium.

1936 Entry into the Luftwaffe.

1937 Commenced pilot-training in the Luftkriegsschule (Air-war School) Berlin-Weder.

1938 Dive-bomber pilot-training course with Stuka-Geschwader 168 in Graz.

1939 Promoted to Leutnant (RAF: Pilot Officer; USAF: Lieutenant)

1941 Initial combat missions with Stuka-Geschwader 'Immelmann 2' on the Eastern Front. Sank the Soviet battleship Marat. German Cross in Gold. 1942 – Knight's Cross of the Iron Cross. Staffelkapitän in III./Stuka-Geschwader 'Immelmann'. 500 combat missions.

1943 1,500 combat missions. His air-gunner companion Oberfeldwebel Hentschel achieved 100 tanks destroyed. Oakleaves and Swords to the Knight's Cross.

1944 2,000th combat mission. Destroyed the 320th tank;17 tanks alone on 26th March 1944. Diamonds to the Knight's Cross with Oakleaves and Swords. Promoted to Oberst- Leutnant (RAF: Wing Commander, USAF: Lieutenant Colonel)

1945 1st January. Rudel, in the meantime Kommodore of SG 2 'Immelmann', as the sole airman received the Golden Oakleaves (according to the citation award list of 29th December 1944, this citation had only been awarded 12 times to the most highly- commended individuals) and was promoted to Oberst (RAF: Group Captain, USAF: Colonel). Notwithstanding severe injuries and an amputation, Rudel flew further Combat missions despite several flying bans imposed on him. 'They can't forbid me to fly, especially when Russian tanks are strolling all over German territory.'

Below: **Hans-Ulrich Rudel (1916-1982) certainly belongs to the bravest airmen of all the armed forces during the Second World War. The picture shows him (left) as a decorated Hauptmann (Captain) together with his gunner Oberfeldwebel (Sergeant) Erwin Hentschel. After 1,490 flights over enemy territory with Rudel in which he destroyed 100 tanks, Erwin Hentschel was drowned on 20th March 1944 during the escape with Rudel from Russian captivity in the icy waters of the Dnjestr river.**

Bottom: **The awards of the most highly decorated airman during the Second World War. At centre: Ribbon and Knight's Cross of the Iron Cross with Golden Oakleaves and Swords and Diamonds – awarded once only. Upper right: Oakleaves with Swords and Diamonds. Centre right: Oakleaves with Swords. Centre left: Oakleaves. Lowest row left: Golden Pilot's Emblem with Diamonds. Lowest row centre: Golden Service Clasp with Diamonds and Pendant for 2,000 Combat Flights – awarded once only. Lowest row right: German Cross in Gold.**

Blohm & Voss P163.01 and P163.02

1942

Blohm & Voss P163.01 with Daimler-Benz DB 613.

Dating from early 1942, the P163 combat aircraft project was designed for the low-level attack, dive-bombing and close-support role. In the documentary submission to the RLM, the project was also described as a so-called Arbeitsflugzeug (literally 'work' aircraft), an up until then almost unknown mission designation. As with all the other aircraft proposals from its Chief Designer Dr-Ing Richard Vogt who was born in Wurttemberg, the P163 had multifarious capabilities.

An unusual feature of the design was the arrangement of the four crew members in two wingtip nacelles. The fuselage, with conventional tail surfaces, housed the powerplant, fuel, and external bombload. The pilot sat in the extensively glazed cockpit together with his navigator/gunner in the port wingtip nacelle, the other two air gunners being located in the starboard nacelle. This wingtip armament arrangement allowed a wide fore and aft field of fire largely undisturbed by the presence of the propeller discs, fuselage, and tail surfaces. A cantilever mid-wing monoplane of all-metal construction, it incorporated detachable wing halves fabricated of steel with wooden flaps and ailerons. The 10m (32ft 9¾in) wide-track mainwheels retracted inwards into the wings, the tailwheel retracting rearwards beneath the rudder. Aft of the double-engine, the all-metal steel-tube fuselage accommodated the likewise steel protected fuel tanks, the rear fuse-

lage and tail surfaces being fabric covered to reduce weight. The two variants differed only in their powerplants, both of which drove six-bladed contraprops of 4.3m (14ft 1¼in) diameter. In addition to the normal MG151/20 armament positions, provision was made for the carriage of one MK114 or five RZ 65 rockets.

Blohm & Voss P163.01 – data

Powerplant
1 x 3,800hp Daimler-Benz DB 613C/D 24-cylinder in-line engine

Dimensions

Span	20.50m	67ft 3in
Length	15.60m	51ft 2¼in
Height, wheels down	6.50m	21ft 4in
Wing area	55.30m²	595.23ft²

Weights

Fuel weight	2,725kg	6,008 lb
Loaded weight, max	15,000kg	33,069 lb

Performance

Max speed	544km/h at 6,000m	338mph at 19,685ft
	610km/h at 8,600m	379mph at 28,250ft
Service ceiling, max weight	8,500m	27,890ft
Absolute ceiling	9,025m	29,610ft
Range, cruising speed		
at sea level	2,050km	1,274 miles
at 6,000m (19,685ft)	2,420km	1,504 miles
Take-off run, grass	490m	1,608ft
concrete	450m	1,476ft
Landing speed	140km/h	87mph

Blohm & Voss P163.02 – data

Powerplant
1 x 4,000hp BMW 803A 28-cylinder twin-row radial engine

Dimensions
Span	20.50m	67ft 3in
Length	15.00m	49ft 2½in
Height	6.50m	21ft 4in
Wing area	55.30m²	595.23ft²

Weights
Empty weight	9,400kg	20,723 lb
Fuel weight	2,850kg	6,283 lb
Loaded weight, max	15,200kg	33,510 lb

Performance
Max speed	570km/h at 6,000m	354mph at 19,685ft
	676km/h at 12,200m	420mph at 40,025ft
Range, cruising speed		
at sea level	2,000km	1,242 miles
at 6,000m (19,685ft)	2,250km	1,398 miles
Service ceiling, max weight	9,150m	30,020ft
Absolute ceiling	10,060m	33,005ft
Take-off run, grass	475m	1,558ft
concrete	435m	1,427ft
Landing speed	145km/h	90mph

Armament (both variants)
Port console	1 x MG151/20 (250 rds) movable (A-Stand)
	plus 2 x MG151/20 (500rpg) twin-mount (B-Stand)
Starboard console	1 x MG151/20 (250 rds) movable (A-Stand)
	plus 1 x MG151/20 (500 rds) movable dorsal turret
	plus 2 x MG151/20 (500rpg) twin-mount (B-Stand)

Bombload
2 x 1,000kg (2,205 lb) beneath fuselage

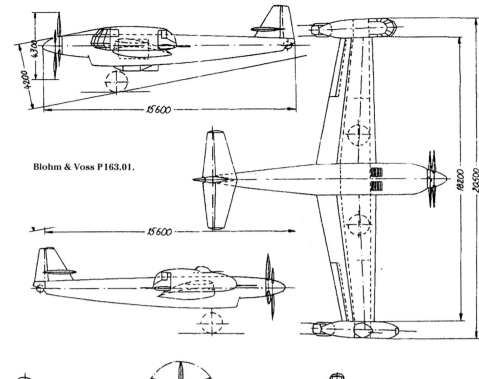

Blohm & Voss P163.01.

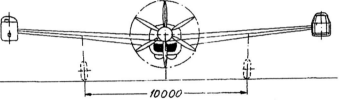

Below: **Blohm & Voss P163.02 with BMW 803.**

Among German aircraft designers, Dr-Ing Richard Vogt was rich in innovations and with his developments was years ahead of aircraft construction internationally. Several of his ideas and proposals are still being employed today in the most modern aircraft designs. Born a Swabian, from humble beginnings he worked his way to the top. It is noteworthy that his office door was always open for everyone. He is shown here at the centre of a group of co-workers on the occasion of a demonstration of the BV 238 large flying boat – the largest aircraft of its type in the world at that time. (See *Luftwaffe Secret Projects: Strategic Bombers 1935-1945* pp.116-117).

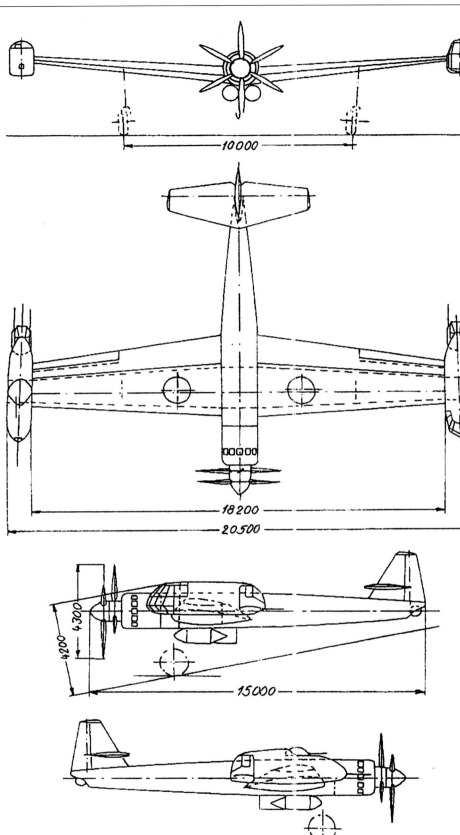

Blohm & Voss P163.02.

Blohm & Voss P 170.01

1942

Another of the most unusual projects emanating from Chief Designer Dr-Ing Richard Vogt was the P 170 Schnellbomber (fast bomber) and Schlachtflugzeug (ground attack aircraft) project of 1942 that featured constant-chord wings and a triple engine arrangement in which, besides the central motor, the other two were located in wingtip nacelles with vertical fins and rudders at their rear extremities. The goal and purpose of this design with its high-performance engines was to achieve equality if not superiority over enemy fighters. To reduce the drag of a conventional fuselage to a minimum and to partly counter the weight of the engines, the two crew members were housed at the rear of the fuselage in which three different cockpit arrangements were investigated during the life of the project. Instead of the usual centre wing section, it had two exchangeable wing halves made of steel covered with 1mm thick steel sheet. The fuel tanks of 2,800kg (6,173 lb) capacity were accommodated in the fuselage behind the central motor, the outer wheels of the full-span wide-track undercarriage retracting into the nacelles behind the powerplants. Of the proposed three BMW 801E engines which produced 2,100hp at

2,700rpm, the two outer units had propellers rotating in opposite directions, calculated to provide an appreciable increase in lift. Because of its high calculated speed, provision for defensive armament had not been firmly decided. Instead of the intended normal bombload of 1,000kg (2,205 lb) and 2,000kg (4,409 lb) at overload, two rocket packs each with six RZ 65 projectiles could be carried.

The design studies submitted to the RLM in October 1942 met with no positive response as there was supposedly no requirement for this project on which work was stopped at the end of the year.[1]

---

[1] Three other Blohm & Voss ground attack projects followed that are not included by the author in the 1943 era of his narrative but only in passing later on. These were:

(1) the asymmetric P 177 two-seat dive-bomber and ground attack aircraft powered by a Jumo 213 engine in the nose of the offset fuselage, the two crew members being housed back-to-back in the starboard crew nacelle. The wide-track mainwheels retracted outwards to rest in bulged bays at the tips of the constant-chord wing of 12m (39ft 4½in) span and 24m² (258.32ft²) area. Unlike the BV 141, it had a symmetrical tailplane. Normal loaded weight was 6,000kg (13,227 lb) and landing weight 5,150kg (11,353 lb). Normal armament consisted of two forward- and two

rearward-firing MG 151/20, provision being made for two additional cannon beneath the outboard wings or for six rocket projectiles. A single SC 500 bomb could be carried beneath the crew nacelle.

(2) the P 178 single-seat dive-bomber, powered by a single Jumo 004 turbojet in a nacelle beneath the starboard wing half, the pilot, fuel, and internal bombload being housed in the otherwise symmetrical fuselage on the port side of the constant-chord inboard wing section, the outer sections tapering on both leading and trailing edges towards the tips. Wingspan and area were identical to the previous project, except that the wide-track mainwheels retracted inwards into the wing roots forward of the tubular main spar. Armament consisted of two MG 151/20s low down in the nose beside the forward cockpit. The SC 500 bomb was enclosed entirely within the fuselage, whereas the SC 1000 protruded slightly below it. A special feature of the design is that a pair of superimposed rocket tubes projected from the fuselage rear beneath the fin and rudder, and may have been intended to shorten the take-off run or to provide acceleration after the bombing attack.

(3) the third proposal, the asymmetric single-seat P 179 dive-bomber powered by a single BMW 801D radial in the fuselage nose, was largely identical in overall appearance to the P 177 but was dimensionally smaller, having a span of 10.4m (34ft 1½in) and wing area 18m² (193.74ft²). Armament consisted of two forward-firing MG 151/20s in the offset starboard crew nacelle, with provision for the carriage of a single bomb beneath it.

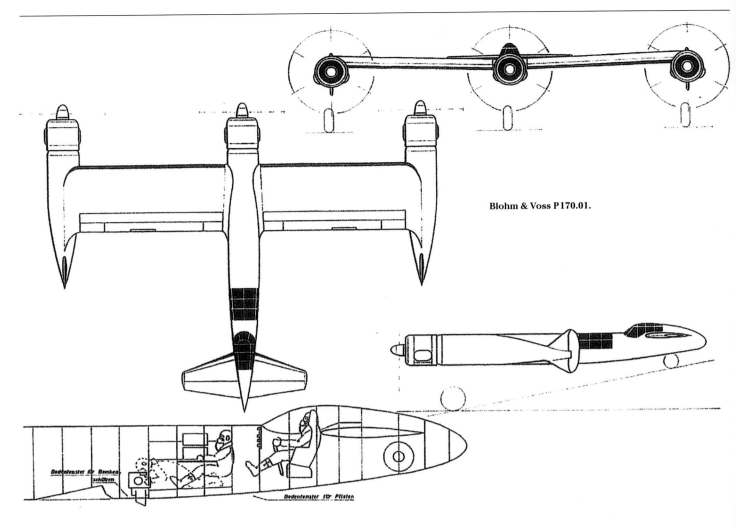

Blohm & Voss P170.01.

Blohm & Voss P170.01 – data of 17.9.1942

Taken from the Blohm & Voss brochure submitted to the RLM.

Powerplant   3 x 1,860hp BMW 801D (or BMW 801E) radials.
Dimensions, weights and performance relate to the BMW 801D.

Dimensions

Span	16.00m	52ft 6in
Length	13.00m	42ft 7¾in
Height	3.65m	11ft 11¾in
Wing area	44.00m²	473.60ft²

Weights

Empty weight	9,100kg	20,062 lb
Loaded weight*	13,300kg	29,321 lb

Performance

Max speed	820km/h at 8,000m	510mph at 26,250ft
Range at 6,000m (19,685ft)*	2,000km	1,242 miles
Service ceiling	10,400m	34,120ft

Bombload   1 x SC1000 or 2 x SC 500 or 4 x SC 250 normal
2 x SC1000 or 4 x SC 500 at overload

*These figures are with a 1,000kg (2,205 lb) bombload.

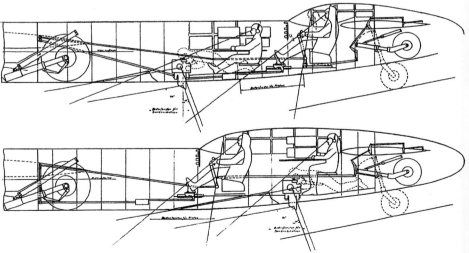

Blohm & Voss P170.01 crew seating proposals.

Focke-Wulf P VII

1943 to 1944

Focke-Wulf P VII full-scale mock-up of 1944.

According to Focke-Wulf drawing 0310226-94, the Focke-Wulf P VII project was laid out as a single-seat Jagdflugzeug (fighter), and in a further variant, as a single-seat fighter and ground attack aircraft powered by a turbojet plus rocket motor. Armed with powerful weapons, the P VII was intended to combat enemy tanks and provide infantry support.

This December 1943 study, submitted to the RLM by the Focke-Wulf Project Office under the leadership of Chief Designer Dipl-Ing Ludwieg Mittelhuber, resembled the de Havilland DH 100 Vampire developed in England at the same time. Like the latter aircraft, it also had its powerplant enclosed in the rear of the fuselage, fed by wingroot air intakes. One variant envisaged the installation of a 5cm MK114A (BK 5 Bordkanone) in the fuselage. Of twin-boom layout, the twin fins and rudders sup-

ported a high-set tailplane and elevator. To provide an increase in performance, an auxiliary liquid-propellant HWK 109-509 rocket motor was situated beneath the Heinkel-Hirth HeS 011 turbojet. Doubts expressed by the RLM concerning the novel arrangement of the wingroot air intakes were dispelled in air-intake test-bed and wind-tunnel investigations. Furthermore, the RLM was inclined to prefer a multi-engined aircraft for this task. It was only when Prof Dipl-Ing Kurt Tank convinced the RLM of the efficacy of the single-engined layout that documentary data for the BMW 003, Junkers-Jumo 004 and Heinkel HeS 011 turbojets were made available.[2] Although the full-size mock-up of the P VII had been inspected at the end of 1944, the RLM showed no inclination to place a construction contract.

Right: Another view of the Focke-Wulf P VII full-scale mock-up of 1944.

Centre right: Focke-Wulf P VII works drawing.

Focke-Wulf P VII – data

Powerplant	1 x 1,300kg (2,866 lb) thrust Heinkel-Hirth 109-011 turbojet and one HWK 109-509 rocket motor	
Dimensions		
Span	8.00m	26ft 3in
Length	10.55m	34ft 7½in
Height	2.35m	7ft 8½in
Wing area	17.00m²	182.98ft²
Weights		
Loaded weight	4,750kg	10,472 lb
Performance		
Max speed	935km/h at 9,000m	581mph at 29,530ft
Service ceiling	13,700m	44,950ft
Armament		
2 x MG 213, 2 x MK 103, 3 x MK 103 upward oblique, 1 x MK 114A (BK 5)		

Several variants of this project existed, each with the same wingspan but of varying lengths, wing areas, rocket fuel weights and armament. Two other proposals, one of 1.2.44 and the other of 5.7.44, each of 8m span and 10.55m length, had wing areas of 15.5m² (166.84ft²) and 14m² (150.69ft²) respectively. Loaded weight for the first was as quoted, but maximum speeds varied from 785km/h (487mph) at sea-level to 825km/h at 9km (516mph at 29,530ft) at mean weight 3,750kg (8,267 lb). Initial climb rate was 20.5m/sec 4,035ft/min); range varied from 470km at 12km (292 miles at 39,370ft) to 890km at 6km (553 miles at 19,685ft). Endurance varied from 46 mins at 615km/h (382mph) to 64 mins at 890km/h at 6km (553mph at 19,685ft). Armament alternatives were two MG151/20s (2 x 175rpg) in fuselage nose plus two MK108s (60rpg) in tailbooms, or one MK103 (80 rds) in fuselage nose plus two MG 213s (2 x 120rpg) in tailbooms, or four MG 213s or two MK108s (60rpg) and two MG151/20s (175rpg). On the wooden mock-up, the target sighting device ahead of the cockpit windscreen was a ZFR 4a for the MK103, or a Revi 16C for the other types of weapons.

² This statement is difficult to comprehend as the Focke-Wulf design offices, like others in the aircraft industry, were obviously aware at that late date of at least preliminary design and performance data on the above three turbojets, since the firm's earliest single-engined jet fighter designs of 1942 and 1943 were planned to use the BMW P3302 and Jumo T1 (the precursors of the 109-003A and 109-004A). Even the P.01-100-series of jet-powered designs from Abteilung L at Messerschmitt dating from spring 1939 show that Dr Alexander Lippisch and his co-workers were aware of the much earlier Junkers, and BMW F9225 turbojet developments that were later abandoned. The dotted outline of the BMW P3302, P3304 and the early HWK rocket motors in the Lippisch projects is clearly recognisable as is the BMW P3304 in the Messerschmitt P1073B of 13th August 1940. The cutaway perspective drawing of 14th March 1944 shows it with the outline of the HeS11V6 prototype. The last known document of this Focke-Wulf jet fighter (usually described as the P VI 'Flitzer' but not designated as such in *Luftwaffe Secret Projects: Fighters 1939-1945*, pp.143-144) was dated 11th December 1944. The HeS 021 turboprop-powered variant is usually referred to as the P VII 'Peterle'.

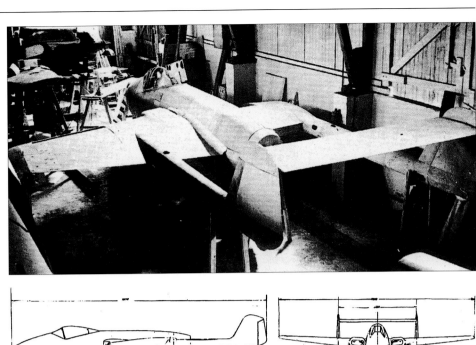

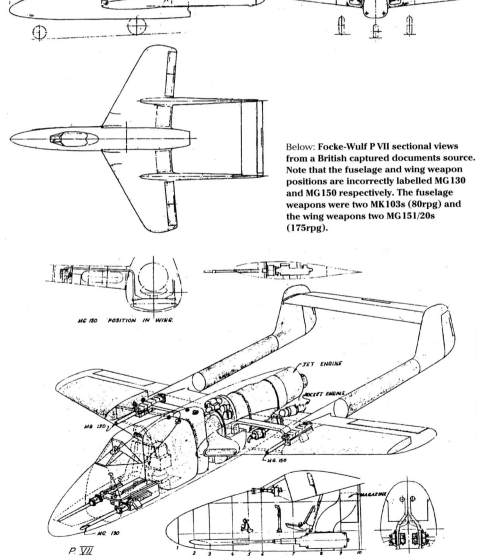

Below: Focke-Wulf P VII sectional views from a British captured documents source. Note that the fuselage and wing weapon positions are incorrectly labelled MG 130 and MG 150 respectively. The fuselage weapons were two MK 103s (80rpg) and the wing weapons two MG 151/20s (175rpg).

Henschel PJ 600/67

Under the leadership of Henschel Chief Designer Dipl-Ing Friedrich Nicolaus, this single-seat canard project dating from 1941/42 was to have been employed as a low-level and ground attack aircraft.

On the basis of the Hs P 87 described earlier, the aim was to produce an aircraft capable of carrying a 2,000kg (4,409 lb) bombload. Dispensing with an undercarriage, when operated from land or ship-based, the aircraft was to take-off with the aid of a catapult but a Mistelschlepp towed take-off was also planned. The project underwent extensive wind-tunnel testing, but despite good measurement results and considerably advanced in design, the RLM ordered work on it to be stopped.[3]

[3] In a brief reference to this aircraft which he calls the Hs P 87 canard of September 1943, Manfred Griehl: *Jet Planes of the Third Reich – The Secret Projects, Vol.1*, Monogram Aviation Publications, Sturbridge, Mass., 1998, p.122, states that most construction details were seized by Soviet forces in 1945, and that wind-tunnel studies which were to have been carried out in late 1944 were cancelled by the Chef TLR as no research vacancy could be made available. Also, as large portions of the Argus pulsejet plants were destroyed during bombing raids, the project was thus doomed. In his data table on p.186, the aircraft is listed as the P 187 day fighter powered by two Argus As 044 pulsejets and armed with four MK108s beneath the fuselage, but other than a similar three-view drawing, no other details are given.

An almost identical three-view drawing to that provided in this volume also appeared in Heinz J Nowarra: *Die deutsche Luftrüstung, Vol.3*, Bernard & Graefe Verlag, Koblenz, 1993, p.38, in whose data

table the author lists the powerplants as two 500kg thrust As 044 pulsejets, span 11.7m, length 8m and height 3.52m, surmising that the armament in the underfuselage bay may have consisted of two MG151/20s. Whether either of the quoted sets of dimensions has any relation to fact is unclear, save that all published drawings, apparently based on a scale model, have a much longer fuselage that does not match in scale with the wingspan.

Henschel PJ 600/67 – data

Powerplant	2 x 410kg (904 lb) thrust Argus As 014 pulsejets	
Dimensions		
Span	12.10m	39ft 8½in
Length	14.80m	48ft 6¾in
Height	2.80m	9ft 2¼in
Performance		
Max speed	810km/h	503mph
Armament	4 x MK108 in underfuselage bay	

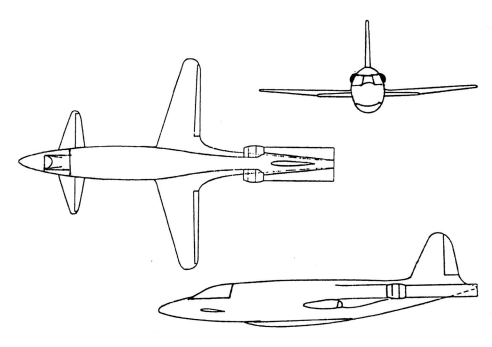

Junkers EF 112

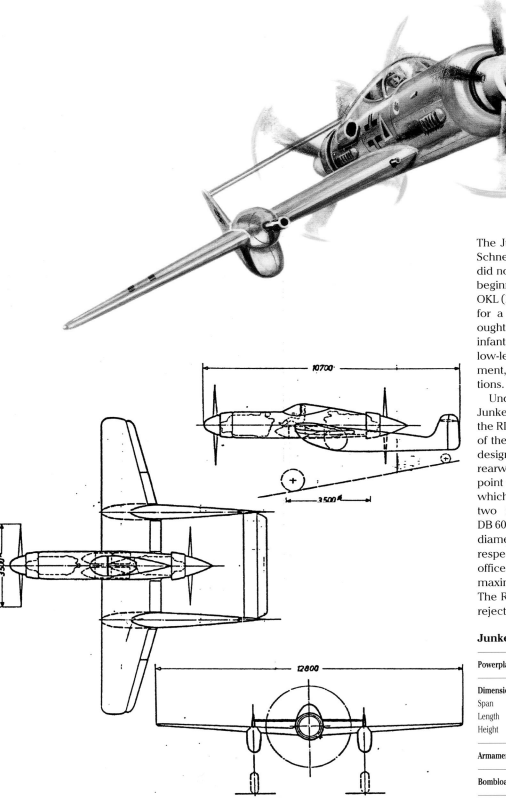

The Junkers EF 112 project for a single-seat Schnellstbomber (super-fast bomber) of 1942 did not pass the drawing board stage. At the beginning of that year, the RLM took up the OKL (Luftwaffe Supreme Command) demand for a Jagdbomber (fighter-bomber) which ought to be capable of supporting advancing infantry by releasing its medium bombload in low-level flight and with its cannon armament, attack enemy positions and concentrations.

Under the date of 11th December 1942, the Junkers-Werke submitted its EF 112 project to the RLM, but it did not meet the expectations of the responsible authority. The EF 112 was designed as a low-wing aircraft housing the rearward-retracting mainwheels of its four-point tail-down undercarriage in twin booms which supported the high-set tailplane. The two fore and aft-mounted Daimler-Benz DB 603G engines each drove 3.5m (11ft 5¾in) diameter airscrews as tractors and pushers respectively, and that according to design office data, were to provide the aircraft with a maximum speed of up to 760km/h (472mph). The RLM, however, showed no interest and rejected the proposal.

Junkers EF 112 – data

Powerplant	2 x 1,460hp Daimler-Benz DB 603G in-line engines	
Dimensions		
Span	12.80m	42ft 0in
Length	10.70m	35ft 1¼in
Height	4.10m	13ft 1½in
Armament	2 x MK103 and 4 x RZ100	
Bombload	2 x 250kg (551 lb)	

Chapter Three

Ground Attack Aircraft 1944-1945

In the period between 22nd June 1941 – the opening of the attack on the Soviet Union and the end of 1943, the German Army and Luftwaffe were primarily occupied with combating air and ground targets on the Eastern front, where newer means needed to be found to overcome the necessity to penetrate the heavily-armoured Soviet tanks and field artillery that were met at the front. Other than conversion packs with heavier-calibre weapons fitted to existing aircraft types such as the Ju 87 and others mentioned earlier on, no completely new Schlachtflugzeuge designs had entered production. With the anticipated Allied invasion of the continent, however, a new impetus was provided for ever newer drawing board projects for low-level and ground attack aircraft, the most proliferous of which were those by Blohm & Voss. The intensity of Allied bombing raids on the homeland as well as the increasing Allied air superiority caused a change in emphasis to the priority production of fighter and fighter-bombers that was furthered by the establishment of the Jägerstab (Fighter Staff), the TLR (Technical Air Armament Board) and the Jägernotprogramm (Fighter Emergency Programme) in 1944. As the opportunity for attacking land, sea, and air targets still presented itself on both the Eastern and Western fronts with newer, precision-aimed and more effective destructive weapons, various jet-propelled projects began to replace the slower piston-engined designs, as reflected in the latest projects worked on until the end of the war.

Blohm & Voss P 192.01

1944

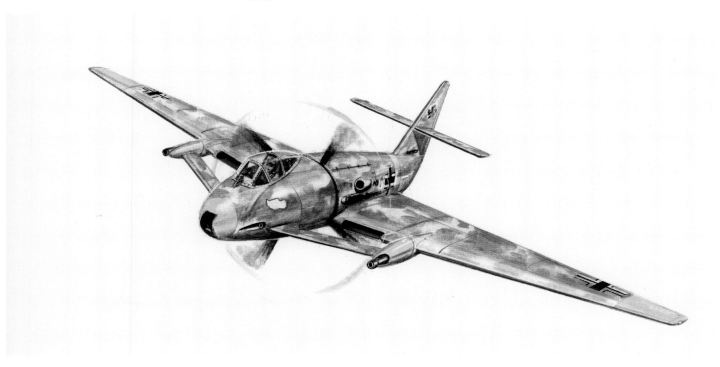

Dr-Ing Richard Vogt's efforts to remove the powerplant from the fuselage nose to improve visibility and provide installation facilities for weapons is reflected in the single-seat P192.01 of February 1944. In an aircraft of symmetrical appearance, Vogt moved the motor and airscrew to the c.g. behind the cockpit which was completely separated from the fuselage proper. This concept was only employed in the C-10 Motorsegler (powered glider) of the FAG Chemnitz, which probably had a degree of influence on the P192.01. This novel arrangement of the engine and propeller was sufficient to cause the RLM to reject the proposal, especially as no data on its efficiency was available.

The project was designed as a Stuka (dive-bomber) and Schlachtflugzeug (ground attack aircraft). The forward fuselage, cockpit and side armament were supported on two outriggers connected to booms projecting from the leading edges of the low-positioned wings. Proposed powerplants were the Daimler-Benz DB 603G for the P192.01 or the more powerful Junkers-Jumo 213E for the P192.02, both with wingroot radiator intakes and with the propeller hubs covered by a ring

which formed the fuselage outer contour at that point. Wing leading edge was unswept, but the trailing edge had marked taper towards the tips. The presence of the propeller necessitated a nosewheel tricycle undercarriage, the nosewheel retracting rearwards and the mainwheels inwards into the wing roots. Normal armament could be considerably strengthened with the addition of a 5cm MK114.

Blohm & Voss P192.01 and 02 – data

Powerplant	1 x 1,900hp DB 603G (P192.01) or	
	1 x 2,150hp Jumo 213E (P192.02)	
Dimensions		
Span	13.20m	43ft 3¾in
Length	11.80m	38ft 8½in
Height	3.10m	10ft 2in
on ground	4.30m	14ft 1¼in
Wing area	26.20m²	282.01ft²
Weights		
Loaded weight	5,700kg	12,566 lb
Armament	2 x MK108 4 x MG151/20 1 x MK114	

The two variants differed somewhat in dimensions and weights, the P192.01 having a constant-chord tailplane and that of the slightly smaller P192.02 equi-tapered towards the tips. In an Air Min Report, span given is 13m (42ft 7⅞in) and wing area 25.4m² (273.4ft²); with 600 litres (132 gals) of fuel and one 500kg (1,102 lb) bomb, all-up weight is 5,735kg (12,643 lb). Armament is given as two MK103s (70rpg) in the wing booms and two MG151/20s (250rpg) in the fuselage nose. Prior to release, the normal bombload, which could consist of two SC 250s or one SC 500, or one SC1000 at overload, was extended to clear the propeller disc. In the slightly smaller P192.02, the DB 603G engine was such a tight fit in the slim fuselage that it was necessary to provide a dorsal bulge to clear the accessories. The performance of both variants, however, was disappointing.

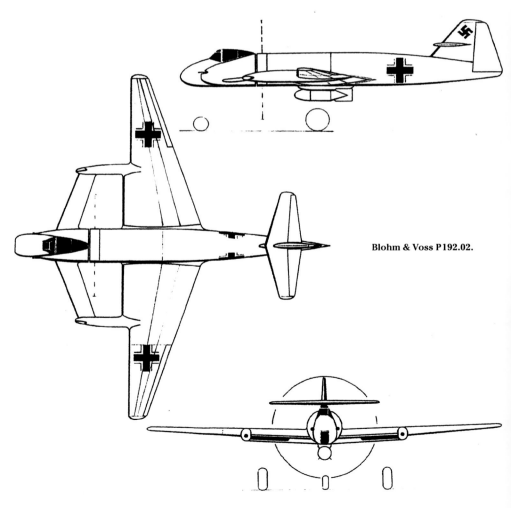

Blohm & Voss P192.02.

Blohm & Voss P 193.01

1944

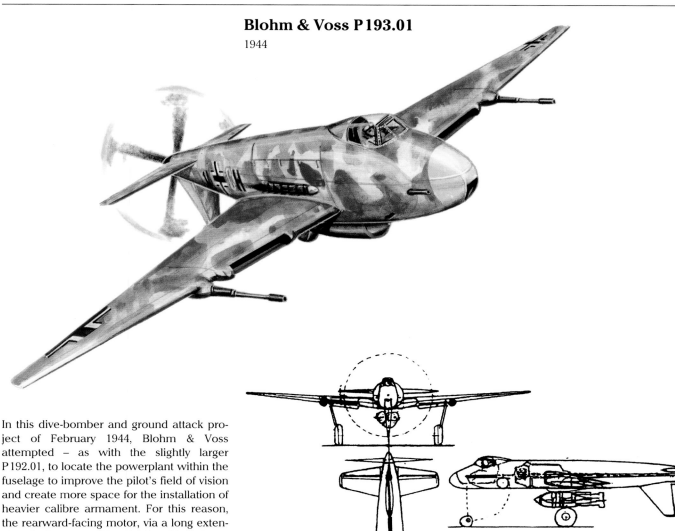

In this dive-bomber and ground attack project of February 1944, Blohm & Voss attempted – as with the slightly larger P 192.01, to locate the powerplant within the fuselage to improve the pilot's field of vision and create more space for the installation of heavier calibre armament. For this reason, the rearward-facing motor, via a long extension shaft, was to drive the four-bladed propellers protected on take-off and landing by the ventral fin and rudder. The low-winged aircraft of monocoque construction had sheet metal skinning which served simultaneously to provide armour protection. The wing planform, identical to its forbear, had an unswept leading edge and tapered 22° forward from root to tip. As before, the radiator air intakes were housed in the extended inboard wing leading edges behind which the mainwheels retracted ahead of the main spar box. The space between the mainwheels served as the suspension point for the underfuselage bombload. But even this improved concept was rejected for paltry reasons by the RLM.

Blohm & Voss P 193.01 – data

Powerplant	1 x 1,750hp Junkers-Jumo 213A-1 in-line engine	

Dimensions		
Span	11.40m	37ft 4⅞in
Length	10.32m	33ft 10¼in
Height, on ground	3.90m	12ft 9½in
Wing area	20.00m²	215.92ft²

Weights		
Loaded weight	5,700kg	12,566 lb

Performance		
Max speed*, at sea level	480km/h	298mph
at 7,000m (22,965ft)	570km/h	354mph
Rate of climb*, at sea level	10.8m/sec	2,126ft/min
at 7,000m (22,965ft)	5.5m/sec	1,082ft/min

Armament	4 x MK103 and 2 x MK108

Bombload	2 x SC 500 or 1 x SC1000

* Both at climb and combat power rating. Absolute maximum was 640km/h (398mph) at that height. Due to the very low climb rates, the project proved unattractive.

Blohm & Voss P194.01

1944

Blohm & Voss P194.01 – data

Powerplants   1 x BMW 801D of 1,700hp at 2,700rpm or 2,250hp with C3 fuel and MW 50 boost, plus 1 x Jumo 004C of 1,220kg (2,690 lb) reheat static thrust or 1 x BMW 003A of 800kg (1,764 lb) static thrust.

Dimensions

Span	15.30m	57ft 10in
Length	11.80m	38ft 8½in
Height	3.64m	11ft 11¼in
Wing area	36.40m²	391.80ft²

Weights

Empty weight, with 003A	6,500kg	14,330 lb
Fuel, BMW 801D	730kg	1,609 lb
BMW 003A	1,000kg	2,205 lb
Loaded weight, 1 x 003A	9,150kg	20,172 lb
1 x 004C	9,330kg	20,569 lb

Performance

Max speed , with TL	640km/h at 2,000m	398mph at 6,560ft
with TL	675km/h at 8,000m	419mph at 26,250ft
with TL plus MW 50	715km/h at 8,000m	444mph at 26,250ft
without TL	540km/h at 6,000m	336mph at 19,685ft
Climb rate		
combat power, at take-off	15.0m/sec	2,953ft/min
with MW 50, at take-off	19.2m/sec	3,779ft/min
Max range, at cruising speed	1,070km at 6,000m	665 miles at 19,685ft
Service ceiling	11,100m	36,420ft
Take-off run	520m	1,706ft
Landing speed	150km/h	93mph

Armament   2 x MK103 (70rpg) and 2 x MG151/20 (250rpg)

Bombload   9 x SC70 or 2 x SC 250 or 1 x SC 500 normal
1 x SC1000 at overload, all in fuselage bomb-bay

All performance figures are quoted at mean take-off weight 7,600kg (16,755 lb). The take-off weight is with 500kg (1,102 lb) bombload; at overload, take-off weight was 9,650kg (21,274 lb).

Developed for a time parallel to the P192 and P193 studies, this asymmetric proposal begun in late February 1944 was laid out to be capable of use as a Zerstörer, Stuka (dive-bomber), Aufklärer (reconnaissance) and Schlachtflugzeug (ground attack) aircraft. As a follow-on to the asymmetric BV141, the P194.01 was conceived, so to speak, as the 'final solution' to the multi-purpose concept. The idea of the asymmetric aircraft ran like an unbroken thread through a number of the firm's projects developed under the leadership of Dr-Ing Richard Vogt, and included the P177, P178, P179 and P204, as well as the BV237 which reached the mock-up stage at the end of the war. The principal advantage of this layout was to provide the pilot of a single-engined aircraft with a much improved field of vision and give the concentrated armament a field of fire unobstructed by the engine.

The arrangement of the powerful motor in the fuselage nose coupled with a turbojet behind or beneath the offset cabin console was an audacious innovation which soon earned the nickname of 'side-car' aircraft. With this, Vogt sought to compensate the inner unsymmetrical airflow pattern associated with a single-engined aircraft by giving it an asymmetrical layout. Flight trials with the BV141 confirmed the theoretical advantages of this layout. On the other hand, the asymmetrical location of the powerplants offered the possibility of counteracting the asymmetric flow pattern caused by the propellers to be reduced along the longitudinal axis by the addition of the offset jet engine, which at take-off, had less than half of the longitudinal moment of the BV141. As a result of the new weight distribution, the lateral distance of the crew cabin from the aircraft's c.g. from 1.8m (5ft 10¾in) on the BV141 was reduced to just 1.45m (4ft 9in) on the P194.01. The arrangement of the two powerplants and enclosure of the bombs in the fuselage, and the fact that unlike a conventional twin-engined aircraft where three drag bodies are necessary, there are only two in this case, so that its high performance at low level exceeds even that of the well-known Do335 Ameisenbär (Ant-eater). The P194.01, unfortunately, did not become a reality, for in many respects it would have been one of the most interesting aircraft of its time.

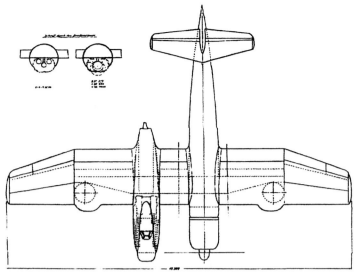

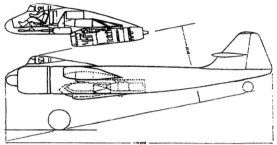

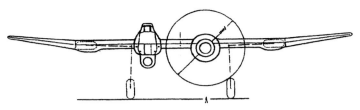

Blohm & Voss P194.01 with one BMW 003A turbojet. The firm's artist inadvertently omitted the empennage surfaces in the head-on view.

Blohm & Voss BV141 prototype.

Stability and Control Processes with Orthodox and Asymmetrical Aircraft

Right: The illustrations show the fluid-mechanical flow patterns and the forces which result on the flying stability between conventional aircraft (upper) and the asymmetrical BV141 (lower). The reaction of the clockwise-rotating airscrew results in the pitching moment (K) which tends to turn the aircraft anti-clockwise around its longitudinal or roll axis. At the same time, the airscrew twist effect (torque) through the c.g. acts on the rudder and turns the aircraft about its vertical or yaw axis. The tendency to enter a left bank can only be offset by trimming of the elevator and ailerons. Similar forces were also experienced with the BV141. Due to the compensating weight of the offset crew nacelle on the right, the c.g. was displaced to a point outside the fuselage. The nacelle weight (P) and its extra drag (W) offset the pitching moment (K) and the rudder force (Q). A balance of forces is thus created and the machine flies in a stable manner.

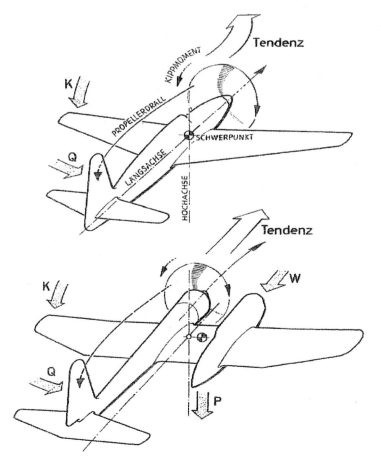

Works illustration of airflow, stability and control patterns on normal and asymmetrical aircraft. Key: Tendenz = tendency, Kippmoment = pitching moment, Propellerdrall = airscrew twist effect, Schwerpunkt = centre of gravity, Längsachse = longitudinal axis, W = drag, K, Q, P = directional forces.

Blohm & Voss P 196.01

1944

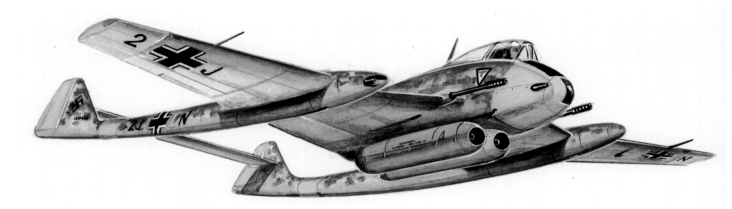

This project for a single-seat dive-bomber and ground attack aircraft dating from April 1944 was noticeable for the wide distance between its twin booms supporting the high-positioned tailplane. Its twin turbojets were housed in a paired nacelle at the rear of the short fuselage projecting ahead of the trapezoidal wing centre section. The mainwheels of the wide-track undercarriage retracted inwards into the wing roots, a single tailwheel retracting into each of the booms beneath the fins. Due to the tail-sitter arrangement on the ground, the pilot was provided with a forward and downward-vision panel in the extreme nose beside which were located the lower pair of forward-firing cannon, the remaining pair being housed in the widened fuselage cheeks beside the cockpit. The internal bombload was accommodated in the forward portion of the twin booms. Despite its favourable layout, this submission by Dr-Ing Richard Vogt failed to meet with the approval of the RLM Technisches Amt.

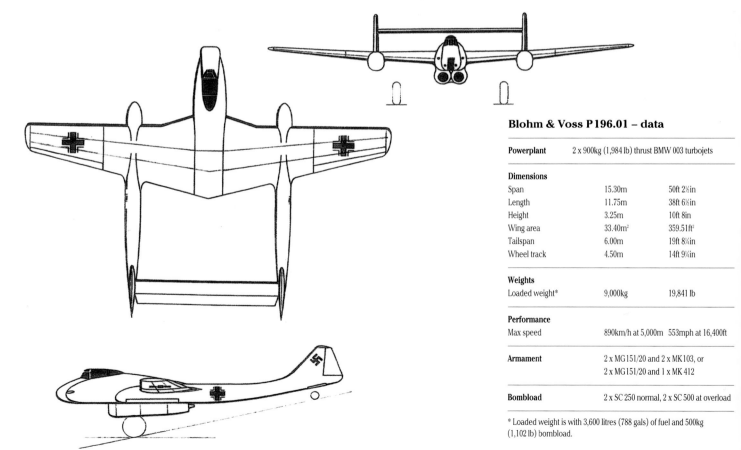

Blohm & Voss P 196.01 – data

Powerplant	2 x 900kg (1,984 lb) thrust BMW 003 turbojets	
Dimensions		
Span	15.30m	50ft 2½in
Length	11.75m	38ft 6½in
Height	3.25m	10ft 8in
Wing area	33.40m²	359.51ft²
Tailspan	6.00m	19ft 8¼in
Wheel track	4.50m	14ft 9¼in
Weights		
Loaded weight*	9,000kg	19,841 lb
Performance		
Max speed	890km/h at 5,000m	553mph at 16,400ft
Armament	2 x MG151/20 and 2 x MK103, or 2 x MG151/20 and 1 x MK 412	
Bombload	2 x SC 250 normal, 2 x SC 500 at overload	

* Loaded weight is with 3,600 litres (788 gals) of fuel and 500kg (1,102 lb) bombload.

Blohm & Voss P 204.01

1944

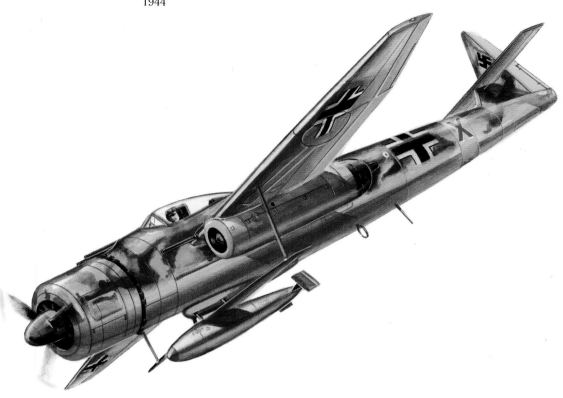

Blohm & Voss P 204.01 with underslung BV 246 glide-bomb.

This single-seat dive-bomber and ground attack project submitted by Dr-Ing Richard Vogt to the RLM in June 1944, like that of the earlier P194.01, also featured a combination of fuselage nose-mounted BMW 801D piston engine and an underslung wing-mounted turbojet, either a Jumo 004 or BMW 003 which could be attached as a Rüstsatz beneath the port wing. In his submission dated 23rd June 1944 to Oberst Siegfried Knemeyer, head of the Aircraft Development Group at the Chef TLR in the RLM, the turbojet was expected to raise maximum speed from 686km/h (426mph) to 730km/h (454mph). Through this increase, as Vogt explained further, the P 204.01 would have approached the performance and operating regime of the Messerschmitt Me 262, its more favourable take-off and landing characteristics in comparison to the Me 262 being the subject of further variants. A mid-wing design of all-metal construction, it was to have been simpler to manufacture and be capable of becoming operational within a very short time so as to be of early assistance at the front. Special features of the design were that in addition to its two fuselage guns mounted above the engine, it had a pair of guns buried completely within the wing firing outside the pro-

peller disc, the wide-track mainwheels retracting outwards towards the tapered wing outboard sections, the constant-chord horizontal tailplane mounted on a special step ahead of the fin and rudder. In addition to its bomb load for the ground attack role, provision was made for the carriage of a BV 246 glide bomb beneath the fuselage, its narrow-chord long-span wings being situated in line with the aircraft's wing leading edge just beneath the turbojet nacelle. This project was the last of a long line of design studies which, with its offset wing-mounted turbojet, counted among the asymmetric layouts.

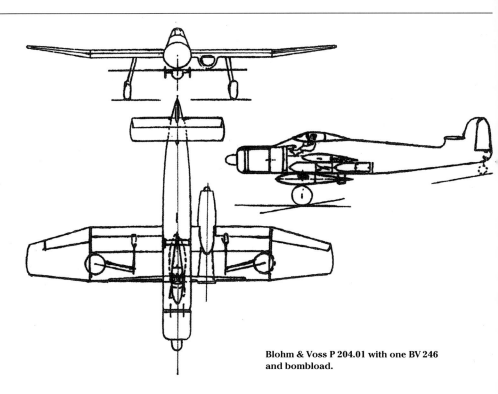

Blohm & Voss P 204.01 with one BV 246 and bombload.

Blohm & Voss P 204.01 – data

Powerplant   1 x 1,700hp BMW 801D and 1 x 900kg (1,984 lb) thrust Jumo 004 or 1 x 800kg (1,764 lb) thrust BMW 003A turbojet

Dimensions

Span	14.30m	47ft 0¼in
Length	12.45m	40ft 10¼in
Height	4.10m	13ft 5½in
Wing area	33.70m²	362.73ft²

Weights

Loaded weight*	8,500kg	18,739 lb
with fuel	1,600kg	3,527 lb
and bombload	500kg	1,102 lb

Performance

Max speeds*, at sea level	650km/h	404mph
with MW 50	755km/h at 8,000m	469mph at 26,250ft
without TL at sea level	550km/h	320mph
without TL	575km/h at 8,000m	358mph at 26,250ft
Rate of climb*		
at take-off	15.0m/sec	2,953ft/min
at 8,000m (1,240ft)	6.3m/sec	1,240ft/min
Range* at sea level	530km	329 miles
at 6,000m (19,685ft)	800km	497 miles
Service ceiling*	9,700m	31,825ft
Take-off run	550m	1,804ft
Landing speed	146km/h	91mph
Landing weight	6,720kg	14,815 lb

Armament   2 x MG 151/20 and 2 x MK 101

Bombload   2 x SC 250 or 1 x SC 500 or 1 x BV 246

* Maximum speeds are at half-fuel mean weight 7,200kg (15,873 lb) with BMW 801D with MW 50 boost and BMW 003A unless otherwise stated. Rate of climb, range and service ceiling are at loaded weight and climb and combat power. Landing weight is at 20% remaining fuel and full ammo less bombload.

Blohm & Voss BV 237

1943 to 1944

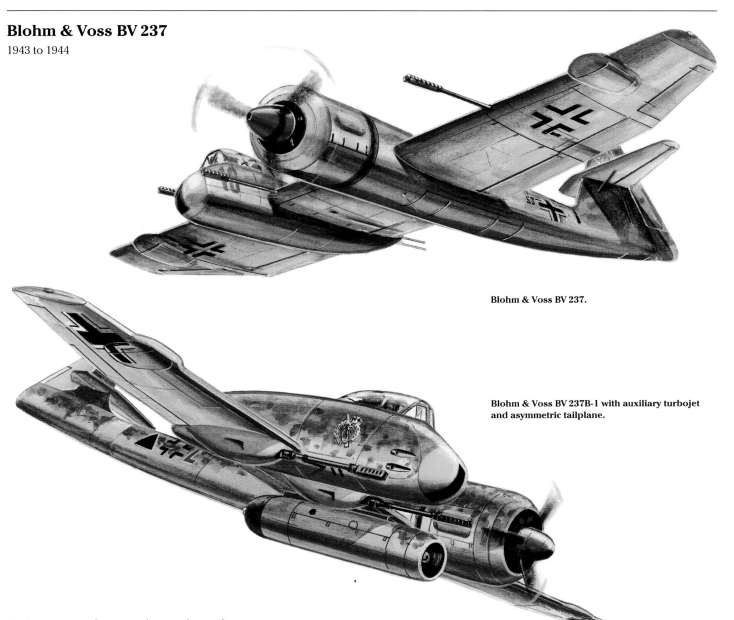

Blohm & Voss BV 237.

Blohm & Voss BV 237B-1 with auxiliary turbojet and asymmetric tailplane.

In the course of progressive work on the development of asymmetric aircraft, the design team headed by Dr-Ing Richard Vogt drew up the design of what became the BV 237 single-seat dive-bomber and ground attack aircraft in early 1943, based directly on the experience gained with the BV 141 whose ancestry is unmistakable. Unlike succeeding projects, this proposal did receive the approval of the RLM and was awarded a construction contract. Although the Blohm & Voss facilities remained undamaged from the effects of severe Allied bombing raids on Hamburg in the summer of that year, a development stop was placed on the firm by the RLM for several of its on-going projects, including BV 237. Detail design of the BV 237 nevertheless continued after a pause, and had reached an advanced stage when work on the similar P 194.01 had started.*

As in standard Blohm & Voss practice, the hollow steel-tube main spar which served simultaneously as a fuel tank, passed through the fuselage up to the position of the outward-retracting mainwheels which rested in bulges in the tapered outboard wing sections. By using a mixture of wood and steel, the proportion of light metal in the aircraft's construction was reduced by over 65% and by further simplification for manufacturing purposes, would have been able to commence series production in mid-1945, following approval from the TLR. By the end of 1944, however, there was no longer an overriding need for a dive-bomber and ground attack aircraft, so that the planned initial 0-series aircraft were not built.

In keeping with some of the earlier asymmetric designs, the BV 237A was to have been powered by a 1,750hp BMW 801D radial, the offset completely armour-plated starboard nacelle housing a crew of one in the Stuka role, and two in the ground attack role. The main fuselage fuel tank, built integral with the forward fuselage, housed 900 litres (198 gals) of fuel that enabled a range of 1,000km (621 miles) to be attained. With an additional 1,000 litres (220 gals) in the hollow main spar, range was increased to 2,000km (1,242 miles). In addition to its underslung wing weapon loads, the proposed BV 237B-1 version was to have had an auxiliary Jumo 004B turbojet mounted in a nacelle between the fuselage and the offset crew nacelle. With this aircraft and the other asymmetric projects previously mentioned, the firm concluded its series of 'side-car' developments, concentrating during the last year of the war – with few exceptions, on pure jet-propelled fighter and fighter-bomber proposals.

* In his German text, the author wrote that the BV 237 'could be regarded as a further development of the BV P 194.' The head of the Konstruktionsabteilung (Design Department) at Blohm & Voss between 1940 and 1945, Hermann Pohlmann, wrote in his book: *Chronik eines Flugzeugwerkes (Chronicle of an Aircraft Company)*, Motorbuch Verlag, Stuttgart, 1979, p.158, that in a telegram of 15th June 1943, the RLM proposed that Blohm & Voss take over the further development and manufacture of the Me 264 as the Messerschmitt firm was fully overloaded with work important for the war effort. Quoting an immediate reply from Dr-Ing Richard Vogt to the RLM GL/C-E2, subject 'Takeover of the Me 264', Dr Vogt stated that the assumption of this task would require an estimated 40,000 hours of additional effort from the Design Office in the following months, and that this figure would in practice be exceeded several times over, requiring at least 50 engineers to be engaged on this for six months. The design office would thus not be available for any other new tasks and that the RLM would have to 'completely relinquish' the BV 237 as well as the P 184 (four-motor long-range reconnaissance aircraft) project which was superior to the Me 264 in all respects. Six weeks later, although the firm's premises remained undamaged from the devastating Allied bombing raids on Hamburg, there came a total stop to previous design and production plans for the BV 222, BV 238, BV 237 and P 184. The Me 264 take-over proposal also fell by the wayside. This clearly confirms that the BV 237 designation and a construction contract existed long before the P 194 took shape on the drawing board. Furthermore, Pohlmann stated on p.81 of his book that although only a few examples of the BV 141 had been built, the asymmetric concept was continued in the design of the Jabo P 177, the Stukas P 178 and P 179, multi-purpose P 194, and again for the Stuka BV 237 which had received construction approval during a Führer Conference at the Obersalzburg. Design and mock-up construction was taken up but due to the war situation or else disagreement in the High Command, work on it was again stopped. In a

final remark, Pohlmann says that 'In its structural composition, the BV 237 was completely in accordance with that of the BV 141.' (ie not the P 194.)

In the project description dated March 1944 for the P 194 submitted to the RLM, the brochure mentions that the 'asymmetrical aircraft ...offers particular possibilities' and that 'We should like to recall the time-consuming but at that time very satisfactory performance of the BV 141 Nahaufklärer (close-range reconnaissance aircraft) and the to a large extent completely-designed BV 237 Stuka project.' – a further confirmation that the BV 237 had existed long before the P 194.

Blohm & Voss BV 237 – data

Powerplant	1 x 1,750hp BMW 801D radial and 1 x 900kg (1,984 lb) thrust Jumo 004B in the BV 237B-1	
Dimensions		
Span	15.30m	50ft 2½in
Length	11.45m	37ft 6¾in
Height	5.20m	17ft 0¾in
Wing area	42.00m²	452.07ft²
Weights		
Loaded weight	6,680kg	14,727 lb
Performance		
Max speed	580-610km/h	360-379mph
As Stuka (dive-bomber)		
Armament	2 x MG151/20 forward-firing	
	2 x MG131 rearward-firing	
Bombload, normal	500kg	1,102 lb
at overload	1,000kg	2,205 lb
Loaded weight, normal	6,250kg	13,778 lb
at overload	6,700kg	14,771 lb
As Schlachtflugzeug (ground attack)		
Armament	2 x MG151/20 forward-firing	
	2 x MG131 rearward-firing	
	2 or 3 MK103 fixed forward	
	1 x MK101 as special Rüstsatz (proposed)	
Bombload, at overload	6 x SC70 inboard of wheel bays	
Loaded weight, normal	6,150kg	13,559 lb
at overload	6,680kg	14,727 lb
Max speed	580km/h at 6,000m	360mph at 19,685ft

Weights, armament and bombload figures are from Translator.

Blohm & Voss BV 237 as Stuka (left) and Ground Attack aircraft (right).

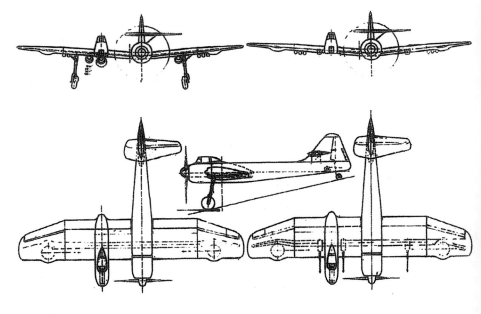

Heinkel 'Lerche'

1944 to 1945

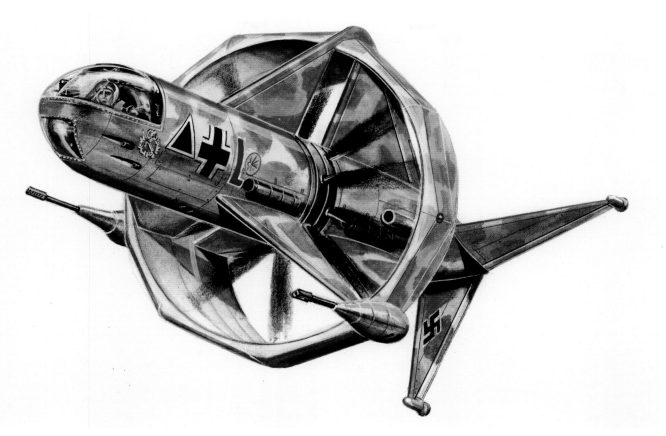

The 'Lerche' (Lark) series of project studies, begun in 1944 in Vienna by Heinkel designers Dr-Ing Kurt Reiniger and Dr-Ing Gerhard Schulz to serve as a single-seat prone-piloted VTOL ground attack aircraft, was of revolutionary concept. The 'Lerche' I to III proposals envisaged use of the aircraft as a light and heavy fighter as also in the ground attack role.

Aside from the numerous unsolved technical and aerodynamic problems that would not have made rapid development or even manufacture possible, by reason of its realistic assessment by its designers, was by no means 'an addled or soft-shelled egg,' as similar principles, based on German documents, were incorporated into prototype aircraft built and flown by the Allies after the war.

In order to increase propeller efficiency, Reiniger and Schulz adopted for the first time the use of an annular wing which served as a shroud for the counter-rotating propellers of the tandem engines. By incorporating adjustable control surfaces on the circular wing, the related thrust and efficiency of the propeller could be enormously increased. In the almost constant-diameter fuselage housing the prone pilot, powerplants, armament and fuel and supporting the three equi-

spaced fins and rudders at the rear, Reiniger and Schulz achieved a smooth aerodynamic form. Its VTOL capability conferred a degree of independence from being confined to a home airfield. The Lerche II, Entwurf C (Draft C) of 25th February 1945 depicts the overall layout and method of operation. A concluding report was dated 8th March 1945.

Heinkel 'Lerche' Start (take-off) and Landung (landing) sequence of 10th March 1945.

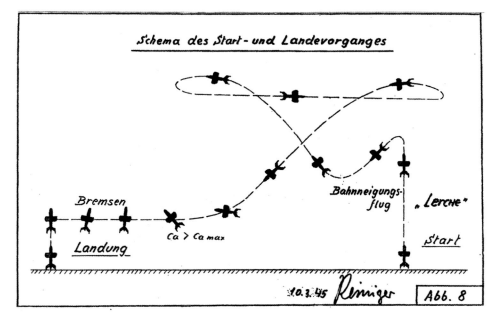

Heinkel 'Lerche II' – data

Powerplants	2 x 2,000hp Daimler-Benz DB 605 (Lerche II drawing) or 2 x 2,400hp DB 603E in-line engines.	

Dimensions

Propeller diameter	4.00m	13ft 1½in
Wing max width	4.55m	14ft 11in
Wing chord	1.50m	4ft 11.1in
Wing area	12.00m²	129.16ft²
Fuselage width	1.25m	4ft 1¼in
Fuselage length	9.40m	30ft 10in
Length, over wheels	10.00m	32ft 9¾in
Surface area	102.80m²	1,106.5ft²

Weights

Fuel weight	600kg	1,323 lb
Loaded weight	5,600kg	12,346 lb

Performance

Max speed	750-800km/h	466-497mph

Armament	2 x MK108 beside cockpit

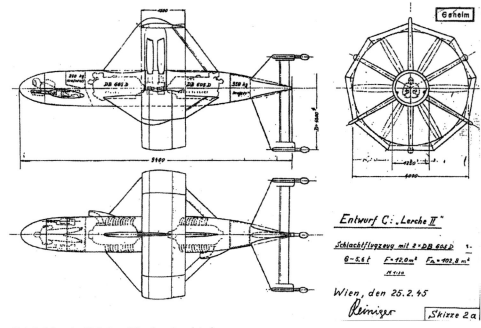

Heinkel 'Lerche II' Entwurf C – drawing dated
25th February 1945.

Henschel Hs 132

1944 to 1945

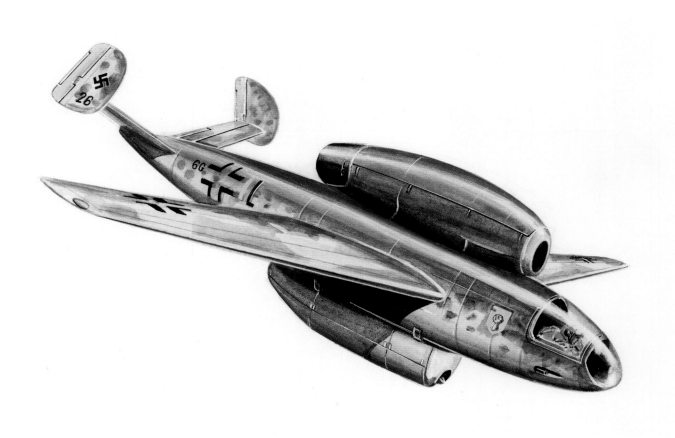

At the beginning of 1944 the Development Department of the Henschel Flugzeugwerke set itself the task of evolving a new dive-bomber and ground attack aircraft capable of attaining the highest horizontal and diving speeds. The P 123 project was officially designated Hs 132 in the RLM GL/C-E2 letter of 31st January 1944 and the firm received an order for the construction of six pre-production aircraft. Detail design, however, begun in May 1944, was 50% complete by 25th July 1944 and construction of component parts commenced.

The Hs 132 consisted of a cantilever mid-wing monoplane of all-metal monocoque construction with an almost circular cross-section fuselage. An armoured enclosure was to afford protection from enemy fire from below for the pilot in the extensively-glazed cockpit, protection from head-on fire being afforded by a 75mm (3in) thick armoured glass panel, itself surrounded by armoured steel plate. The dihedral tailplane and twin endplate fins and rudders was made of light metal, the wood-covered wings stressed for 12g loads.

The first two prototypes were to have been powered by a BMW 003 and Jumo 004 turbo-jet respectively. The Hs 132D variant, specially conceived for the low-level and ground attack role, had a larger wingspan of 9.1m (29ft 10½in), wing area 16m² (172.22ft²) and was to have been powered by the more powerful HeS 011 turbojet of 1,300kg (2,866 lb) static thrust.

At an Aircraft Development Committee meeting held on 21st/22nd November 1944, the Hs 132, also described as a fighter, was to have been the subject of discussion on the quantity to be produced. The Hs 132 V1, in process of completion in the spring of 1945, was expected to commence flight trials in May 1945. At the collapse, however, it was almost 95% complete, the V2 being 80% and the V3 being 75% complete, with the V4 to V6 prototypes in final assembly before being captured by the advancing Soviet troops.

Henschel Hs 132. The prototype was not fully completed when the Henschel plant was overrun by Soviet troops. G Heumann impression

Henschel Hs 132 – data

Powerplant	1 x BMW 003A (Hs 132A), 1 x Jumo 004B (Hs 132B), 1 x HeS 011A (Hs 132C) and 1 x HeS 011 (Hs 132D).	
Dimensions		
Span	7.20m	23ft 7½in
Length	8.90m	29ft 2½in
Height	2.95m	9ft 8in
Wing area	14.80m²	159.30ft²
Wheel track	4.32m	14ft 2in
Weights		
Loaded weight	3,400kg (7,496 lb) with 1 x 500kg bomb	
Performance		
Max speed, without bomb	800km/h at 4,000m	497mph at 13,125ft
	780km/h at 6,000m	485mph at 19,685ft
with bomb	710km/h at 4,000m	441mph at 13,125ft
Range, at 10,000m (32,800ft)	1,120km	696 miles
Service ceiling	10,500m	34,450ft
Endurance, with bomb	1 hr 20mins	
Weapons load		
Hs 132A Stuka	1 x SC 500 or 1 x SD 500 bomb	
Hs 132B Ground Attack	2 x MG 151/20 (250rpg)	
	and a 250-500kg bombload	
Hs 132C Ground Attack	2 x MG 151/20 (250rpg) plus	
	2 x MK 108 (60rpg) and a 1000kg bomb	

Bombload
A single semi-recessed underfuselage bomb could be carried, ranging from an SC/SD 250 to a maximum of one SC/SD 1000 bomb.

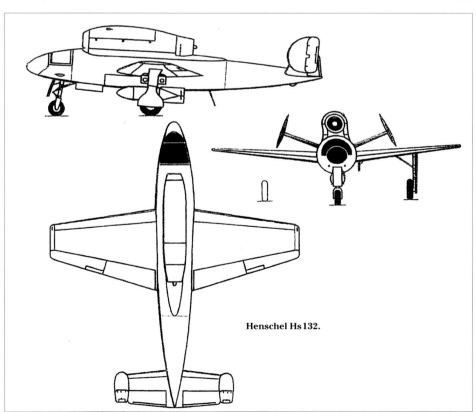

Henschel Hs 132.

Hütter Low-level & Ground Attack Project

1945

Among the proposals submitted by Dipl-Ing Wolfgang Hütter to the EHK or Entwicklungshauptkommission (Development Main Committee) in March 1945 was a project for a single-seat Erdkampf (close-support) and Schlachtflugzeug (ground attack) aircraft. Of mixed wood and metal construction, the shoulder-wing design with its spindle-shaped fuselage and butterfly tail was to have been powered by two Argus As 044 pulsejets attached to the fuselage and wing leading edges. According to the construction description, Hütter was able to reduce the destructive frequency-produced pulsejet vibrations to such an extent that no damage was sustained to the wings and fuselage. Take-off was to have been by the Starrschlepp (rigid pole-tow) method or with the aid of a take-off ramp on its wide retractable skid which would be extended for landing. Armament was to consist of two MK 108 cannon on either side of the fuselage nose and a battery of R4M Bordraketen (air-to-air rockets) beneath each wing.

Hütter Project – data

Powerplants	Two Argus As 044 pulsejets	
Dimensions		
Span	7.35m	24ft 1½in
Length	6.85m	22ft 5¾in
Weights		
Airframe weight	680kg (1,499 lb) w/o As 044 or weapons	
Loaded weight	not known	
Performance		
Endurance	40 mins	
Armament	2 x MK 108 24 x R4M or RZ 100	

Junkers EF126

1944 to 1945

At the beginning of 1944, the Junkers Flugzeugwerke evolved the design of a single-seat lightweight low-level attack aircraft to combat enemy tanks and provide infantry support. Laid out in both high- and mid-wing configurations, the EF126 had wings fabricated of wood, the latter third being fabric covered. The two wing halves containing spanwise fuel tanks, were attached to the fuselage by a continuous tube similar to that on the V-1 flying bomb. The circular cross-section fuselage, housing a large fuel tank and spherical pressure cylinder for the pulsejet, could be made either of steel or wood, adapted to the available materials at the manufacturing centres. The mid-wing variant had a central fin and rudder which served as a

rear support for the dorsal pulsejet. The shoulder-wing variant had twin endplate fins and rudders, the pulsejet supported at the rear on a central pylon.

Take-off, with the aid of two 1,200kg (2,646 lb) thrust solid-propellant RATO units, was on a jettisonable undercarriage, the landing being made on extensible skids although provision had been made for a normal tricycle undercarriage. The EF126 did not progress beyond a mock-up which was captured by the Soviets.* With the help of captured German personnel, the EF126 was built and flown under Russian supervision, whereby the Junkers works pilot Flugkapitän Joachim Mathies was fatally injured in a crash.

Top: **Junkers EF126 shoulder-wing variant.**

Above: **Junkers EF126 mid-wing version.**

* The EF126 'Elli', designed as a ground attack aircraft, was also referred to in the November 1944 EHK conference as a Zerstörer (heavy fighter), powered by one or two Argus As 014 pulsejets. Tested in the wind-tunnel at the beginning of 1945, an order for 20 aircraft placed on 20th January was cancelled by the TLR in March 1945.

Junkers EF126 – data (mid-wing version)

Powerplant	1 x 475kg (1,047 lb) static thrust Argus As 044 pulsejet	

Dimensions

Span	6.65m	21ft 9¾in
Length, fuselage	7.80m	25ft 7in
Length, overall	8.46m	27ft 9¾in
Height, skids up	1.90m	6ft 2¾in
Wing area	8.90m²	95.80ft²

Weights

Loaded weight	2,800kg	6,173 lb
with RATO	2,970kg	6,548 lb

Performance

Max speed, at sea level		
clean	780km/h	485mph
with external load	680km/h	423mph
Range, 100% thrust	300km	186 miles
60% thrust	350km	217 miles
Endurance, 100% thrust	23mins	
60% thrust	45 mins	
Take-off run, with RATO	270m	886ft

Armament	2 x MG 151/20 (180rpg) or 2 x MK 108	

Underwing loads	2 x AB 250 (2 x 108 SD 2 bombs) containers	
	12 'Panzerblitz' rocket projectiles	
	12 R4M rocket projectiles	

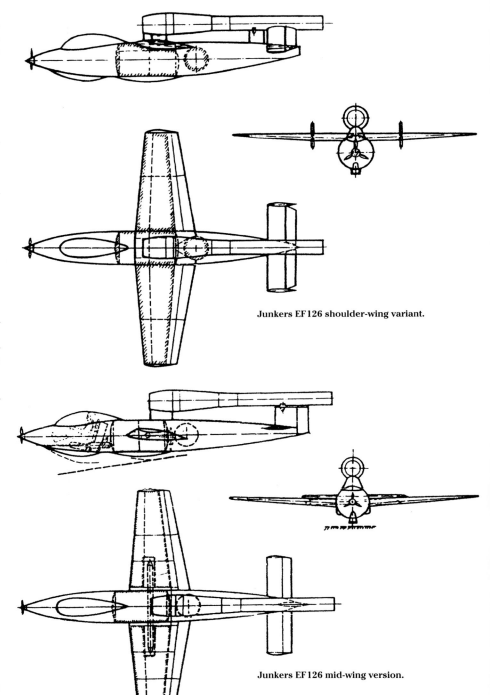

Junkers EF126 shoulder-wing variant.

Junkers EF126 mid-wing version.

An EF126 model with and without undercarriage in the Junkers wind-tunnel.

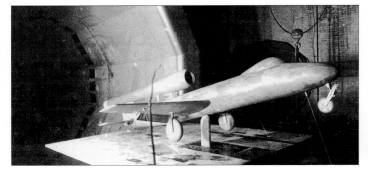

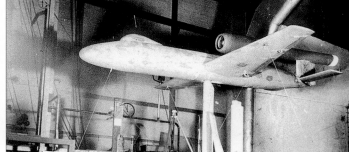

Argus-Junkers Ground Attack Project

1945

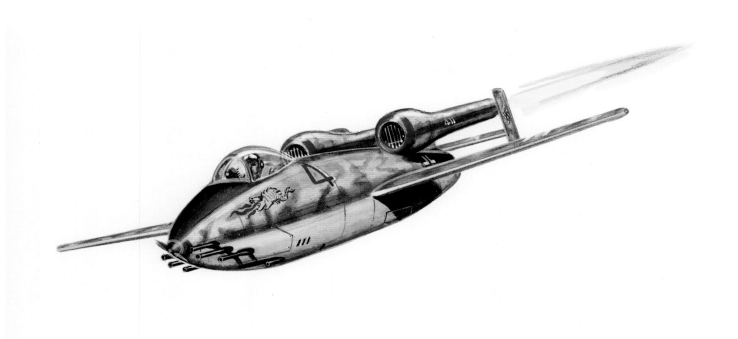

Argus-Junkers Ground Attack Project with two As 044 pulsejets and six nose cannon.

As a further development of the Junkers EF 126 infantry-support and ground attack aircraft, this joint proposal of the Argus and Junkers firms was to have been powered by two 475kg (1,047 lb) static thrust As 044 pulsejets – a further development of the As 014. This single-seat undesignated mid-wing aircraft differed from its forbear in that the pulsejets were located at the fuselage sides and supported on pylons at the rear above the tailplane which had twin endplate fins and rudders. Equipped with a fuselage-mounted retractable tricycle undercarriage and with the aid of two solid-propellant RATO units, it was capable of deployment from unhardened forward airfields to fulfil its ground attack function, and could also be used as a fighter.

The co-operation between Argus and Junkers was aimed above all at reducing the amount of critical materials and manufacturing time in which its simple design and construction in Baukasten (building-block) form suited it primarily for series-production by small firms in the last months of the war. As a result of the destruction of the at Junkers and those of the parts manufacturers as well as the continual advance of Russian forces, the production plan was no longer possible to carry out.

It was presumably this proposal that was mentioned by Prof Heinrich Hertel of Junkers at the EHK conference in November 1944 as a very cheap and simple fighter and ground attack aircraft powered by two Argus As 014 pulsejets. Its range of 300-350km (280-310 miles) and endurance of 40-50 minutes, however, was considered to be completely inadequate, so that the project's realisation was considered impracticable. No objection was raised at the time to its further development for experimental purposes but with no intention to place it in quantity production.

Argus-Junkers Ground Attack Project – data

Powerplant	2 x 475kg (1,047 lb) static thrust Argus As 044 pulsejets	
Dimensions		
Span	6.65m	21ft 9¾in
Length, fuselage	7.80m	25ft 7in
overall	8.46m	27ft 9in
Height, wheels up	1.30m	4ft 3¼in
wheels down	1.80m	5ft 11in
Wing area	8.90m²	95.80ft²
Weights		
Loaded weight	2,980kg	6,570 lb
Performance		
Max speed	810km/h	503mph
Armament	2 x MK 108 or 1 x MK 214 in fuselage nose	
Underwing loads	24 x R4M or 2 to 4 x RZ 100 rocket projectiles	

**Argus-Junkers Ground Attack Project –
drawing dated 21st April 1945.**

**Argus-Junkers Ground Attack Project with
two wing-mounted cannon.**

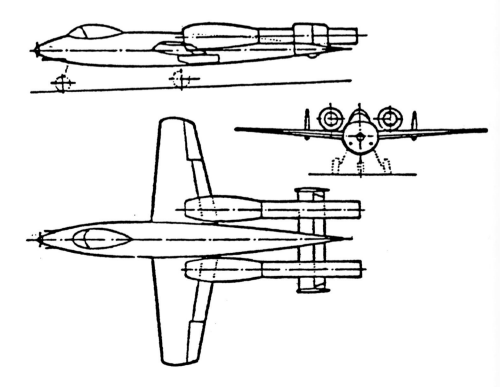

Chapter Four

The Development of the Twin-Engined Heavy Fighter 1935 - 1939

The term Kampfzerstörer (Kampf = battle or combat, Zerstörer = destroyer) was a concept which the Luftwaffe had adopted from the Navy that was intended to reflect the fighting power of the strategic fighter whose origins dated back to the 1920s. An important prerequisite for the creation of such an aircraft was the introduction of the 'all-metal monoplane with retractable undercarriage' that was to operate in the speed regime which justified the tactical employment of the Zerstörer as a superior combat vehicle.

For defensive purposes, the Kampfzerstörer with its powerful armament was expected to be decisive in destroying enemy bombers, and above all, prove itself superior to single-engine fighters. It was expected to penetrate deep into enemy territory, engage in combat with enemy bombers as well as fighters sent up to oppose the approaching bomber force, and then retire at high speed on the return flight – a truly optimistic notion.

In the course of a study undertaken in summer 1934 by the Luftwaffe leadership, ideas were crystallised for the tactical employment of new types of weapons in an offensive type of air war. Deviating from the then current viewpoint that strategic bombing attacks should be undertaken by sufficiently-armed bomber formations that would fight their way through to the target with their own weapons, the latter role was to be accomplished by the newly-defined Kampfzerstörer, in other words, to clear the path ahead for the bombers approaching their targets. This was to have been achieved using heavily-armed multi-seat twin-engined aircraft the size of a medium bomber which, instead of carrying a jettisonable load, would possess a heavy concentration of cannon armament.

In the late autumn of 1934, the RLM issued a requirement for a Kampfzerstörer to several aircraft firms: AGO, Dornier, Focke-Wulf, Gotha, Heinkel, Henschel and Messerschmitt, who were requested to submit proposals that, dependent upon approval, would lead to the construction of three prototype aircraft. Initial evaluation of the projects by the RLM Technisches Amt (Technical Office)

in early December 1934 led to the preliminary decision that the Henschel and Focke-Wulf proposals appeared to be the most promising of the contestants whose submissions conformed extensively to the RLM requirements in respect of both dimensions and the arrangement of the weapons. Messerschmitt on the other hand, with its completely novel project layout, had little chance of securing the RLM's interest as it had envisaged a two-seat fighter that was smaller and lighter but therefore faster than an equivalent bomber design. With the prevailing viewpoint concerning the usefulness and tactical employment of the planned heavy Kampfzerstörer, their proposal was regarded as impracticable. Generalluftzeugmeister Ernst Udet, however, exercised his influence with the RLM Technisches Amt to include the Messerschmitt Zerstörer project. Thus, as the third contestant in the competition, Messerschmitt was also awarded a construction contract.

The three successful tenders received the RLM designations Fw 57, Hs 124 and Bf 110 respectively. For the powerplant, the RLM suggested the use of the newly-developed Daimler-Benz DB 600, but allowed the Junkers-Jumo 210 as an alternative in the event the DB 600 would not be available in sufficient numbers. For development of the armament to be carried, the Mauser and Rheinmetall firms were made responsible, and before the end of 1934, prototype development of the three selected types had commenced.

Designed under the leadership of Dipl-Ing Wilhelm Bansemir, the Fw 57 Kampfzerstörer was the largest of the three submissions selected by the RLM Technisches Amt and was the first attempt by Focke-Wulf to build an all-metal aircraft. With a wingspan of 25m (82ft), the low-wing monoplane, however, was too heavy: the wing turned out five times heavier than the calculated design weight. Its armament was to consist of four forward-firing 20mm cannon in the fuselage nose and two 20mm cannon in a traversible dorsal turret (B-Stand) at the rear of the three-crew compartment. The Fw 57 V1, V2 and V3

prototypes were intended to be powered by the DB 600 in-lines which, with a loaded weight of 8,300kg (18,298 lb) without armament, had a maximum speed of 405km/h (251mph). But the flying characteristics of the Fw 57 V1 proved unsatisfactory in every respect. A few weeks after its maiden flight in early 1936, it became damaged in an emergency landing and had to be written off. The second and third prototypes underwent flight trials for several months but were decreed unusable and eventually scrapped.

Headed by Dipl-Ing Friedrich Nicolaus at Henschel, the Hs 124 design was laid out as an all-metal aircraft with twin fins and rudders and was powered by two Junkers-Jumo 210 engines. The Hs 124 V1 had a movable electrically-operated 20mm Mauser cannon in the nose and an open MG weapon position in the dorsal fuselage, manned by the radio operator. This weapon arrangement was regarded as purely provisional since, in the opinion of the RLM Technisches Amt, the armament was insufficient to meet the Kampfzerstörer specifications. The Hs 124 V2 was equipped with two movable 20mm nose-mounted cannon, served by a prone gunner via an optical range sight. The first to meet the specified RLM armament requirements was the Hs 124 V3, which instead of the nose weapon mounts on the V1 and V2, had a solid nose with four fixed MG 17 and two MG FF weapons.

Because of delays experienced and non-availability of the more powerful Daimler-Benz engines, the construction programme was not increased beyond the three prototypes, and as delays were also experienced with weapon deliveries, the Hs 124 V3 was used as a trainer for Zerstörer crews. At the beginning of 1939, the Hs 124 V3 was fitted with air-to-surface rockets and employed on weapons trials, in the course of which it demonstrated its excellent manoeuvrability. In mock combat, it often out-manoeuvred the Messerschmitt Bf 109 fighter.

The first prototype of the third contender, the Bf 110, made its maiden flight on 12th May 1936, eighteen months after the conclusion of project work. Following initial difficulties that became evident during the flight-test phase, its maximum speed was found to be only slightly higher than single-seat fighters. In order to make the Bf 110 competitive with single-seat fighters, despite its lightweight type of construction, the Messerschmitt concern placed a limit on fuel capacity. Although capable of carrying a larger quantity of fuel, the increased flying weight would have caused a considerable detriment to speed and manoeuvrability. It was solely the powerful offensive armament, initially consisting of four MGs and two cannon, that made it superior to all other pre-war fighter aircraft and constituted a deadly opponent.

During the course of manufacturer's trials, continual difficulties were experienced with the powerplants which partially led to flight-testing being interrupted for long periods. In the upper speed region, the Bf 110's handling characteristics were good, but in the medium and low-speed range, it was poor in terms of acceleration, turns, and stability. Its strong tendency to swing violently on take-off and

landing was improved by alterations to undercarriage wheel alignment, but even though lessened, remained a unique feature of the Bf 110 throughout its career. Due to delivery problems by Daimler-Benz, the Bf 110 V2 prototype was only able to effect its first flight on 26th October 1936. Works trials were conducted at the beginning of 1937 in Augsburg, the prototype then being delivered to the E-Stelle Rechlin. Despite its known deficiencies, it was clearly shown that the Bf 110 was superior to the Hs 124, so that the Technisches Amt decided in favour of the Bf 110 as Kampfzerstörer and Messerschmitt was given a contract for the manufacture of a pre-production batch of machines. Production of the Bf 110A-0 commenced at the beginning of 1937, and despite the negative judgements from the E-Stellen and frontline service units, it remained in service in this role until 1945 with a production total of 6,000 aircraft.

Besides the three contenders referred to in the abovementioned RLM competition, a number of other pre-war Zerstörer projects are described below.

Dornier Do 29

1934 to 1935

Dornier Do 29 in flight. Note the six MG 15s in the fuselage nose.

One of the unsuccessful contenders to the RLM Kampfzerstörer requirement of late 1934, the Do 29 design was submitted in February 1935. Other than minor differences in having a lower-positioned wing, fully-glazed two-man cockpit enclosure raised above the fuselage contours and the solid nose housing a total of six fixed forward-firing MGs and an MG position at the rear of the cockpit (B-Stand), it could not deny its direct ancestry

to the earlier Do 17 of 1934. Dimensionally, it was about the same size as the Junkers Ju 85, the forerunner of the Ju 88. Following completion of a mock-up, the proposal did not receive RLM approval for prototype construction and design work on the project was ordered to be stopped at the beginning of 1936.

Dornier Do 29 – data

Powerplants	2 x 840hp Bramo 323A radial engines	
Dimensions		
Span	17.56m	57ft 7⅛in
Length	15.35m	50ft 4¼in
Height	4.60m	15ft 1in
Wing area	c.55.00m²	592.00ft²
Weights		
Loaded weight	4,270kg	9,414 lb
Performance		
Max speed	460km/h	286mph
Armament (proposed)	6 x MG15 in fuselage nose	
	1 x MG15 in dorsal (B-Stand)	

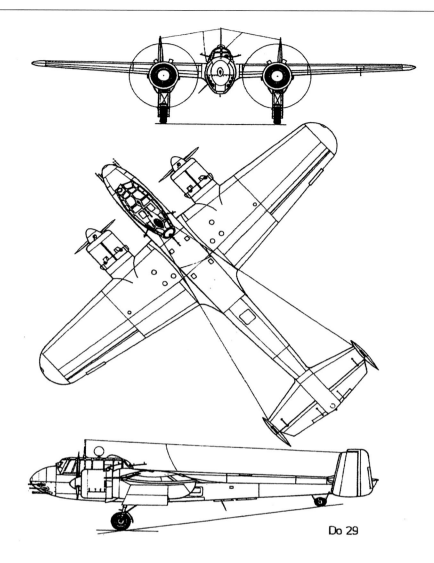

Do 29

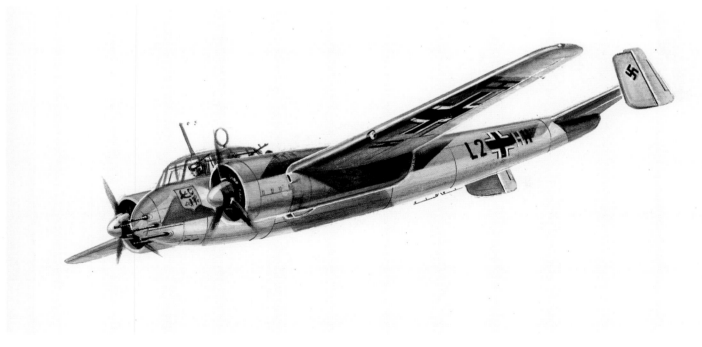

Focke-Wulf Fw 57

1934 to 1936

Focke-Wulf Fw 57 V3.

Focke-Wulf Fw 57.

In response to the RLM requirement of late autumn 1934, under the design leadership of Dipl-Ing Wilhelm Bansemir, Focke-Wulf submitted their proposal for a three-crew twin-engined Kampfzerstörer of all-metal construction a few months later.

A cantilever low-wing monoplane and the first all-metal aircraft by the company, the Fw 57 was an extremely modern aircraft, but due to the installation of the lower-powered Junkers-Jumo 210 engines, was not able to attain the calculated performance and in-flight handling.

Following the completion of three prototypes, the first of which flew in May 1936, fur-

ther work was terminated later in the year by the RLM, as in the meantime, there appeared little possibility of obtaining the originally intended DB 600 and DB 601 for installation. This was not the fault of the engine manufacturer but due solely to the incomprehensible development and construction restrictions imposed by responsible individuals in the RLM, whose directives hindered the engine firms from carrying out development and series-production of high-performance engines. Although the dorsal fuselage turret was fitted on all three prototypes, none had been tested with the proposed nose and turret armament.

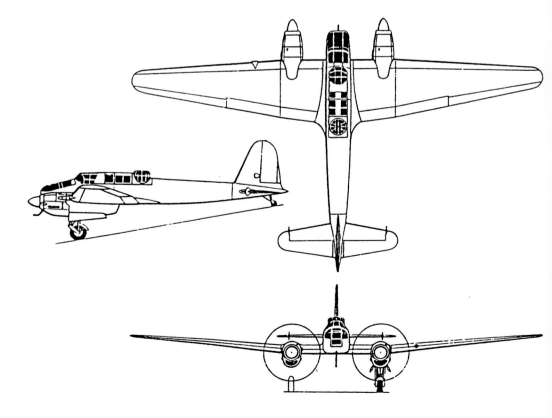

Focke-Wulf Fw 57 V3 with Daimler-Benz DB 600 engines.

Focke-Wulf Fw 57 V2 perspective view with revised fin and rudder.

Focke-Wulf Fw 57 V1. Note that the planned 20mm cannon have not been installed in the Mauser dorsal turret.

Focke-Wulf Fw 57 – data

Dimensions

Span	25.00m	82ft 0¼in
Length	16.40m	53ft 9¾in
Height	4.10m	13ft 5½in
Wing area	73.50m²	791.13ft²

Weights

Empty weight	6,800kg	14,991 lb
Loaded weight, unarmed	8,300kg	18,298 lb

Performance

Max speed, at sea-level	365km/h	227mph
at 3,000m (9,840ft)	405km/h	251mph
Range	1,550km	963 miles
Service ceiling	9,100m	29,855ft

Armament (proposed)	4 x MG17 fixed in nose and
	2 x 20mm in dorsal turret

Henschel Hs 124 V3

1934 to 1939

Designed by Dipl-Ing Friedrich Nicolaus of Henschel, the Hs 124 was the parallel development of the Focke-Wulf Fw 57. The two-seat cantilever mid-wing all-metal design featured a three-part wing, twin fins and rudders and a hydraulically retractable undercarriage.

With the Hs 124, Henschel took part in the 1934 RLM competition for a Kampfzerstörer. Only three prototypes of the design which could be used as a light bomber, long-range reconnaissance or heavily-armed fighter were completed. The Hs 124 V3, laid out for the Zerstörer function and completed in 1936, differed from the earlier transparent-nosed V1 and V2 prototypes in having a solid nose intended to house four to six MG 17 and MG FF weapons. Due to delays with the Daimler-Benz DB 601 engines, these could not be installed and had to be substituted with two of the weaker 850hp Junkers-Jumo 210C. The resultant loss in performance led to the RLM losing interest in the type. The Hs 124 V3, nevertheless, continued to be used until 1939 as a training aircraft for Zerstörer crews.

Top: **Henschel Hs 124 V3 with four fixed nose weapons.**

Centre right: **Henschel Hs 124 V3 with Jumo 210 engines.**

Right: **Henschel Hs 124 with BMW 132 radials.**

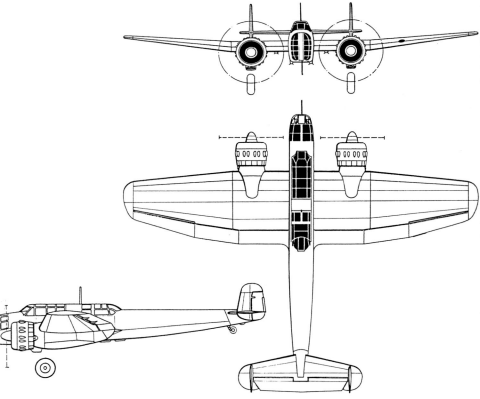

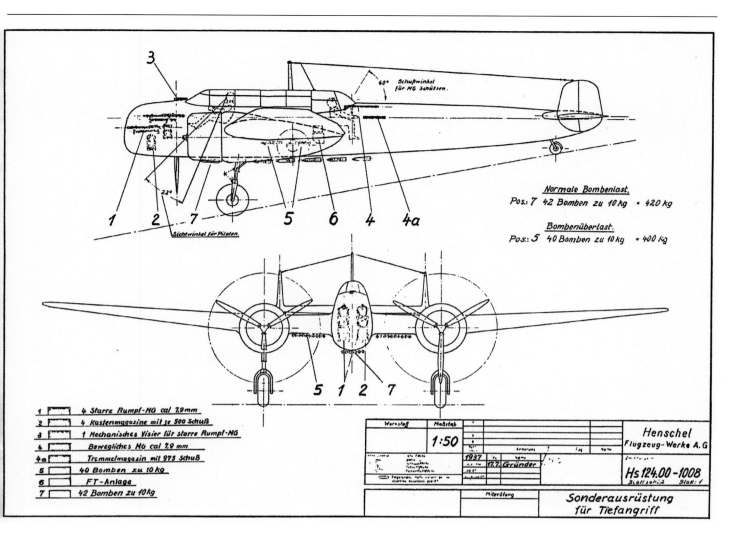

The drawing contains the following annotations:

Schußwinkel
für MG Schützen.

Sichtwinkel für Piloten.

Normale Bombenlast.
Pos.: 7 42 Bomben zu 10 kg = 420 kg

Bombenüberlast.
Pos.: 5 40 Bomben zu 10 kg = 400 kg

	4 Starre Rumpf-MG cal 7,9 mm
1	4 Starre Rumpf-MG cal 7,9 mm
2	4 Kastenmagazine mit je 500 Schuß
3	1 Mechanisches Visier für starre Rumpf-MG
4	Bewegliches MG cal 7,9 mm
4a	Trommelmagazin mit 975 Schuß
5	40 Bomben zu 10 kg
6	FT-Anlage
7	42 Bomben zu 10 kg

Werkstoff   Maßstab   1:50

Henschel
Flugzeug-Werke A.G

1937   17.7. Gründer

Hs 124.00-1008

Sonderausrüstung
für Tiefangriff

Henschel Hs 124 V3 – data

Powerplants 2 x 850hp Junkers-Jumo 210C in-line engines.
Alternatives planned were two DB 601 in-lines or two BMW 132Dc radials.

Dimensions

Span	18.20m	59ft 8½in
Length	14.50m	47ft 7in
Height	3.75m	12ft 3½in
Wing area	54.60m²	587.70ft²

Weights

Empty weight	4,200kg	9,259 lb
Loaded weight*	6,947kg	15,315 lb

Performance

Max speed*	435km/h at 3,000m	270mph at 9,840ft
Range at 3,000m (9,840ft)	1,860km at	1,156 miles at
	337km/h	209mph
at 6,000m (19,685ft)	2,450km at	1,522 miles at
	300km/h	186mph
Service ceiling	7,900m	25,920ft

Armament (proposed) 4 x MG17 in nose, 1 x MG in dorsal turret

Bombload 40 x 10kg (22 lb) underwing bombs, normal or
42 x 10kg (22 lb) bombs beneath fuselage, overload

* Loaded weight is with 1,340kg (2,954 lb) fuel, 420kg (926 lb) bombload and 75kg (165 lb) ammunition; maximum speed is with two BMW 132Dc radials.

Above: **Henschel Hs 124 works drawing of 17th July 1937 showing the projected and bombload positions in the low-level attack role.**

Messerschmitt Bf110

1935 to 1939

Messerschmitt Bf110B-1.

Messerschmitt Bf110 V1 in flight.

Upon receipt of the RLM contract to build three prototypes, the Bf110 V1 lifted off from Augsburg airfield on 12th May 1936 on its maiden flight. In contrast to the other two contestants of the RLM requirement of 1934 – the Fw 57 and the Hs 124, the Bf110 V1 was powered by two Daimler-Benz DB 600 engines. With a loaded weight of 5,000kg (11,023 lb), it attained a level speed of 505km/h at 3,300m (314mph at 10,830ft). Numerous difficulties experienced with the engines in the course of works trials led to extensive interruptions to the test programme. As a result of delays in engine deliveries, the Bf110 V2, powered by two DB 600A-1 engines, only accomplished its first flight on 24th October 1936, followed by the Bf110 V3 on 24th December 1936. On the basis of improved performance with these powerplants, the RLM Technisches Amt awarded a contract for a pre-production batch of the Bf110A-0 series. During flight tests with the Bf110, however, further problems were encountered with the DB 600A-1 which resulted in termination of the flight trials.

In the meantime, Daimler-Benz had developed the direct fuel-injection DB 601 motor which with the same 33.9 litre cylinder capacity, delivered 1,100hp at take-off. The Luftwaffe Commander-in-Chief Hermann Göring put pressure on placing the Bf110A-0 into immediate production with this powerplant, but because of delivery difficulties, this could not be effected. As an interim solution for the manufacture of this two-seat aircraft that had meanwhile been decided upon, Messerschmitt had to install the weaker 610hp Junkers-Jumo 210 in the Bf110A-0s of 5,600kg (12,346 lb) loaded weight that resulted in a performance reduction in all areas of the flight regime. Maximum speed sank from 430km/h (267mph) to 380km/h (236mph) at 3,800m (12,470ft). Armament was four MG17 in the nose and an MG15 in a dorsal turret.

It was only at the end of 1938 that sufficient quantities of the DB 601A-1 became available, so that their installation in the Bf110B-1 whose fuselage was lengthened from 12m (39ft 4⅛in) by another 30cm (11¾in), could commence. Upon the outbreak of the Second World War on 1st September 1939, the Zerstörer squadrons had 95 Bf110B-1s on hand.

Above left: **Messerschmitt Bf 110C-4 in flight.**

Above right: **Messerschmitt Bf 110C-1 with clipped wingtips.**

Centre and below left: **Messerschmitt Bf 110 in North Africa.**

Bottom left: **Messerschmitt Bf 110D-2.**

Messerschmitt Bf 110B-1 – data

Powerplants	2 x 1,100hp Daimler-Benz DB 601A-1 in-line engines	
Dimensions		
Span	16.25m	53ft 3¾in
Length	12.30m	40ft 4¼in
Height	4.13m	13ft 6½in
Wing area	38.70m²	413.32ft²
Weights		
Loaded weight	5,600kg	12,346 lb
Max speed	540km/h at 6,000m	356mph at 19,685ft
Armament	4 x MG 17 in nose (upper)	
	2 x MG FF in nose (lower)	
	1 x MG 15 in fuselage (B-Stand)	

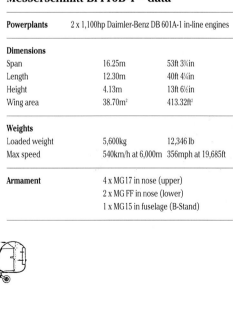

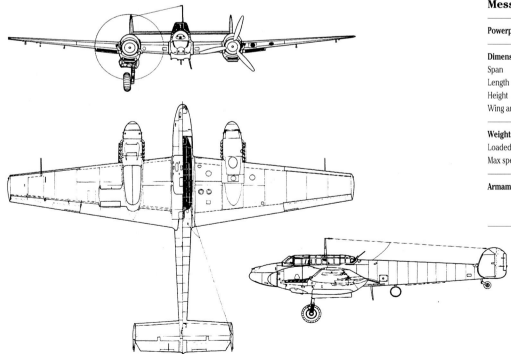

Focke-Wulf Fw 187

1935 to 1939

Focke-Wulf Fw 187 V6.

Focke-Wulf Fw 187 V6 preparing for take-off.

In 1935 Dipl-Ing Kurt Tank, Technical Director at Focke-Wulf, commenced studies on a twin-engined single-seat fighter which was expected to be greatly superior in performance to existing fighters, but at the time, no RLM requirement existed for such an aircraft.

Early in 1936, however, an exhibition of new secret weapons was held at the Henschel plant in Berlin-Schönefeld at which aircraft, engines and armament developments were appraised. Focke-Wulf was represented by its abovementioned design that captured the interest of the high-ranking visitors. The new

aircraft, when powered by two DB 600 motors, would have been capable of a maximum speed of 560km/h (348mph) and thus exceed the top speed of the Messerschmitt Bf 109 fighter. Despite the favourable impression which it and its designer created, the RLM Technisches Amt for paltry reasons was averse to awarding a construction contract.

Kurt Tank thereupon submitted revised, extremely detailed plans to Oberst Freiherr Wolfram von Richthofen, Chief of the RLM C-Amt (Development Department) in order to dissipate all existing doubts. Richthofen recognised the advantages of such an aircraft, and in long drawn out negotiations, secured a construction contract for three prototypes of what became the Fw 187. Development work was begun in 1936 under the design leadership of Obering Rudolf Blaser who carried out detail design of the single-seat fighter. A significant change was that the three prototypes would be powered by the weaker Jumo 210 engine as all production DB 600s were earmarked for other priority projects. Already in the summer of 1937 Kurt Tank, taking-off in the Fw 187 V1 on its maiden flight, reached a speed of 525km/h at 4,000m (326mph at 13,120ft). Despite its Jumo 210D of only 680hp at take-off and its loaded weight of 4,560kg (10,053 lb), the Fw 187 was faster than the single-engined Bf 109 and He 112 fighters.

From 1937 until May 1938, the Fw 187 V1 underwent extensive flight-testing. To increase performance, the Jumo 210Ds were exchanged for the more powerful 670hp

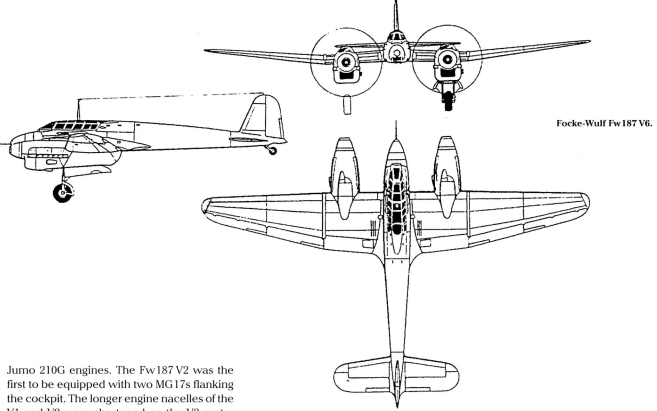

Focke-Wulf Fw 187 V6.

Jumo 210G engines. The Fw 187 V2 was the first to be equipped with two MG 17s flanking the cockpit. The longer engine nacelles of the V1 and V2 were shortened on the V3 prototype that allowed the installation of one-piece landing flaps. Whilst the V3 was still equipped with the Jumo 210D, the two-seater V3 had DB 600A-1s and more powerful armament – the two MG 17s being supplanted by two MG FF cannon. It was also the first to be fitted with rearward-firing armament. In further trials with the DB 600A-1-powered variant, a maximum level speed of 635km/h (395mph) was attained.

Series production of the meanwhile-approved Fw 187A-0 began in 1939, but its early cancellation by the RLM put paid to this aircraft. Kurt Tank, however, was granted permission to modify the aircraft into a two-seat Zerstörer in order to conduct comparative trials with the Bf 110. Three further prototypes therefore appeared: the Fw 187 V4 and V5 with the Jumo 210G and the V6 with the DB 600A-0, serving as prototypes for the Fw 187A-0 series whose production was soon to be terminated. The A-0 series was not to be delivered to the Luftwaffe and resulted in Kurt Tank employing them as factory protection machines in Bremen, thus proving itself under operational conditions. Dipl-Ing Kurt Mehlhorn, designer and test pilot at Focke-Wulf, had the V6 modified into a single-seater and with it, shot down several Allied bombers. The incomprehensible directives and construction bans from the highest authorities in the RLM are still unexplainable even today.

A short time later, when the RLM despairingly sought a twin-engined fighter which performancewise closely approached that of the new jet fighters in order to regain the commencing Allied air superiority, the stupidity of the directives was obvious. Had sufficient numbers of the Fw 187 been available at the time, it is highly probable that they could have brought about a turning point in the air war over Germany.

Focke-Wulf Fw 187A-0 – data

Powerplants	2 x 700hp Junkers-Jumo 210G in-line engines	
Dimensions		
Span	15.30m	50ft 2½in
Length	11.10m	36ft 5in
Height	3.85m	12ft 7½in
Wing area	30.40m²	327.21ft²
Weights		
Empty weight	3,700kg	8,157 lb
Loaded weight	5,000kg	11,023 lb
Performance		
Max speed	530km/h at 4,200m	329mph at 13,780ft
Range	900km	559 miles
Service ceiling	10,000m	32,810ft
Armament	4 x MG 17 in nose (upper)	
	2 x MG FF in nose (lower)	

Left: **Focke-Wulf Fw 187 V6. A mechanic is assisting the pilot with take-off preparations.**

Centre left: **A Focke-Wulf Fw 187 in flight.**

Below: **A head-on view of the Focke-Wulf Fw 187 cockpit and weapons positions.**

Bottom: **An Fw 187 of the Focke-Wulf Works Protection Unit.**

Arado E 500

1935 to 1936

The Arado E 500 four-crew Zerstörer project drawn up in 1935, incorporated an advanced armament concept that had been put forward by Dipl-Ing Walter Blume and Kurt Bornemann. This consisted of a pair of turret-mounted 20mm Rheinmetall-Borsig Lb. 202 cannon which could be rotated through a full 360°, the weapons also being able to be trained from the horizontal through a full 90° to fire vertically upward in the fuselage dorsal B-Stand, a similar ventral turret being capable of the same degree of rotation horizontally and vertically downward. To provide the dorsal and ventral gunners – the latter lying prone in the under-fuselage pannier – an almost unrestricted field of fire, the tail surfaces were placed on the outboard portion of the twin fins and rudders. Seated slightly behind and to the right of the pilot, the forward gunner operated the fixed forward-firing armament in the nose of the short central fuselage nacelle. The two rotatable weapon turrets were located directly behind the pilot and forward gunner, the fuselage space behind the turrets serving to house the fuel tanks and internal bombload. The main load-carrying element of the E 500 project was the tapered box main spar extending from one wingtip to the other and joined to the equally load-carrying twin booms which housed not only the two Daimler-Benz in-line engines, but also the rearward-retracting main-wheels of the conventional undercarriage. The RLM considered the armament idea worthy of further development up to the prototype stage, but following the construction of a full-scale mock-up in co-operation with the Rheinmetall firm, work on the project was stopped by the Technisches Amt as it was not prepared to further the early introduction of electrically-operated rearward-firing movable turrets.

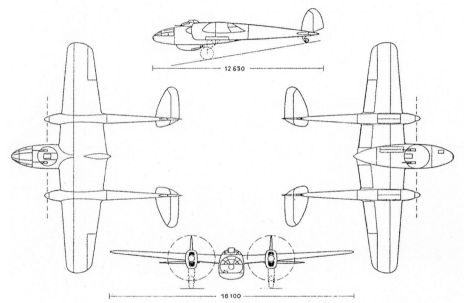

12 650

18 100

Stand: Dezember 1935

Arado E 500 – drawing dated December 1935.

Arado E 500 full-scale mock-up showing the dummy Rheinmetall Lb. 202 twin mounts.

Arado E 500 – data

Powerplants	Two Daimler-Benz 600-series in-line engines	
Dimensions		
Span	18.10m	59ft 4¼in
Length	12.65m	41ft 6in
Height*	4.00m	13ft 1½in
Wing area*	c.50.00m²	518.18ft²
Bombload (probable)*	2 x 250kg (551 lb)	

* Figures estimated from the works drawing. In his original brief data table, the author listed span as 17.32m (56ft 10in) and length as 15.20m (49ft 10½in), stating that nothing was known concerning the powerplants or other dimensions. In a written reply to my queries, the author stated that his dimensions relate to a later variant with more powerful engines – Translator.

AGO FP-30 (Ao 225)

1937 to 1938

As early as 1937 the Aerowerke Gustav Otto at Oschersleben participated in the RLM requirement for a heavy Kampfzerstörer with its FP-30 submission. It differed from the other firms' projects in featuring wing-mounted propellers driven remotely from two engines inside the fuselage. Besides AGO, Arado's entry was the Ar E 561. The point of origin for this project was the multi-seat Ao 192 touring aircraft whose sleek lines and good flying qualities were among the best of its class. The two powerful DB 601 engines coupled nose to nose and located in the wingroot-fuselage junction, were arranged to drive three-bladed airscrews at the extremities of long, slim nacelles in the wings. In another variant of this design, a 24-cylinder DB 606 coupled unit was to have provided the power in place of the DB 601s. A partial full-size mock-up of the FP-30 (Ao 225) was built to study in detail the remote-drive installation for various power-plants. Problems experienced with the design of remote drive mechanisms and angular bevelled gearing as well as strong vibration tendencies led the AGO designers to

consider the use of two separate wing-mounted engines. Because of the complexity of the remote drive system and its difficulty to manufacture, the RLM rejected the proposal on which work was terminated in favour of the Me 210 and Ar 240. The RLM number 225 was later reassigned to the Focke-Achgelis Fa 225 autogyro, and only a few technical details of the Ao 225 have survived.

AGO Ao 225 – data

Powerplants	2 x 1,100hp DB 601 or
	1 x 2,400hp 606 in-line engines

Dimensions		
Span	21.05m	69ft 0¾in
Length	19.10m	62ft 8in

Armament	4 x MG 131 in fuselage nose, plus
	1 x MG 151 in dorsal turret (B-Stand)

Partial full-size mock-up of the AGO FP-30 (Ao 225).

AGO Ao 225 inverted model in the DVL wind-tunnel.

Gotha Zerstörer Projects
1937 to 1938

Gotha P 8-01.

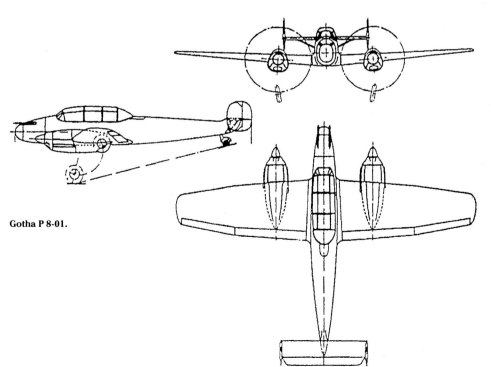

Gotha P 8-01.

In 1938 the Gothaer Waggonfabrik (Gotha Wagon Works) submitted to the RLM three Zerstörer projects that had been drawn up under the design leadership of Dipl-Ing Egwin Leiber. As opposed to the proposals of other firms, by virtue of their design and performance, these lay in the 'lightweight' category. In the view of Gotha Chief Designer Dipl-Ing Albert Kalkert, the proposals would have been useful for providing homeland and regional defence, since their potential opponents – the French Potez 63 and the Polish PZL 38, were also meagre performers. His argument, however, was not taken notice of by the RLM as it had no requirement for a 'lightweight Zerstörer' and rejected the proposals without further comment. It was only several years later, during the last year of the war, that Kalkert's notion that similar aircraft of this type were to assert themselves as so-called Nahkampfflugzeuge (close-combat aircraft).[1]

[1] According to published accounts, as early as 1935 the GWF had submitted their P 3.01 project to meet the 1934 RLM requirement for a Zerstörer. Like the similar Arado E 500, it was of the high-wing twin-boom layout housing a crew of two or three and powered by two DB 600 engines (with the Jumo 211 as alternative) in the fuselage driving the wing-mounted airscrews via extension shafts. Span was 16m (52ft 6in), length 11.1m (36ft 5in) and height 2.5m (8ft 2½in). The slightly larger P 3.02 variant with cranked wings and the same powerplants had a span of 17m (55ft 9¼in), length 12.8m (42ft) and height 2.9m (9ft 6¼in), but no other details are available.

Gotha P 8-01 – data

Powerplants	2 x 240hp Argus As 10C in-line engines	
Dimensions		
Span	11.00m	36ft 1in
Length	8.67m	28ft 5¼in
Height	2.80m	9ft 2¼in
Armament	2 x MG15 in fuselage nose	

Gotha P14-02 – data

Powerplants	2 x 465hp Argus As 410 in-line engines	
Dimensions		
Span	12.36m	40ft 6½in
Length	8.96m	29ft 4¼in
Height	2.80m	9ft 2¼in
Armament	2 x MG15 in fuselage nose	

Gotha P14-02.

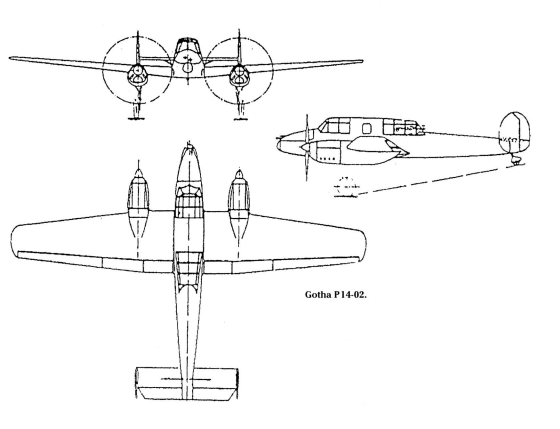

Gotha P14-02.

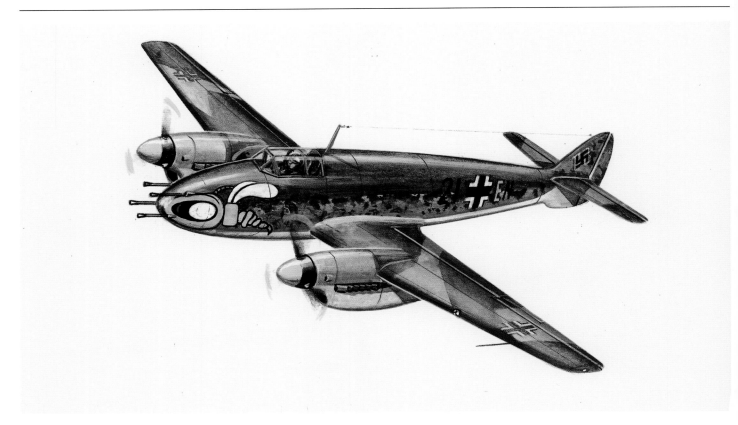

Gotha P 20 – data

Powerplants	2 x 240hp Argus As 10C or	
	2 x 465hp As 410 in-line engines	
Dimensions		
Span	12.15m	39ft 10¼in
Length	8.00m	26ft 3in
Armament	4 x MG 15 in fuselage nose	

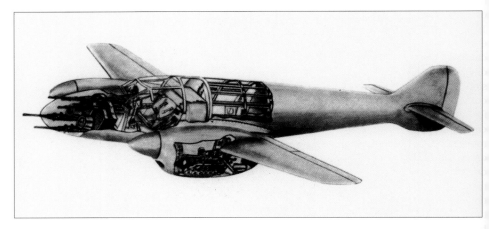

Top: **Gotha P 20.**

Above: **Gotha P 20 works perspective illustration dated 14th February 1938.**

Arado E 561

1936 to 1937

In response to the new RLM requirement issued at the end of 1936 for a multi-seat heavy Zerstörer capable of reaching speeds of 550-600km/h at 6,000m (342-373mph at 19,685ft), Arado submitted its E 561 proposal early in the following year. Featuring a low cantilever wing, twin endplate fins and rudders and a crew of three, it gave the impression, like the AGO Ao 225, of being a conventional twin-engined aircraft, except for the buried side-by-side DB 600 engines canted at 44° to each other driving opposite-rotating wing-mounted propellers via remote drive shafts as seen in the accompanying drawing. Besides the remote drive mechanisms, it had annular radiators in the wing nacelles. Subsequent analysis showed that this type of remote drive was unsuitable for high-power engines due to lack of component strength which resulted in distortions, and that manufacture of such a drive was beyond the capabilities at that time. The Junkers shaft-drive system intended to be used was only suitable for low-powered engines, so that Arado withdrew the E 561[2] as the RLM did not show any further interest in it.

[2] This project has been described in other published accounts as the E 561 or E 651 or even under both project numbers by the same author. In the book by Jorg Armin Kranzhoff: Arado – Geschichte eines Flugzeugwerks, Aviatic Verlag, Oberhaching, 1995, p.107, it is described in the text and accompanying three-view drawing as the E 651. In a later more comprehensive work by the same author, namely: Die Arado Flugzeuge – Vom Doppeldecker zum Strahlflugzeug, Bernard & Gräfe Verlag, Bonn, 2001, pp.190-191, the same two-view drawing as illustrated here bears a very blurred but readable Arado designation E 651 under which the aircraft is described. Unlike the colour illustration in this volume which shows the three crew members sitting in tandem, the drawing clearly shows that the second crew member was seated slightly behind to the right of the pilot and above the powerplant. The third crew member, seated above the rear of the powerplants and facing aft, took up a prone position in the rear fuselage when operating the ventral rearward-firing armament. In the event of an engine failure during the critical take-off phase, the remaining engine was able to provide power to drive both sets of propellers. All three undercarriage wheels were retractable.

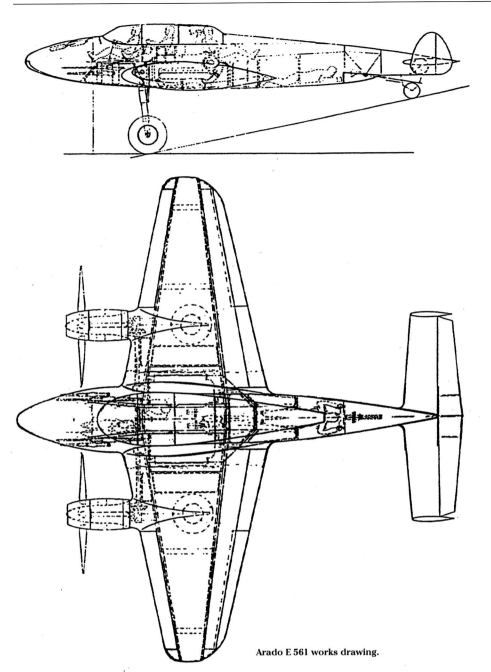

Arado E 561 – data

Powerplants	Two Daimler-Benz DB 600 in-line engines	
Dimensions		
Span*	18.30m	60ft 0½in
Length*	17.50m	57ft 5in
Armament	4 x guns in nose and 2 x MG 81Z in each fuselage dorsal (B-Stand) and ventral (C-Stand) position	

* Both dimensions from the author appear unduly high, as this gives a nosewheel diameter of 1.55m (5ft 1in) and a propeller diameter of around 5.1m (16ft 8¾in) – much too high for a 1937 engine. Taking the outstretched rear prone gunner as a scale, this gives a span not greater than 11m (36ft 1in), fuselage length no more than 10m (32ft 9¾in) and mainwheel diameter 1.10m (3ft 7¼in) – the same as those used on the Do17 and Ju 88A aircraft of similar vintage. But this is pure conjecture; actual dimensions probably lay somewhere in between.

Arado E 561 works drawing.

Lippisch-Messerschmitt P 04-106

1939

Shortly after Dr Alexander Lippisch commenced work at Abteilung L (for Lippisch) at the Messerschmitt AG on 2nd January 1939, the P 04-106 project for a tailless Kampfzerstörer was drawn up under the leadership of Dipl-Ing Rudolf Rentel. Powerplants were to have been two DB 601Es driving pusher propellers via extension shafts. Calculations indicated a maximum speed of 510km/h (317mph), making it a superior internal competitor to the Bf110.[3]

Other than the tailless layout, a notable feature of the design was the retractable tailwheel which was extended telescopically some 1.9m (6ft 2¾in) when at rest on the ground. The crew of two were housed beneath a long cockpit canopy, the rear gunner operating the rearward-firing movable pair of MG 131s and the pilot the four fixed forward-firing MG 151s in the fuselage nose. The entire project was rejected without explanation by the RLM and development work on it was terminated.[4]

[3] The Bf110C-1, in Luftwaffe service in 1939 and powered by two 1,050hp DB 601A-1s, had maximum speeds of 475km/h at 0km (295mph at sealevel), 525km/h at 4,000m (326mph at 13,120ft) and 540km/h at 6,000m (336mph at 19,685ft). Dimensions were: span 16.25m (53ft 3¾in), length 12.07m (39ft 7¼in), height 4.13m (13ft 6¼in) and wing area 38.4m² (443.32ft²). Loaded weight was 6,028kg (13,289 lb) normal, and 6,750kg (14,880 lb) at overload. Armament was two MG FFs (180rpg), four MG 17s (1,000rpg) plus one flexible MG 15 (175 rds).

[4] In the book by Alexander Lippisch: *Ein Dreieck Fliegt (A Delta Flies)*, Motorbuch Verlag, Stuttgart, 1976, p.73, is a three-view drawing of 26th August 1939 of the P 04-107a Schulflugzeug (basic trainer), identical in dimensions and overall layout but powered by two 450hp Argus As 410 in-lines mounted as pushers. Another three-view drawing beneath it dated 8th December 1939, powered by two 1,200hp DB 601Es, is labelled P 04-106 but described in the text as the P 04-108 Zerstörer and Bomber, suggesting that it was a later variant.

A further tailless project of the same basic configuration developed by Dr-Ing Hermann Wurster drove pusher propellers via extension shafts. In place of the telescopic tailwheel, it had a nosewheel tricycle undercarriage and a tall single fin and rudder in place of the outboard twin fins; a drawing designated P 04-114 dates from May 1941. Lippisch describes this two-seater as a Schulflugzeug powered by two 240hp Argus As 10C in-lines – rather underpowered for an aircraft with a span of 16.8m (55ft 1¼in), length 5.86m (19ft 2¾in) and height 3.575m (11ft 8¾in). Weight and performance figures are not known.

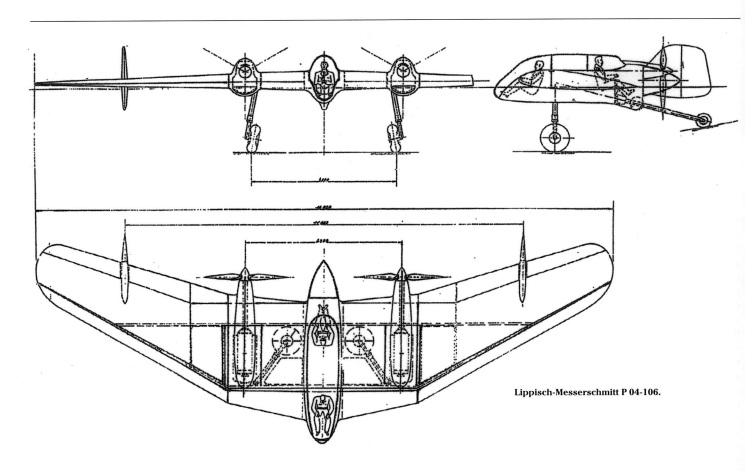

Lippisch-Messerschmitt P 04-106.

Lippisch-Messerschmitt P 04-106 – data

Powerplants	2 x 1,200hp DB 601E 12-cylinder in-line engines	
Dimensions		
Span	16.00m	52ft 6in
Length	5.83m	19ft 1½in
Height	3.15m	10ft 4in
Performance		
Max speed	510km/h	317mph
Armament	4 x MG151 in fuselage nose, plus	
	2 x MG131 at fuselage rear	

Chapter Five

Fast Bomber Projects 1941-1943

During this period of the war, when the first axial-flow turbojet prototypes of BMW (P.3302) and Junkers (Jumo T1) were undergoing their initial test-bed runs, the success of this new motive power was still too early to predict. As will be seen from the projects described below, designers still concentrated on utilising conventional powerplants, but in their projected higher-powered variants. Nevertheless, the first tentative steps to combine the advantages of jet propulsion of considerably higher fuel consumption with the more economic reciprocating engine began to be made in mixed-power proposals by Arado and Dornier in order to attain higher speeds. Pure jet-propulsion studies begun during the previous period by Arado, Lippisch and Messerschmitt were continued but led to no production orders for these new types by the RLM.

Arado Ar 240

The Ar 240 counted among the most modern aircraft designs of its time that employed all the newly-developed manufacturing and assembly techniques, which in turn led to considerable problems and delivery delays during manufacture. Towards the end of 1938 when Arado submitted proposals to the RLM for a fast multi-purpose aircraft taking into account its possible operational use as a Zerstörer and Schlachtflugzeug (ground attack aircraft), the RLM decided in favour of the twin-engined design of conventional layout and construction.

On 2nd April 1938, the Arado Flugzeugwerke received an RLM contract for the construction of three prototypes of the E 625 with a view to examining its possibilities as a successor to the Messerschmitt Bf110. Development proceeded under the RLM designation Ar 240 in the hands of Dipl-Ing Wilhelm van Nes who had taken over the leadership of the Development Department at Arado. The focal point of development lay in diversity:

the Ar 240 was to be suitable for various operational uses, reach a high speed and with various types of defensive and offensive armament, possess considerable fighting power. The design was laid out in such a way that the most powerful engines could be installed without extensive structural alterations being necessary for powerplant attachment. This possibility proved itself of particular advantage during operational trials with the Ar 240, when various types of engines were installed. The required high speeds at various combat altitudes also rendered the installation of a special high-altitude pressurised cabin necessary and as wing area was reduced to a minimum, resulted in a very high wing loading. During final assembly of the initial prototypes, Arado received a further contract for a total of ten experimental machines.

In co-operation with the Daimler-Benz AG,

the Ar 240 powerplants were enclosed within an optimised aerodynamic cowling but this reduced the cooling surface area to a minimum and thus made necessary the use of encased propeller hubs in order to reduce engine overheating.

Following receipt of Rechlin test reports, the RLM was impressed with the Ar 240's performance and became interested in a continuation of development work on it. The planned large-scale production was to have been assumed by AGO in Oschersleben in order to make available production capacity at Arado for further aircraft developments such as the Ar 234. These production plans, however, were not realised since due to the war, appreciable difficulties were experienced in materials supplies that led to a drastic curtailment of aircraft types within the entire aircraft industry and termination of Ar 240 production at the end of 1942.

Arado Ar 240 V2 with BMW 801 radials.
G W Heumann

Arado Ar 240 with 7.5cm BK 7.5 Bordkanone (aircraft cannon).

Between the maiden flight of the Ar 240 V1 on 10th May 1940, almost two years elapsed before the first pre-production Ar 240A-01 and A-02 flew, in October 1942. The design appeared to have been plagued from the outset by numerous problems, partly associated with its innovative tailcone dive-brakes, the fixed wing outboard slots, pressure cabin, remotely-controlled gun barbettes, double-slotted trailing-edge flaps, periscopic sight, etc. Several alterations were made to its dimensions and structure to cure aerodynamic instability, so that the Ar 240 V3 featured major changes. Although markedly improved over the V1 and V2, the flying characteristics left much to be desired. Delivered to a special reconnaissance unit for operational evaluation, the aircraft was flown by Oberst Siegfried Knemeyer on unarmed sorties over England where its high speed and altitude capabilities avoided interception. Although preparations for production for the first 40 Ar 240A-0s were at an advanced stage at AGO in 1942, several variants had been planned from 1939 with various powerplants. Typical of expected maximum speeds for the Ar 240A were: July 1939 – 682km/h (424mph), November 1941 – 625km/h (388mph), and April 1942 – 610km/h (379mph). Proposed powerplants were the DB 601E, DB 603A-2, DB 603G, DB 605, DB 614, DB 627, Jumo 213E and BMW 801J. The Ar 240A-0 of 26th Sep-

tember 1942 had an empty weight of 6,800kg (14,991 lb), loaded weight 8,700kg (19,180 lb), maximum speed 623km/h at 6km (387mph at 19,685ft), range 2,000km (1,242 miles) and service ceiling 9,300m (30,510ft). Proposed to fulfil the role of day or night fighter, fast bomber or unarmed reconnaissance, the variants stretched up to the Ar 240F fighter. Various bombloads up to 500kg (1,102 lb) and armament combinations, too numerous to mention, could be fitted.

Experience gathered with the Ar 240 led to development of the new Ar 440 model. Structural alterations to the Ar 240 V10 served as the prototype Ar 440 V1, followed by three further prototypes whose components were taken from the Ar 240 production line. At the end of October 1943, however, the RLM stopped the entire Ar 240/Ar 440 programme.

The Ar 240 ultimately counted among the finest developments of the German aircraft industry. Its superb performance and high quality of manufacture made it a top-notch product. Noteworthy in this connection is the co-operation between Arado and Rheinmetall-Borsig on the installation of heavy Bordwaffen (aircraft armament). A report dated 1st September 1944 by the latter firm mentions the installation of the L 40 of 7.5cm (2.95in) calibre, but does not indicate whether flight trials with this weapon actually took place.

Arado Ar 240 side-view with BK 7.5.

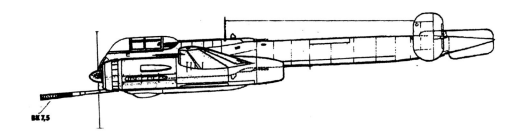

Arado Ar 240 V3 with DB 603 engines.

Arado Ar 240A-02 – data

Powerplants	2 x 1,175hp Daimler-Benz DB 601E engines	
Dimensions		
Span	14.34m	47ft 0½in
Length	12.81m	42ft 0¼in
Height	3.95m	12ft 11½in
Wing area	31.00m²	333.67ft²
Weights		
Loaded weight	9,450kg	20,833 lb
Performance		
Max speed	685km/h	426mph
Range	2,800km	1,740 miles
Service ceiling	10,500m	34,450ft

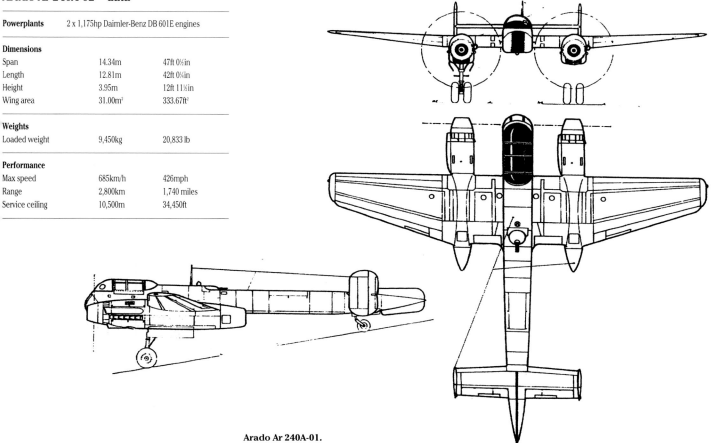

Arado Ar 240A-01.

Arado Ar 240 TL

1942

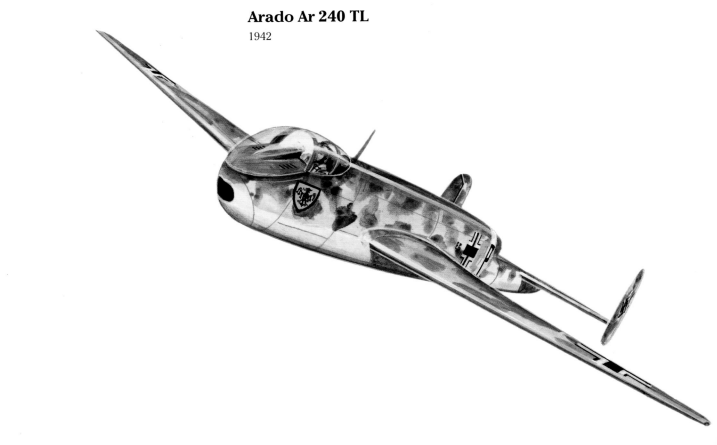

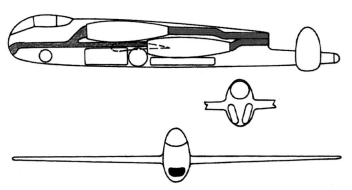

Arado Ar 240 TL sectional side-view showing superimposed turbojets fed by a common cusped fuselage nose intake, fuselage fuel tanks, and undercarriage bays.

Among the numerous projects from Arado Chief Designer Dr-Ing Walter Blume were also studies for a Zerstörer and Nachtjäger (night fighter) which Blume in 1942 designated the Ar 240 TL (for Turbinen-Luftstrahltriebwerk = turbojet).

The new turbojets in 1942 gave the German aviation industry a decisive impulse for revolutionary developments, and Blume also utilised these propulsion units for his TL variant. Externally it resembled, and leads to the conclusion that it used components of the two-seat Ar 240 in featuring a fully-glazed nose crew compartment, mid-wings and twin fins and rudders, but with a redesigned fuselage housing the two turbojets internally so as to reduce drag to a minimum and thus attain high speeds. As can be seen in the accompanying sketches, various turbojet layouts were studied, mounted either directly above each other or stepped. In one study, both turbojets were fed from a common air intake in the fuselage nose beneath the cockpit, and in others, via lateral intakes at the fuselage sides. The high thermal stresses imposed on the fuselage by the turbojets may have presented an insoluble problem. From available documents, no details of dimensions and performance can be ascertained.[1,2]

These studies and original sketches[3] by Walter Blume who concerned himself with these propulsion problems, provides the reader with an insight into the wide diversity of designs undertaken at that time.

[1] Not mentioned by the author is that subsequent to the Arado E 370 twin-jet proposals drawn up by Dipl-Ing Wilhelm van Nes and Emil Eckstein in 1941, which led to the RLM order for a full-scale mock-up of the Ar 234A in February 1942 and for 6 prototypes in April, the design office in late 1941 had also studied a number of mixed-power (piston engine + TL) projects, including a variant of the Ar 240A-0.

A drawing dated 4th November 1941 shows an Ar 240A-0 powered by two 1,350hp DB 601E motors with an auxiliary Jumo 004, the latter based on Junkers data of 4th November 1941. For the sake of simplicity, the complete TL nacelle was to be mounted beneath the forward fuselage such that the lower weapons bays, rear-view periscope and tail surfaces would not be affected by its presence. Dimensions were: span 14.3m (46ft 11in), length 12.8m (42ft), horizontal height on ground 4.1m (13ft 5½in) and wing area 31m² (33.67ft²). Empty weight was 7,700kg (16,975 lb); fuel for the DB 601Es was 1,500 litres (330 gals), for the TL 550 litres (121 gals), and loaded weight 9,700kg (21,385 lb) – 1,000kg (2,205 lb) more than the Ar 240A-0 without the turbojet. By using the TL for for short periods to escape enemy fighters, the reconnaissance aircraft would have had maximum speed raised from 620km/h (385mph) to 703km/h (437mph) at 6km (19,685ft) at normal combat power, and to 764km/h at 6km (475mph at 19,685ft) at emergency power. If used for a total of 30 mins, the TL would have reduced normal range from 1,800km (1,118 miles) at normal combat power to around 1,300km (808 miles). The aim here was to use the same fuel for all three engines to avoid the necessity for separate fuel tanks and fuel switches. The project was later shelved in favour of the pure TL-powered aircraft.

[2] Further Arado mixed-power projects around the turn of 1941/42 were the E 480 and E 490, each with either two DB 609, DB 614 or DB 627s plus one TL unit. Compared with conventional fighters, maximum speed at low-level was 113km/h (70mph) higher and 77km/h (48mph) higher at 10km (32,810ft) altitude. As the pure TL fighter could reach an estimated 780km/h (485mph) at altitude for a lower airframe weight and showed a lower speed deterioration at sea-level and medium altitudes, the mixed-power proposals were not pursued.

[3] The side and head-on views dated 6th September 1942 had a wingspan and length of c.15.3m (50ft 2½in), wing area 27.5m² (296ft²), fuselage height 2.55m (8ft 4½in), fuselage width 1.31m (4ft 3½in), fuel capacity 2,300 litres (506 gals) and loaded weight 9,300kg (20,503 lb). Its equipment was based on the Ar 240F fighter which had multiple forward-firing MG151 and MG131 positions.

Arado E 530
1942

Also in 1942, as a comparison to the two-seat Ar 440, Arado put forward the single-seat E 530 Zerstörer and Schnellbomber (fast bomber) powered by two Daimler-Benz DB 603G engines. Like the Ar 240 and Ar 440, the tailsitter had twin mainwheels retracting rearwards behind the powerplants. The pilot was seated in the port fuselage in a pres-surised cockpit as the aircraft was designed to reach very high altitudes and speeds. The fuselages were joined by constant-chord low wing and tail surfaces; the wings tapering outboard on both leading and trailing edges towards the wingtips. The Kampfzerstörer version was to have had an underwing weapons container housing a 5cm cannon and two MG151/20s, the bomber carrying an external 500kg (1,102 lb) bomb at this position. Details of the aircraft's type of construction are not known.

The DB 603G, which had a single-stage supercharger, had a normal rated altitude without GM-1 boost of 7,000m (22,965ft). With GM-1 (nitrous oxide), this was raised to 9,500m (31,170ft) and produced an additional 95kg (209 lb) exhaust thrust, the aircraft carrying a sufficient quantity to allow GM-1 operation for 30 minutes. In a comparison with the Ar 440 equipped with the same powerplants, despite its twin fuselages the E 530 had estimated speeds both with and without GM-1, 40km/h (25mph) faster than the Ar 440 at all altitudes between 8,500m (27,890ft) and 11,300m (37,070ft). At combat power without GM-1, maximum speed at 8,500m was 705km/h (438mph) or 728km/h (452mph) at emergency power. With GM-1, speeds were raised to 760km/h at 11,200m (472mph at 36,745ft) and 780km/h at 11,300m (485mph at 37,070ft) at combat and emergency power respectively. Initial climb rate at 11.5m/sec (2,264ft/min) compared with only 7.5m/sec (1,476ft/min) for the Ar 440, due to its lighter all-up weight and lower wing loading.

Despite the higher speed advantages, Walter Blume realised that the cockpit had the drawback of a poorer forward view for ammunition and bombload release due to the nose-mounted engine compared with the forward cockpit of the Ar 440 that could attain frontline service much earlier. Although submitted to the RLM, the E 530 was rejected for 'lack of need'.

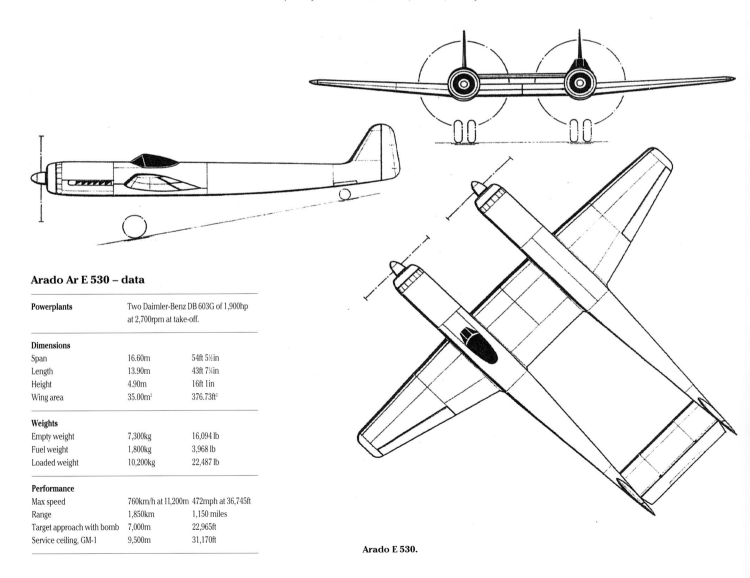

Arado Ar E 530 – data

Powerplants	Two Daimler-Benz DB 603G of 1,900hp at 2,700rpm at take-off.	
Dimensions		
Span	16.60m	54ft 5½in
Length	13.90m	43ft 7¼in
Height	4.90m	16ft 1in
Wing area	35.00m²	376.73ft²
Weights		
Empty weight	7,300kg	16,094 lb
Fuel weight	1,800kg	3,968 lb
Loaded weight	10,200kg	22,487 lb
Performance		
Max speed	760km/h at 11,200m	472mph at 36,745ft
Range	1,850km	1,150 miles
Target approach with bomb	7,000m	22,965ft
Service ceiling, GM-1	9,500m	31,170ft

Arado E 530.

Arado E 556

1943

One of the proposals drawn up by the Arado design office and submitted to the RLM in 1943 was the two-seat E 556 Zerstörer project, a further development of the Ar 440 whose components were to be embodied into the design in order to speed up its readiness for production. In its overall layout, the project incorporated Dr-Ing Blume's ideas, but had little similarity with the Ar 440. A mid-wing design with twin fins and rudders, it had a completely new fuselage, and with the new 16-cylinder DB 609 engines and single MK 108 cannon in the tail, the E 556 was a completely new aircraft capable of meeting all possible operational tasks, but failed to win RLM approval. A decisive factor which doomed the project was the termination of manufacture of the DB 609 by the RLM for unexplained reasons, although this unit developed a superb performance.

Arado E 556 – data

Powerplants	Two Daimler-Benz DB 609s, each of 2,660hp at 2,800rpm at take-off	
Dimensions		
Span	16.35m	53ft 7.7in
Length	14.60m	47ft 10.8in
Weights		
Loaded weight	12,800kg	28,219 lb
Performance		
Max speed	680km/h at 6,000m	423mph at 19,685ft
Armament	4 x MK 108 in fuselage nose, plus 1 x MK 108 in fuselage tail	

Daimler-Benz DB 609 works document.

Daimler Benz	Motorenmuster DB 609 A-F Kraftstoff C3					Blattzahl: I Blatt: 1
Höhe km	Leistungsstufe	PS	U/min	Ladedruck ata	Kraftstoffverbrauch g/PSh	l/h
0	Start- und Notleistung	2660	2800			
0	Steig- und Kampfleistung	2270	2500			
0	Höchstzul. Dauerleistung	1950	2300			
6,6	Notleistung	2450	2800			
8,7	Steig- und Kampfleistung	1980	2500			
8,0	Höchstzul. Dauerleistung	1780	2300			
	Höchste Dauersparleistung					
10,0	Notleistung	1680	2500			

Untersetzung: A u. D 1:1,93   C u. F 1:2,40
           B u. E 1:2,14

Vergleichsgewicht: 1150 kg ± 3 %

Abzuführende Wärmemenge bei Steig-u. Kampfleistung:
a. aus Schmierstoff 75 000 kcal/h in rd 6,5 km; 64 000 an Boden und 60 000 in Volldruckhöhe
b. aus Kühlstoff 720 000 kcal/h
c. aus Ladeluft 100 000 kcal/h in Volldruckhöhe (Zwischenkühlung)

Sonstige Vermerke:
DB 609 A-C = linkslaufend
DB 609 D-F = rechtslaufend

Merkmale und Eignung:
Grundmuster 16 Zylinder in Λ Form mit Dreigang-Grenslader, mit Ladeluft-Kühlung um 50° auf 80° C, über am Motor angebauten Wärmeaustauscher.

Wenn diese Ladeluftkühlung nicht verwendet wird, so verringern sich die Höhenleistungen um 10 – 12 %.

Neue Geräteanordnung und 2 Zündmagnete.

Arado E 556 plan-view.

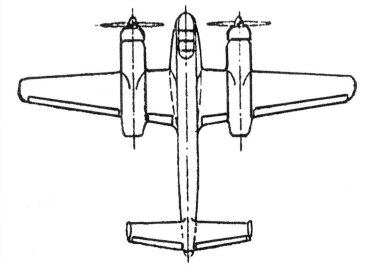

Arado E 560

1943

On the occasion of a conference held at the Arado plant at Landeshut on 18th August 1943, Chief Designer Dr-Ing Walter Blume presented to RLM representatives his designs for a series of combat aircraft, intended as Schnellbomber (fast bomber), Zerstörer (heavy fighter), Nachtjäger (night-fighter), Schlechtwetterjäger (bad-weather fighter) and Aufklärer (reconnaissance) aircraft. The studies carried the company designations TEW 16/43, signifying T = Technik und Konstruktion (technology and design), EW = Entwurf (study), 16 = the design series, and 43 = the year 1943, followed by a two-digit mission category. In this particular case, -22 signified a Flakträger (anti-aircraft weapons carrier), -24 a Zerstörer, and -26 a Schnellbomber. Besides 'TEW', the collective project number E 560 was used.

Starting point for these projects was the Ar 440 which, equipped with high-performance piston engines as well as turbojets, was to achieve an enormous improvement in flight performance. Blume held the view that a 'fast two-seater' represented the best basis for such a multi-purpose aircraft armed with large-calibre weapons. By using a common airframe similar to the Ar 440, he sought to reduce materials and manufacturing costs in order to obtain simplification in series production.

At the apex of the most progressive of Blume's plans was the TEW 16/43-24 scheme for a Kampfzerstörer which in its design layout, formed the basis for further development (eg the TEW 16/43-26 Schnellbomber). The Kampfzerstörer was to receive very powerful armament and for an interim period was to replace the Me 210. Other than three MK 103 and four MK 213s as fixed cannon and two

MK 213s as remote-controlled tail weapons, it was also to be equipped with a weapon firing vertically downwards. This so-called 'Glockenschwengel' (bell-clapper) – a 5cm cannon, was intended to be used against ground targets whilst flying over them. Rheinmetall-Borsig was to manufacture the weapon and special gun mounts according to Blume's plans. Due to considerable problems associated with loading and target sighting mechanisms, development of the weapon was broken off. As powerplants for the Zerstörer, Schnellbomber and Flakträger, the more powerful Junkers-Jumo 012 turbojets, upon entering series production, would replace the improved Jumo 004C turbojets.

The Flakträger TEW 16/43-22 was to combat enemy bombers with the aid of rocket-propelled shells and guided missiles. Blume further maintained that the design studies exhibited could be realised in the shortest possible time, provided that the reciprocating and jet engines currently under development would soon mature to reach production status. The planned mock-up of the Kampfzerstörer did not materialise as the RLM, without giving any reason, ordered all development work to be stopped.[4,5]

[4] An early E 560 design study of October 1942, powered by 4 Jumo 004s, had a span of 18m (59ft 0¾in), length 19.1m (62ft 7in), height 5.4m (17ft 8½in) and tailspan 6.6m (21ft 7¾in). Armament was two MG 151s (200rpg) in fuselage nose, two MG 151s (200rpg) firing rearwards and one MG 151 in the tail remote-controlled by periscope, plus two ETC 1000 racks with combined bomb jettison and rear-vision sight and corrective calculator.
[5] The 'TL 1500' studies of February 1942 envisaged a multi-purpose swept wing twin-jet powered by two 1,600kg (3,527 lb) thrust turbojets. Span and

length were 16.2m (53ft 1¾in), wing area 46.6m² (501.59ft²), empty weight 8,370kg (18,464 lb) and loaded weight 16,000kg (35,274 lb). Another, powered by two 2,200kg (4,850 lb) thrust turbojets, had span 19m (62ft 4in), wing area 64m² (688.87ft²), empty weight 12,310kg (27,142 lb), loaded weight 16,000kg (35,274 lb) and take-off run 1,200m (3,937ft).

Further E 560 variants, powered by four HeS 011 or four Jumo 004C turbojets, had span 18m (59ft 0¾in), wing area 48m² (516.66ft²) and take-off weight 17,000kg (37,478 lb) with various armament, bombloads and performances. Maximum speed ranged from 790km/h (491mph) with the Jumo 004C to 935km/h (581mph) with the HeS 011 at sea-level, with ranges of 2,300km (1,429 miles) and 1,920km (1,193 miles) respectively.

Basic Arado E 560 / TEW – data

Powerplants	Two 2,200kg (4,853 lb) thrust turbojets (February 1943 status)	
Dimensions		
Span	16.20m	53ft 1¾in
Length	18.00m	59ft 0¾in
Height	3.00m	9ft 10in
Wing area	46.00m²	489.28ft²
Wing sweep	25° at ¼ chord	
Weights		
Loaded weight	16,000kg	35,274 lb

Arado E 560 (TEW 16/43 study).

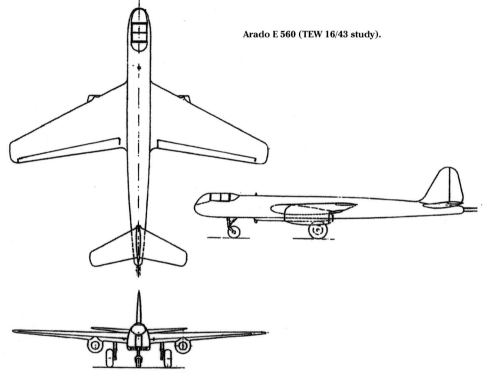

Arado E 560 chart of armament and bombs for the various roles.

	Schnellbomber	Zerstörer	Nachtjäger	Schlechtwetterjäger	Zwangs – Aufklärer
	3 2 MK 213 ferngesteuert mit je 300 Schuß	1 3 MK 103 starr mit je 100 Schuß	1 2 MK 108 starr mit je 100 Schuß	1 3 MK 103 starr mit je 100 Schuß	3 2 MK 213 ferngesteuert mit je 300 Schuß
	4 Periskopvisier	4 MK 213 starr mit je 300 Schuß	3 MK 213 starr mit je 300 Schuß	4 MK 213 starr mit je 300 Schuß	4 Periskopvisier
	2 Vor-Rückblickvisier m.BZA-Anlage	2 Vor-Rückblickvisier	2 Vor-Rückblickvisier	2 Vor-Rückblickvisier	
	7 Lotfe	3 2 MK 213 ferngesteuert mit	5 2 MK 108 starr unter 70° mit je 100 Schuß	3 2 MK 213 ferngesteuert mit je 300 Schuß	
		4 Periskopvisier	3 MK 213 starr unter 70° mit je 300 Schuß	4 Periskopvisier	
	6 Schloßlafetten für 1 x 2500 kg 2 x 1000 " 4 x 500 " 10 x 50 " 700 x 1 "	6 Schloßlafetten für 2 x 250 2 x 500 4 x 50	Nachtjäger-Leitanlage	Schlechtwetter-jäger-Leitanlage	8 2 Reihenbildner jeder Größe

Arado E 654

1941

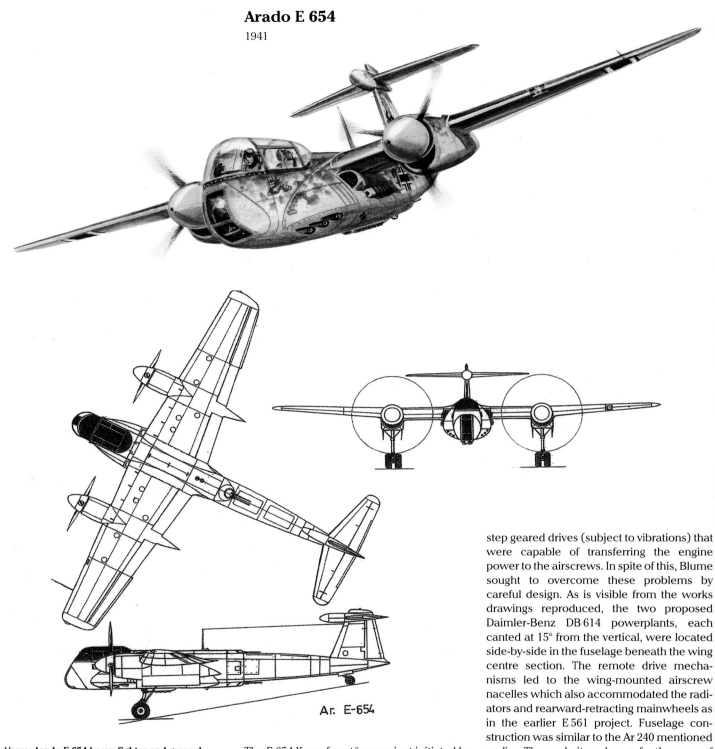

Ar. E-654

Above: **Arado E 654 heavy fighter and ground attack aircraft.**

The E 654 Kampfzerstörer project initiated by Arado Chief Designer Dipl.Ing. Walter Blume and employing buried engines driving wing-mounted propellers was a controversial design dating from the autumn of 1941. Remote drives of this type, experimented with by Junkers and Dornier, brought satisfactory results when short distances away from the engine were involved. There was, however, no practical experience with multi-

step geared drives (subject to vibrations) that were capable of transferring the engine power to the airscrews. In spite of this, Blume sought to overcome these problems by careful design. As is visible from the works drawings reproduced, the two proposed Daimler-Benz DB 614 powerplants, each canted at 15° from the vertical, were located side-by-side in the fuselage beneath the wing centre section. The remote drive mechanisms led to the wing-mounted airscrew nacelles which also accommodated the radiators and rearward-retracting mainwheels as in the earlier E 561 project. Fuselage construction was similar to the Ar 240 mentioned earlier. The cockpit enclosure for the crew of two was located far forward near the nose, the gunner operating the weapon positions on the dorsal and ventral fuselage by means of a periscopic sight. The pilot controlled the six MK 108 cannon in the fuselage sides ahead of the engine compartment. From the handwritten documents prepared by Walter Blume, he termed the E 654 the 'Scorpion' on account of its porcupine-like streamlined aerodynamic body at the junction of the high-mounted tailplane and vertical fin. As had happened with the E 561, this project also suffered not only from material and delivery problems in relation to the remote drive system, but also the lack of RLM interest in this type of propulsion idea. Derived from later versions of the DB 603, both the DB 614 and DB 627 were each designed to produce 2,000hp at take-off. The DB 614 was abandoned in 1942.

Arado E 654 – data

Powerplant	2 x 2,000hp Daimler-Benz DB 614 or DB 627 in-line engines	
Dimensions		
Span	14.34m	47ft 0½in
Length	12.81m	42ft 0¼in
Height	3.95m	12ft 11½in
Wing area	31.30m²	336.90ft²
Armament	6 x MK108 in fuselage cheeks, plus 4 x MG131 in dorsal (B-Stand) and ventral (C-Stand) positions	

Below: **Arado E 654 works drawing of 8th October 1941 showing weapons and engine nacelle data.**

Bottom: **Arado E 654 works drawing of 1st October 1941 showing the proposed engine installation.**

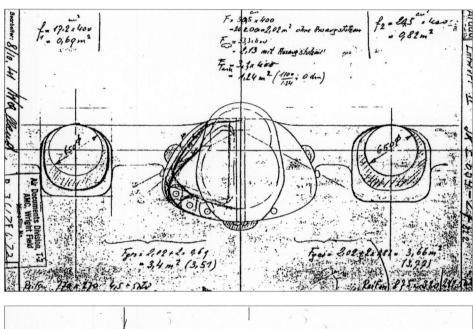

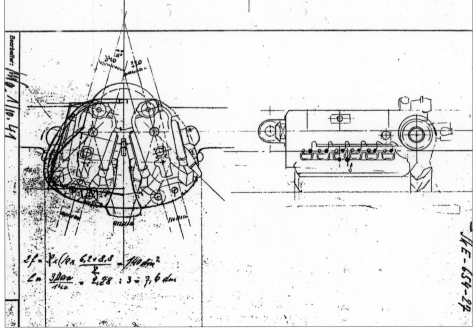

Dornier Do 217N – Interim Solution for a Heavy Night Fighter

1942 to 1943

Lack of suitable night fighters in 1942 forced the RLM to employ the Do 217, which upon conversion of the series-produced Do 217E-2, undertook night fighting duties as the Do 217J. For this purpose, the standard glazed crew compartment of the Do 217E-2 was replaced by one with a redesigned solid nose housing four MG17 and four MG FF guns in the Do 217J-1. Lack of intercept radar resulted in the installation of the FuG 202 'Lichtenstein B/C 1' search radar and aerial array in the nose, this variant becoming the Do 217J-2 powered by two 1,850hp BMW 801ML radials and weighing 13,180kg (29,057 lb) at take-off.

This was followed shortly after by a new night fighter model, the Do 217N, whose airframe and engines were the same as the Do 217M-1 with the exception of the fuselage nose portion which was taken over from the Do 217J-2 together with its armament. As an interim solution, the four MG FFs were replaced by four MG151/20 cannon that had a higher velocity and better trajectory. Simultaneously, the dorsal (B-Stand) and ventral (C-Stand) armament positions were removed, the latter covered by a long wooden fairing which extended the fuselage undersurface towards the rear. With the installation at mid-fuselage of four MG151/20s to fire 70° obliquely upwards as a standard Rüstsatz 22, the series-produced aircraft bore the designation Do 217N/R22. At a take-off weight of 13,500kg (29,762 lb), it attained a maximum speed of 500km/h at 6,000m (311mph at 19,685ft). A further electronic target identification system installed was the FuG 350 'Naxos', tuned to

the frequency of the H2S radar of RAF pathfinder aircraft.* With this equipment, the series-built model was known as the Do 217N-2, and with a loaded weight of 13,200kg (29,101 lb), had a maximum speed of 515km/h at 6,000m (320mph at 19,685ft). Series production of the Do 217 ended in 1943 after a total of 1,730 machines had been built, consisting of 1,366 bombers and 364 night fighters.

*Other special equipment (nò Rüstsatz = field conversion pack number) included the FuG 25 providing the flak batteries with identification and acted as a direction indicator for the 'Himmelbett' controller, and the FuG 227 'Flensburg' for homing onto the tail-warning 'Monica' fitted in RAF bombers.

Dornier Do 217N-2 – data

Powerplant   Two Daimler-Benz DB 603LA of 1,750hp at take-off (normal) or 2,150hp with MW 50, or 2 x 1,560hp BMW 801 ML-2 radials.

Dimensions		
Span	19.00m	62ft 4in
Length	18.10m	59ft 4½in
Height	5.00m	16ft 5in
Wing area	56.60m²	609.22ft²

Weights		
Loaded weight	13,500kg	29,762 lb

Performance		
Max speed	435-500km/h	270-310mph
Range	1,750km	1,087 miles
Service ceiling, without R22	8,900m	29,200ft

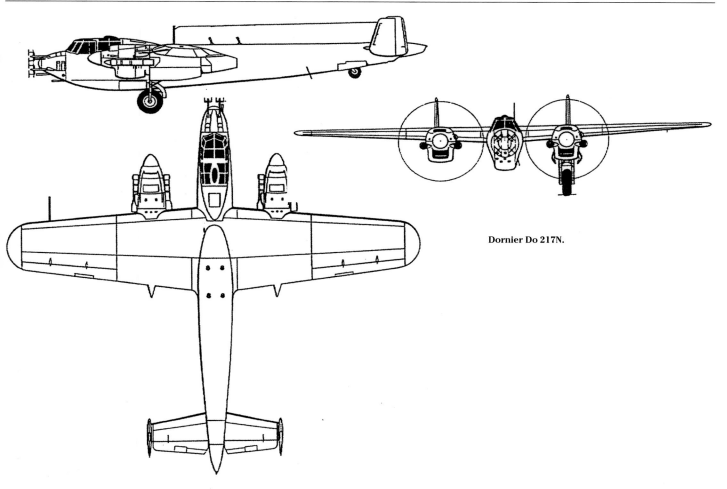

Dornier Do 217N.

Dornier Do 217N night
fighter without night
search devices.

Dornier Do 217N with
FuG 212 'Lichtenstein
C-1' antenna.

Dornier P 232/2

1943

Dornier P 232/2 with fuselage lateral air intakes for the turbojet.

The Dornier P 232/2 project* was initiated in May 1943 as a single-seat fast bomber, heavy fighter and low-level attack aircraft. Its overall configuration was derived from the earlier Do 335 'Pfeil' (arrow), nicknamed the 'Ameisenbär' (ant-eater), where the rear reciprocating engine was replaced by a Jumo 004C turbojet which could be attached to the rear fuselage at the standard attachment points. When not in use, the two lateral air intake scoops for the turbojet could be retracted into the fuselage to reduce drag.

The combination of airscrew and turbojet enabled the advantage of both types of propulsion unit to be realised: long ranges when the piston engine was used alone, and a high cruising speed especially at low altitude with both engines in operation. In the latter case, calculations gave a maximum continuous cruising speed of 646km/h (401mph) – an increase of 85km/h (53mph) over the Do 335, and a range of 1,250km (777 miles). With the TL engine switched off, range was 3,500km at 530km/h (2,175 miles at 329mph). According to the construction brochure of 28th May 1943, the tactical capabilities of the P 232/2 at low-level were ideal. Since turbojet installation resulted in reduced weight, the expendable load could be increased to 1,000kg (2,205 lb). Fixed armament was to consist of a single MK 103 firing through the propeller shaft and two

MG 151/20 (200rpg) above the engine, replaced later by two MK 108s.

A variant of the P 232/2 was the P 232/3 of September 1943, where the lateral air intakes were replaced by a single intake scoop above the rear fuselage. When only the forward engine was in operation, the intake could be closed by a spring-operated flap to reduce drag. This air intake location, selected in various German projects, was also adopted abroad after the war in civil and military aircraft. Work on the project was stopped in late autumn due to the firm's concentration on the Do 335.

* This had been preceded by the similar P 231/3 'mixed-power' project that had a DB 603G in the fuselage nose and a rear Jumo 004C also fed by lateral air intakes .

Dornier P 232/2 – data

Powerplants	1 x 1,900hp Daimler-Benz DB 603G and 1 x 1,200kg (2,646 lb) reheat thrust Junkers-Jumo 004C turbojet.		

Dimensions			
Span	13.80m	45ft 3¼in	
Length	14.00m	45ft 11¼in	
Height	4.50m	14ft 9¼in	
Wing area	38.50m²	414.40ft²	

Weights			
Empty weight	5,370kg	11,839 lb	
Fuel capacity	2,550 litres	561 gals	
Loaded weight	8,450kg	18,629 lb	

Performance			
Max speed at sea level	660km/h	410mph	
at 8,700m (28,545ft)	808km/h	502mph	
Service ceiling, gross max wt	13,200m	43,310ft	
Take-off run	740m	2,428ft	

Armament	1 x MK103 and 2 x MG151/20 or 2 x MK103		

Bombload	500kg (1,102 lb) normal or 1,000kg (2,205 lb) at overload		

Dornier P 232/3 – data

Powerplants	1 x 1,900hp Daimler-Benz DB 603G and 1 x 1,200kg (2,646 lb) reheat thrust Junkers-Jumo 004C turbojet.		

Dimensions			
Span	13.80m	45ft 3¼in	
Length	13.80m	45ft 3¼in	
Height	5.60m	18ft 4½in	
Wing area	33.50m²	360.58ft²	

Weights			
Empty weight	5,100kg	11,243 lb	
Fuel capacity	1,980 litres	436 gals	
Loaded weight	7,750kg	17,086 lb	

Performance			
Max speed at sea level	675km/h	419mph	
at 8,700m (28,545ft)	838km/h	528mph	
Service ceiling, gross max wt	13,300m	43,435ft	
Take-off run	580m	1,902ft	

Armament	1 x MK103 and 2 x MG151/20 or 2 x MK103		

Bombload	500kg (1,102 lb) normal or 1,000kg (2,205 lb) at overload		

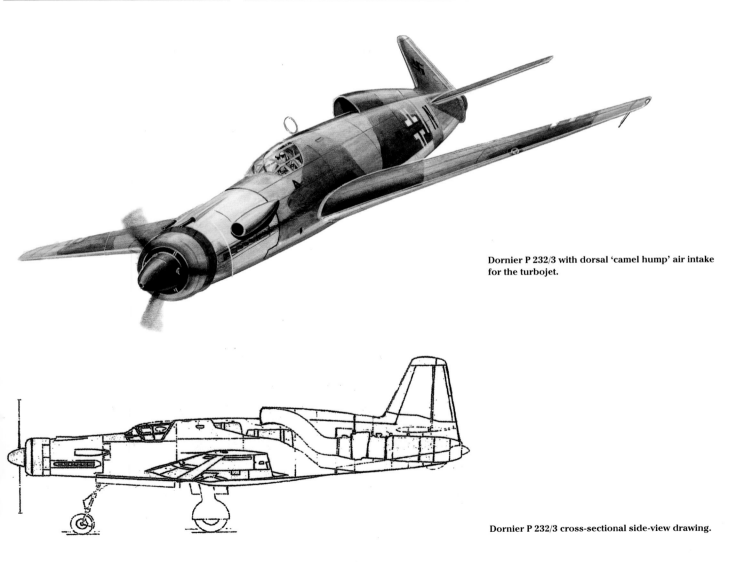

Dornier P 232/3 with dorsal 'camel hump' air intake for the turbojet.

Dornier P 232/3 cross-sectional side-view drawing.

Heinkel P1055.01-16 –
From Record Aircraft to Kampfzerstörer

1941 to 1942

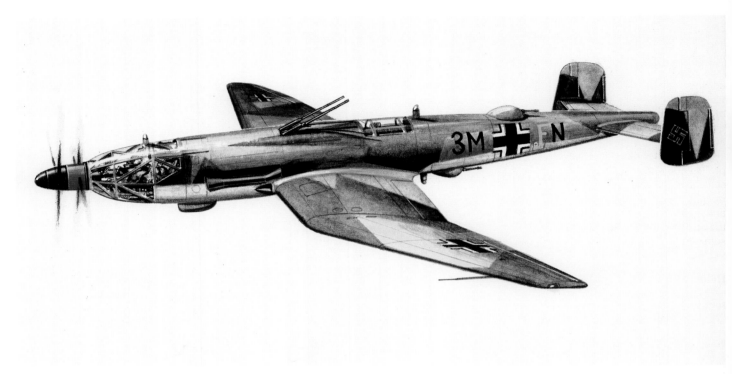

In November 1940, the Heinkel-Werke submitted to the RLM its Kampfzerstörer proposal whose design and layout can be viewed as a further development of the He 119 'world record' aircraft. From documents of the Ernst Heinkel Studien GmbH, Vienna, the P 1055.01-05 of 24th November 1940 was also to have had a coupled 24-cylinder DB 606 of 2,700hp in the fuselage behind the pilot and forward gunner, the second gunner seated in the fuselage facing aft. Other drawings of July 1942 describe a Kampf- and Begleitzerstörer (combat and escort heavy fighter) which corresponded to an enlarged He 119 (P 1055.01-01). Besides the twin fins and rudders and new wings, the project featured a very tall nose-wheel tricycle undercarriage (proposed by Dipl-Ing Rudolf Lusser) not incorporated up to that time in any other German aircraft.[7]

[7] Except for the 'very tall' nosewheel needed for propeller ground clearance, several aircraft had been designed from the outset between 1939 and 1942 with a tricycle undercarriage, among them the Arado Ar 232, Ar 233, Ar E 340, Ar E 470; Dornier P 231, P 232; several Focke-Wulf projects; Göppingen Gö 9; Heinkel He 219, He 280; Henschel P.54, P.72, P.75; Junkers Ju 287; Lippisch P.04-114, P.11(TL), and the Messerschmitt Me 264, Me 265 and Me 329, to mention but a few.

The later P 1055.01-16 three-crew project for a Kampf- and Begleitzerstörer, was to have been powered by a DB 613C/D with which a speed of 720km/h at 9,000m (447mph at 29,530ft) was expected. The P 1055.01-05 and P 1055.01-16 were both mid-wing aircraft but with the Daimler-Benz powerplant located ahead of the wings instead of astride the wingroot forward section on the He 119. The engine extension shaft, located on the fuselage axis and which ran through the cockpit between the two crew members, was some 10cm (4in) lower to enable the engine to be attached to the fuselage bulkheads – unlike the He 119 where the motor was fastened to the wing main spar. This engine relocation resulted in a more forward c.g. position and allowed an improvement in engine and drive shaft access and maintenance. In the performance envelope, noticeable differences existed in engine power at rated altitude. Estimated maximum speeds ranged from 650km/h (404mph) to 720km/h (447mph). The use of auxiliary Argus pulsejet aids fitted as Rüstsätze (standard equipment packs) would have brought about further performance increases. The RLM agreed with the Heinkel suggestion to employ exchangeable outer wing sections which varied the wing area from 35m² (376.73ft²) to 45m² (484.36ft²). The extraordinarily strong armament intended for the P 1055.01-16 Zerstörer escort variant earned it the nickname of 'Waffenigel' (weapon porcupine). The RLM also demanded armour-protected retractable weapon positions as well as protection from enemy fire for the intended annular radiator located behind the crew compartment around the entire fuselage periphery. The possibility was also to be investigated of installing raised 'teardrop' hoods for the pilot and forward air gunner. In addition to the two dorsal weapon turrets with their triple and quadruple weapon mounts, a ventral weapon stand was also envisaged; all weapons positions to have had automatic sighting and firing mechanisms.

The RLM was basically in agreement with project documentation submitted by Heinkel since all performance criteria such as speed, range and armament fulfilled requirements, and besides the provision of a pressure-cabin the RLM held the load factor of five to be sufficient.

Some weeks later, however, to his astonishment Ernst Heinkel learned that the RLM had lost interest in the project. Enquiries made with the Luftwaffe leadership resulted in rejection without any reason being given. But that was not all: a directive had even forbidden Heinkel to continue development work on the P 1055.01-16. Exactly why this revolutionary advanced development was suppressed and hindered by the highest authorities is a secret that still remains unsolved to this day. Was it shortsightedness, sabotage, or simply mere ignorance? All in all, the developments by the Heinkel-Werke were not the only victims of grave misdecisions, for the entire aero-engine industry had suffered from the same causes. The eternal to and fro of often senseless manufacturing bans which were partially retracted in 1944/45 when large parts of the aviation industry already lay in ruins is simply illogical, if one is to exclude the term sabotage.

Heinkel P 1055.01-16 – data

Powerplant	One turbo-supercharged 3,800hp Daimler-Benz DB 613C/D driving 4.4m (14ft 5¼in) diameter contraprops	
Dimensions		
Span	19.25m	63ft 2in
Length	15.35m	50ft 4¼in
Height	4.95m	16ft 3in
Wing area	45.00m²	484.36ft²
Weights		
Mean weight	c.11,000kg	24,250 lb
Performance		
Max speed	720km/h at 9,000m	447mph at 29,530ft
Service ceiling	11,000m	36,090ft
Armament	3 x MG151 in mid dorsal turret	
	4 x MG131 in rear dorsal turret	
	4 x MG131 in ventral turret	

Heinkel P 1055.01-16 sectional side-view.

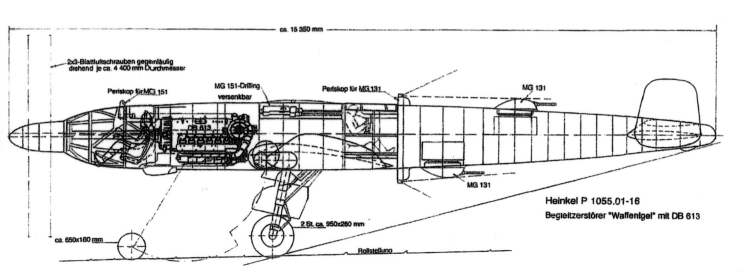

Heinkel P 1055.01-16
Begleitzerstörer "Waffenigel" mit DB 613

Heinkel P1065

1942

Heinkel P1065.01-19 (Ic).

Heinkel P1065.01-19 (Ic) sectional side-view.

In response to an RLM specification of June 1942 for a Kampfzerstörer and so-called Arbeitsflugzeug (army co-operation aircraft) as a replacement for the Bf110, Ju88/Ju188 and Do217, Heinkel participated with the P1065 project. The requirement was for a three-seat bomber and Zerstörer able to attain a speed between 600km/h (373mph) and 700km/h (435mph) and a range from 400km (248 miles) to 1,000km (621 miles). The RLM-suggested Arbeitsflugzeug designation, as mentioned earlier, had already been used in the First World War for a number of aircraft which, depending on the type of equipment fitted, could be employed in the bomber, fighter, or reconnaissance roles. Whilst the Reichswehr (former Defence Ministry) adopted the term Mehrzweckflugzeug (multi-purpose aircraft), the RLM reverted to the original term.

Heinkel submitted a host of studies and designs, all under the P1065 project designation, differing considerably in their design layouts. The requirement centred around a medium class of Arbeitsflugzeug with series production envisaged for 1945/46. By conscious reduction of the range to a value that had been shown to be sufficient for operational needs, a medium-weight combat aircraft was to be evolved using engines that had attained production maturity such as the BMW 801E, DB 609, DB 619, Jumo 222 and BMW 803 which had considerable improvements over the aircraft it was to replace. This included simpler manufacturing techniques calling for minimum maintenance and extensive use of exchangeable materials such as steel and wood. Powerful defensive and offensive armament was to be housed in manually-operated weapons positions, arranged so that there were no 'dead angles' within the field of fire.

The Heinkel project studies encompassed designs with various powerplants and installation possibilities. Thus, the P1065-Ia bomber and heavy fighter was to have had the BMW 801E as well as the Jumo 222C as in the P1065-Ic, depending on the mission requirement. The P1065-IIc variants envisaged an asymmetrical low-level and ground attack aircraft with offset crew compartment and powered by a single 28-cylinder BMW 803 driving contraprops, whereas the P1065-IIIb and P1065-IIIc fast bombers were to use the 28-cylinder DB 619 (coupled DB 609) similar to the P1055.01-16 described above, housed in the fuselage behind the nose crew compartment. The use of various types of powerplants resulted in varying performance figures, all of which were based on the maximum engine power developed. Similar to the P1055.01-16, the P1065 projects also included the use of auxiliary propulsion units in the shape of Argus As 014 and As 044 pulse-jets to supplement the airscrew power, as mentioned in a document of 15th June 1942.

All of these projects that were far advanced in the planning and development stage did not meet with RLM concurrence, and even with other developments, Heinkel had little luck with the RLM. Scoffers had maintained that Prof Dr-Ing Ernst Heinkel, with his Swabian pigheadedness and his 'un-Germanic appearance' had thrown away his chances of favourable decisions by certain 'responsible dunderheads' in the RLM.

Heinkel P1065 – data

Model	P1065-Ia	P1065-Ib	P1065-Ic	P1065-IIc	P1065-IIIb
Crew	Three	Three	Two	Two	Three
Powerplant	Jumo 222C	Jumo 222C	DB 609	BMW 803	DB 619
Take-off power	2 x 2,600hp	2 x 2,600hp	2 x 2,270hp	2 x 3,500hp	1 x 4,540hp
Dimensions					
Span	23.00m	23.00m	23.00m	20.40m	20.30m
	75ft 5½in	75ft 5½in	75ft 5¼in	66ft 11¼in	66ft 7¼in
Length	15.40m	15.40m	15.40m	19.50m	14.40m
	50ft 6¼in	50ft 6¼in	50ft 6¼in	63ft 11¾in	47ft 3in
Wing area	51.00m²	56.50m²	57.00m²	57.50m²	45.00m²
	548.96ft²	608.16ft²	613.55ft²	618.93ft²	484.38ft²
Weights					
Loaded weight	15,220kg	16,890kg	17,050kg	15,500kg	14,870kg
	33,560 lb	37,242 lb	37,595 lb	34,177 lb	32,788 lb
Performance					
Max speed at sea level	550km/h	572km/h	525km/h	519km/h	560km/h
	342mph	355mph	326mph	323mph	348mph

Heinkel P1065.01-20 (Ia).

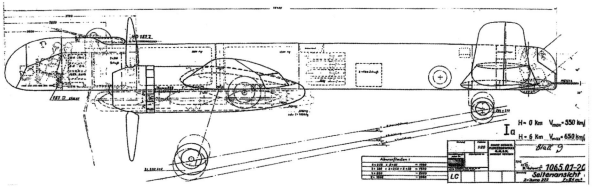

Heinkel P1065.01-20 (Ia) sectional side-view.

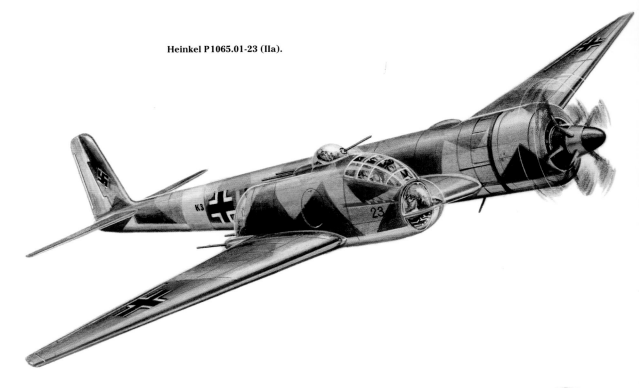

Heinkel P1065.01-23 (IIa).

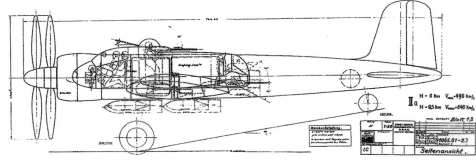

Heinkel P1065.01-23 (IIa) sectional side-view.

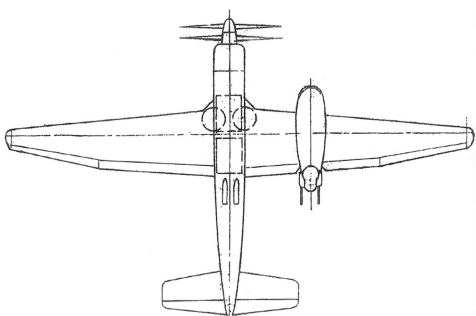

Heinkel P1065.01-23 (IIa) plan-view.

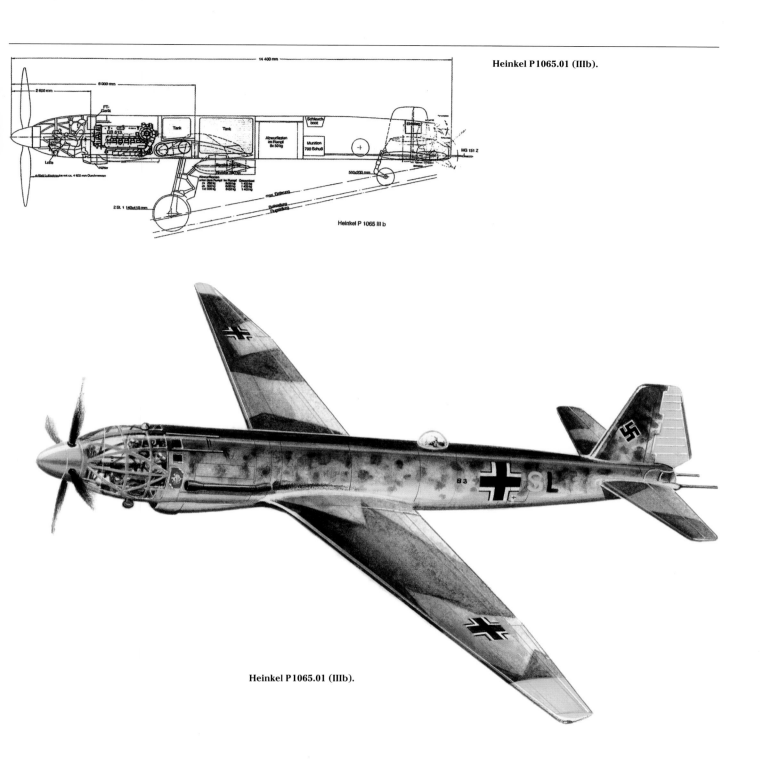

Heinkel P 1065.01 (IIIb).

Heinkel P 1065 III b

Heinkel P 1065.01 (IIIb).

Hütter Fernzerstörer
1942

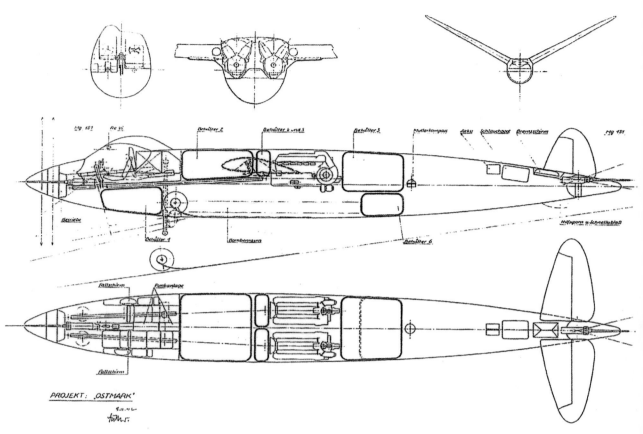

PROJEKT: 'OSTMARK'

Sectional views of the Hütter long-range Zerstörer project. Drawing key (upper row): MG151; Revi (gunsight); fuel tanks Nos 2, 4 and 5, 3; compass; accumulator; dinghy; brake parachute; MG151. (Lower row): gearing; fuel tank No.1; bomb bay; fuel tank No.6; quick-release tailwheel. Plan-view: parachutes, and radio installation (between crew seats).

Under the date of 9th December 1942, Wolfgang Hütter submitted to the RLM a further project designated the 'Ostmark' (Ost = east, Mark = frontier, limit, boundary), consisting of a three-seat twin-engined long-range Zerstörer of all-metal construction. Proposed propulsion units were the DB 601A-1 installed side-by-side at mid-fuselage, each engine driving one set of contra-rotating four-bladed airscrews via extension shafts and so arranged that in the event of one engine being shut down, the remaining one would provide the power to drive both sets of propellers, the crew of three located near the fuselage nose above the drive shaft. The fuel, housed in six fuselage tanks, enabled the aircraft to have a penetration radius of 3,000km (1,864 miles). As an alternative to the 1,000kg (2,205 lb) bombload, a BT 1000 Bomb-Torpedo could be housed internally in the long bomb bay. Armament consisted of a heavy Bordkanone firing through the propeller boss.* Additional weapons for attacking air and ground targets were contained in underwing nacelles. No construction was undertaken as the RLM rejected the project, and no reliable details concerning dimensions and other data are available.†

* The drawing shows an MG151 firing through the propeller boss and another MG151 at the fuselage rear between the 30° butterfly tail surfaces.

† Based on the known dimensions of the DB 601A-1, measurements from the drawing give the following approximates: fuselage maximum width 1.8m (5ft 10¾in), overall length 15.2m (49ft 10½in), horizontal height 3.72m (12ft 2⅜in) and tail span 4.84m (15ft 10½in). The wing root chord of no more than 1.5m (4ft 11in) indicates that the aircraft had a typical Hütter sailplane-type high aspect ratio wing as on the later Hü 211 of 1944. The narrow-track mainwheels retracted into the fuselage ahead of the bomb compartment.

Left: **Prof Dr-Ing Ulrich Hütter (1910-1990) at work on an aero-engine. Together with his brother Wolfgang, he became well-known through the design of high-performance sailplanes (eg the Schempp-Hirth Gö 4 and Hi 20 'Mose' of 1937, and for his development of dive-brakes used by the DFS in 1938). During the war, the brothers developed various projects of the so-called 'Ostmark' series, among them the long-range Zerstörer and various long-range reconnaissance and bomber aircraft.**

Lippisch P 09 – A Rocket-powered Schnellbomber

1942

The project designations P 09 to P 13 inclusive, were confusingly used more than once to cover different design layouts for aircraft powered by reciprocating engines, turbojets, rocket motors and finally, ramjets. This single-seat fast bomber proposal, like the earlier turbojet-powered P 09 fighter project of November 1941, was also designed by Dipl-Ing Rudolf Rentel and submitted to the RLM in May 1942. Unlike the earlier swept wing tailless design, this P 09 Schnellbomber featured a delta wing with a leading edge sweep of 30° and unswept trailing edge. Dimensionally smaller, it was powered by three Walter HWK liquid-propellant rocket motors – the two more powerful units housed in the rear fuselage beside the single vertical fin and rudder, with the third lower-thrust motor, presumably used for the cruising condition, exhausting directly beneath the tail. The deep fuselage centre section of 2m (6ft 6¾in) width and some 1.85m (6ft 0¾in) depth, housed the two large fuel tanks and variable internal bomb load that could consist of an SB 1000 bomb, an LT 350 Luft-Torpedo (air-launched torpedo), or a PC 1000RS rocket-propelled Panzer-bombe (armour-piercing bomb) weighing 978kg (2,156 lb) and 2.2m (7ft 2½in) in length plus two forward-firing guns, one on each side of the cockpit floor. The outboard portions of the wings had fixed leading-edge slots, and ahead of the constant-chord inboard flaps, each wing had upper and lower pairs of dive brakes. Like the Me 163B, the three-point undercarriage consisted of a semi-enclosed retractable tailwheel and a pair of landing skids situated 1.4m (4ft 7in) apart instead of the single skid as on the Me 163B, the take-off being with the aid of a wheeled dolly.

Lippisch P 09 – data

Powerplants	2 x 1,500kg (3,307 lb) thrust HWK and one lower-thrust HWK liquid-propellant rocket units	

Dimensions		
Span	10.00m	32ft 9⅝in
Length	7.40m	24ft 3¼in
Height, skids down	3.40m	11ft 2in

Performance		
Max speed	1,000km/h	621mph
Range	300km	186 miles
Service ceiling	12,500m	41,010ft

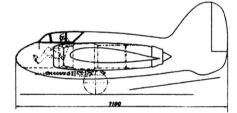

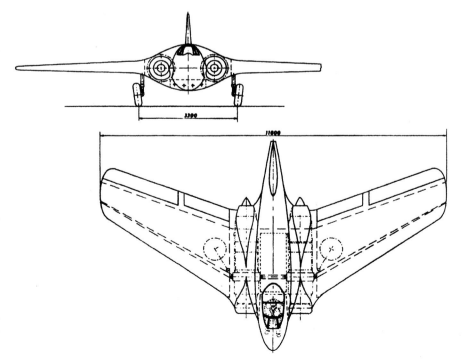

Lippisch P 10-108

1942

The P10-108 Schnellbomber and Zerstörer project of May 1942 differed from the earlier P10 project of November 1941 already described, in being powered by a coupled Daimler-Benz DB 606 engine consisting of two 12-cylinder DB 601s installed side-by-side at the c.g. in the upper fuselage behind the cockpit and driving four-bladed pusher airscrews of 4m (13ft 1½in) diameter via an equally long extension shaft. Like the P 09 described above, it had a fixed forward-firing armament of two MGs but housed in the extended wingroot contours beside the cockpit, and to improve visibility for take-off, landing and bomb aiming, forward and downward-vision panels were provided in the nose. As shown in the drawing, the internal bay beneath the powerplant could accommodate bombloads of up to 1,000kg (2,205 lb).

Dipl-Ing Hubert and Dr-Ing Wurster have been named as project leaders for the P10-108 which appeared in RLM documents as the Me 334, but this designation cannot be confirmed to date with 100% certainty. At that time, Dr Lippisch was still engaged at the Messerschmitt AG. The P10-108 did not pass the drawing board stage as the RLM did not make any funds available for construction of a mock-up.

Lippisch P 10-108 – data

Powerplant	1 x 2,400hp Daimler-Benz DB 606 driving four-bladed pusher airscrews	
Dimensions		
Span	18.00m	59ft 0⅝in
Length	9.85m	32ft 3⅜in
Height, level with u/c down	6.00m	19ft 8¼in
Armament	2 x MG151/20 in nose, 2 x MK108 in tail	
Bombload	1,000kg (2,205 lb)	

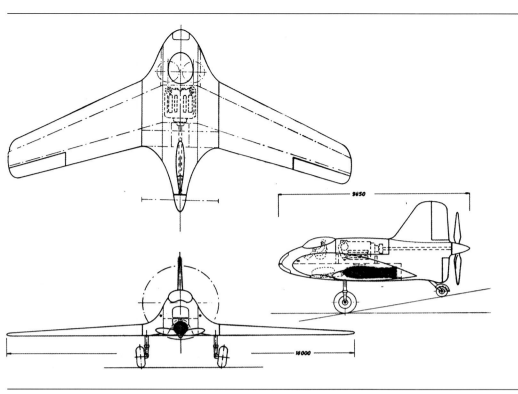

Lippisch P 11-92

1942

The Lippisch P11 series of Schnellbomber and Zerstörer designs undertaken by team member Dipl-Ing Hendrick, dates from September 1942. Like the earlier P 09 and P 10 of November 1941, it was of a 30° swept wing tailless configuration and powered by two Jumo 004 turbojets buried in the thickened wing roots, but had a tricycle nosewheel undercarriage in place of the skids and tail-wheel. The nosewheel retracted rearwards beneath the pilot's feet, the 3.8m (12ft 5½in) track mainwheels diagonally outwards into the wings behind the main spar. New in this proposal was the provision of a clear-vision teardrop canopy for the two crew members seated in tandem in the pressurised cockpit and auxiliary flaps ahead of the trailing edge controls which stretched right across beneath the fuselage centre section. In addition to the internally housed SC1000 bomb, armament consisted of two fuselage MK108s and two wingroot-mounted MG151/20 cannon. The design also had two built-in RATO units on either side of the fin and rudder. The project, however, ended on the drawing board in November 1942 as the RLM refused to finance the construction of a mock-up.

Lippisch P11-92 – data

Powerplants 2 x 750kg (1,653 lb) thrust Junkers-Jumo 004 turbojets and two Rheinmetall-Borsig solid-propellant RATO units.

Dimensions

Span	13.00m	42ft 7¾in
Length	7.50m	24ft 7¼in
Height, wheels down	3.00m	9ft 10in

Armament 2 x MG151/20 in wing roots, 2 x MK108 in fuselage

Bombload 1,000kg (2,205 lb) internally

Lippisch P11-92 – drawing dated 13th September 1942.

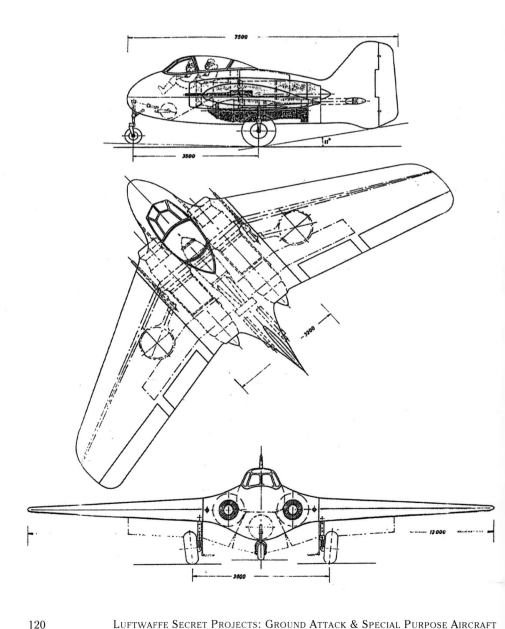

Lippisch P11-105

1942

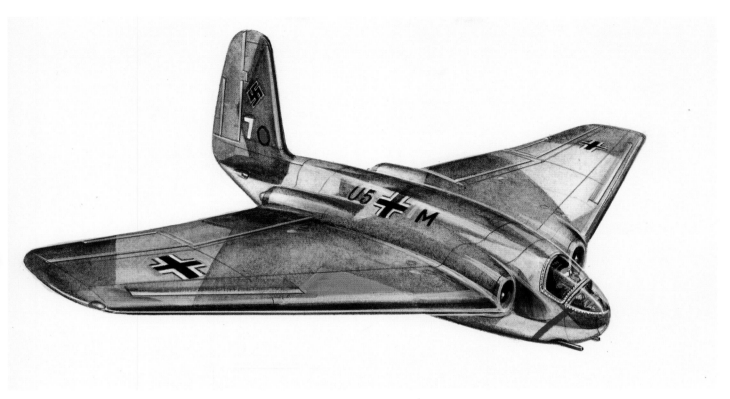

Like the earlier P11-92 proposal, design of the P11-105 Schnellbomber and Zerstörer 30° swept wing tailless aircraft was also headed by Dipl-Ing Hendrick. The initial design study of 13th November 1942 was laid out for a crew of two, but this was reduced to one in the revised design of 1st December 1942. Unlike earlier studies, the cockpit blended fully into the fuselage nose contours. Although the wingspan was slightly reduced, wing root and tip chords were both increased, the former to 3.8m (12ft 5½in) at the junction with the turbojet housing. The four wing tanks and forward fuselage tank held a total of 2,200kg (4,850 lb) of fuel for the Jumo 004 turbojets.

In the P11-105, the tricycle undercarriage track was reduced to 2.9m (9ft 6½in), the nosewheel rotating to lie flat beneath the pilot's seat. As in the earlier project, an SC1000 bomb was accommodated internally. An unusual feature of this proposal was that the heightened vertical fin incorporated rectangular stabilising surfaces of total area 3m² (32.29ft²) on each side that extended hydraulically to the horizontal position to take advantage of the increased moment arm and increase maximum lift coefficient. This type of foldable auxiliary tail surface had previously been tested on P01 models in the wind

tunnel. To shorten the take-off run, two solid-propellant RATO units were housed beneath the fin and rudder. Although the P11-105 fulfilled RLM requirements, only a few weeks after design details had been submitted, the project was turned down by the Technisches Amt without giving a reason.

Lippisch P11-105 – data

Powerplants	2 x 760kg (1,675 lb) thrust Junkers-Jumo 004 turbojets and 2 Rheinmetall-Borsig RATO units.	
Dimensions		
Span	12.65m	41ft 6.0in
Length	8.14m	26ft 8½in
Height	4.00m	13ft 1½in
Wing area	37.30m²	401.48ft²
Weights		
Empty weight	4,005kg	8,829 lb
Fuel weight	2,200kg	4,850 lb
Loaded weight	7,500kg	16,535 lb
Performance		
Max speed	900km/h	559mph
Armament	4 x MK108	
Bombload	1 x SC1000	

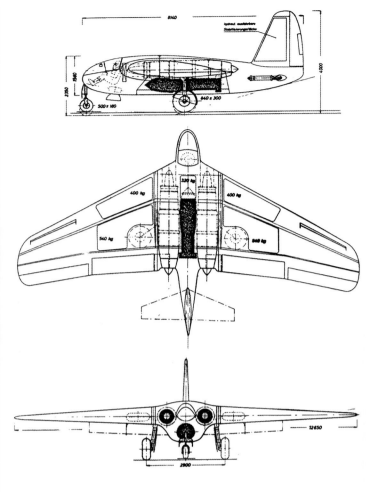

Top and above: **Lippisch P11-105 model showing the fixed outboard wing leading-edge slots, trailing-edge controls and the rectangular extendable tail surfaces outlined on the fin.**

Above: **Lippisch P11-105 – drawing dated of 2nd December 1942.**

Lippisch P11-105 cross-sectional side-view.

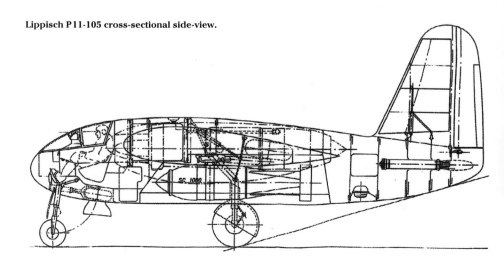

Lippisch P 13
1942

This single-seat twin-engined Zerstörer and Schnellbomber, evolved under the design leadership of Dipl-Ing J Hubert and dating from November 1942, was one of the most interesting if also controversial project studies of the highly-talented Dr Alexander Lippisch whilst at the Messerschmitt AG. Of all-metal construction, its wing planform featured compound leading- and trailing-edge sweep angles, and like preceding projects, housed the wide-track mainwheels and most of the fuel in four wing tanks which, together with the smaller 260-litre fuselage tank, held a total of 1,480 litres (326 gals). Powerplants consisted of two DB 605B engines installed fore and aft of the cockpit, the nose unit driving propellers of 3m (9ft 10in) diameter, the rear engine driving 2.9m (9ft 6¼in) diameter propellers via an extension shaft; the underwing radiators were located at the wingroots behind the mainwheel bays and 300 litres (66 gal) fuel tanks. For take-off and landing, the tailwheel extended telescopically beneath the ventral fin. As opposed to preceding designs, the bombload was carried partially recessed beneath the fuselage. Despite the advantages of this type of layout remarked upon by Lippisch in his project description, the RLM showed no interest in it.

Lippisch P 13 – data

Powerplants	2 x 1,475hp Daimler-Benz DB 605B engines driving tractor and pusher propellers.	
Dimensions		
Span	12.80m	42ft 0in
Length	9.40m	38ft 2in
Height	5.10m	16ft 8¾in
Performance		
Max speed	750km/h	466mph
Bombload	1 x SC 500	

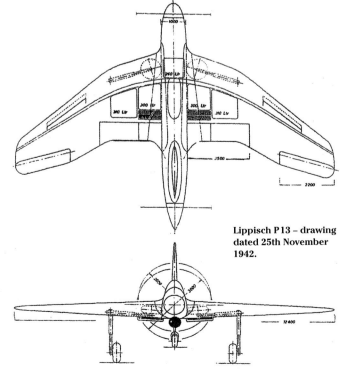

Lippisch P 13 – drawing dated 25th November 1942.

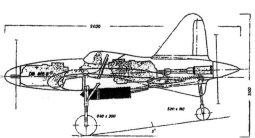

Messerschmitt Me 265

1942

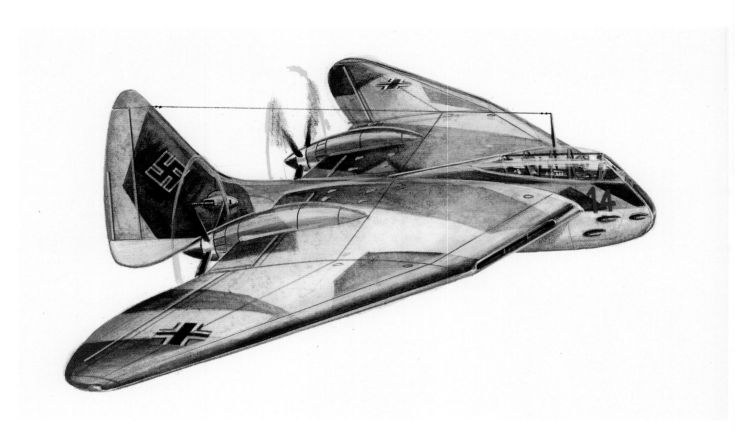

The two-seat Me 265 Zerstörer project, developed under the leadership of Dipl-Ing Walter Stender from the beginning of 1942[7] and also known as the Lippisch P10, was to have filled the gap caused by the discontinuation of the Me 210. Convinced of its advantages, Lippisch vigorously advocated the tailless configuration in this case too. The Me 265 was so configured that despite its completely new swept wing and empennage, it incorporated much of the existing Me 210 fuselage. Powerplants were two wing-mounted DB 603 engines driving pusher propellers.

Despite having advanced to a mock-up and flying model, the Me 265 was stopped in the advanced design stage in autumn 1942 after it had been established that the meanwhile improved Me 210 in the form of the Me 410 using large quantities of available parts from its forbear, was able to be manufactured more quickly.

[7] In his book *Ein Dreieck Fliegt (A Delta Flies)*, Lippisch states that it was preceded by the earlier unrelated P10 Schnellbomber project of 20th May 1942 designed by Dr-Ing Hermann Wurster. A low-wing tailless single-seater with a wing sweep of 25° at ¼ chord, it was powered by a 2,700hp DB606 (coupled DB 601s) located at the fuselage c.g. and driving pusher propellers behind the tail via an extension shaft, the propeller being protected by a dorsal and ventral fin and rudder and tailwheel. Span was 18m (59ft 0¾in), length 9.85m (32ft 3⅜in) and height 6m (19ft 8¼in). Armament comprised two forward-firing weapons in the thickened wing roots and an SC1000 bomb could be carried in the internal bomb bay beneath the coupled power-plant.

Messerschmitt Me 265 model (January 1943).

Messerschmitt Me 265 – drawing dated 1942.

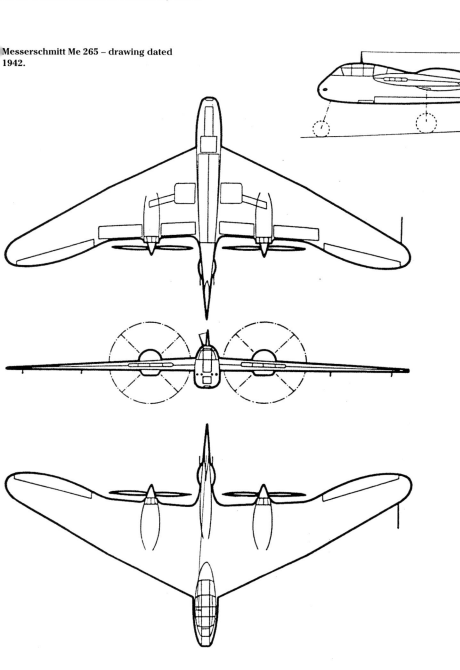

Messerschmitt Me 265 – data

Powerplants	2 x 1,750hp Daimler-Benz DB 603s driving pusher propellers	
Dimensions		
Span*	17.40m	57ft 1in
Length	10.00m	32ft 9¾in
Height	4.20m	13ft 9½in
Wing area	45.00m²	484.36ft²
Weights		
Empty weight	6,300kg	13,889 lb
Loaded weight	11,000kg	24,251 lb
Performance		
Max speed	675km/h at 5,400m	419mph at 17,720ft

* In a drawing of 4th November 1942, Me 265 span is 18.45m (60ft 6½in), height 4.6m (15ft 1in) and wing area 53m² (570.47ft²). Armament and bombload are not stated, but like the Me 210A-1, probably consisted of two MG17s (1,000rpg) and two MG151/20s (350rpg) in the fuselage nose and two MG131s (450rpg) in the rearward-directed FDL 131 barbettes, plus eight SC 50, two SC 250 or two SC 500 bombs internally.

As a comparison, the Me 210 had span 16.34m (53ft 7¼in), wing area 36.2m² (389.64ft²), empty weight 7,070kg (15,586 lb) and loaded weight 9,460kg (20,856 lb). Maximum speed was 563km/h at 5,400m (350mph at 17,720ft), range 1,850km (1,150 miles) and ceiling 8,900m (29,200ft).

Messerschmitt Me 329

1942 to 1945

Detail design of the Me 265 was still in progress when Dr Alexander Lippisch together with Dr-Ing Hermann Wurster began design work on the two-seat P 10 (later Me 329) Zerstörer project, both proposed as comparative studies with the Me 410.

Unlike the Me 265 which employed much of the slender oval cross-section fuselage of the Me 210 with its tandem-seated crew of two, the Me 329 had a wider fuselage of circular cross-section housing the two crew members side-by-side beneath a bubble canopy. Of tailless layout with a broad-chord tapered mid-wing, its two DB 603G engines in wing nacelles drove four-bladed pusher propellers via extension shafts. Its estimated maximum speed of 685km/h (426mph) at rated altitude was much higher than the Bf 110 and Me 210. Armament carried was also formidable: in addition to the nose-mounted four MK 108s (with provision for an MK 114), it had two MG 151/20s in the nose and tail, the latter movable through a 90° arc and remote-controlled by a rear-view periscope. The internal bomb bay could house an SC 1000 bomb.

In the winter of 1944/45, a full-scale motorless glider was flight-tested in tow behind a tow-craft in Rechlin, but due to the adverse war situation, work on the Me 329 was terminated at the beginning of 1945.[8]

Messerschmitt Me 329 – data

Powerplants	2 x 1,750hp Daimler-Benz DB 603 engines driving 3.40m (11ft 1¾) diameter pusher airscrews.	
Dimensions		
Span	17.50m	57ft 5in
Length	7.715m	25ft 3¾in
Height	4.74m	15ft 6½in
Wing area	55.00m²	592.00ft²
Weights		
Empty weight	6,950kg	15,322 lb
Loaded weight	12,150kg	26,786 lb
Performance		
Max speed	685km/h at 7,000m	426mph at 22,965ft
Range	2,520km	1,506 miles
Service ceiling	12,500m	40,010ft
Armament	2 x MG 151/20 in fuselage nose, 4 x MK 108 in fuselage nose, 1 x MK 114 in fuselage nose, 1 x MG 151/20 in fuselage tail	
Bombload	1,000kg (2,205 lb) internally	

[8] Lippisch has stated that the P 10 (Me 329) stemmed from the earlier Lippisch P 04 Zerstörer and Schnellbomber of 1st August 1942 designed by Dr-Ing Wurster. Photographs of the full-scale Me 329 mock-up were included in the Messerschmitt brochure to the RLM. According to other sources the project, begun in March 1942 and favoured by Prof Willy Messerschmitt, was designed to fulfil the roles of heavy-, escort- and night-fighter, dive-bomber, ground attack, and reconnaissance. Powerplants proposed were two DB 603G or two Jumo 213s. Normal forward-firing armament was four MG 151/20s with provision for two MK 103s in the fuselage nose, plus the remote-controlled MG 151/20 tail barbette, the aircraft capable of carrying a maximum bombload of 2,500kg (5,512 lb). Wind-tunnel and flying models were also tested, with and without the vertical fin and rudder. In a comparison requested on 26th August 1942 by Willy Messerschmitt between the Lippisch P 10 and Me 329, the Messerschmitt design engineers Woldemar Voigt and Hans Hornung in a report dated 1st December 1942 established that for the same mission capability and landing speed of 170km/h (106mph) for each aircraft, the following estimates pertained:

	Lippisch P 10 (Nov 42)	Me 329 (Apr 42)	Me 410 (Nov 42)
Dimensions			
Span	18.45m	17.50m	16.34m
	60ft 6½in	57ft 5in	53ft 7¼in
Length	10.00m	7.715m	12.60m
	32ft 9¾in	25ft 3¾in	41ft 4in
Height	3.70m	4.74m	3.70m
	12ft 1¾in	15ft 6½in	12ft 1¾in
Wing area	53.00m²	55.00m²	36.00m²
	570.47ft²	592.00ft²	387.49ft²
Weights			
Loaded weight	11,000kg	12,150kg	10,680kg
	24,251 lb	26,786 lb	23,545 lb
Performance			
Max speed at	682km/h	685km/h	672km/h
7,000 m (22,965ft)	424mph	426mph	418mph
Range	2,480km	2,520km	2,020km
	1,541 miles	1,541 miles	1,355 miles
Service ceiling	12,100m	12,500m	10,900m
	39,700ft	40,010ft	35,760ft

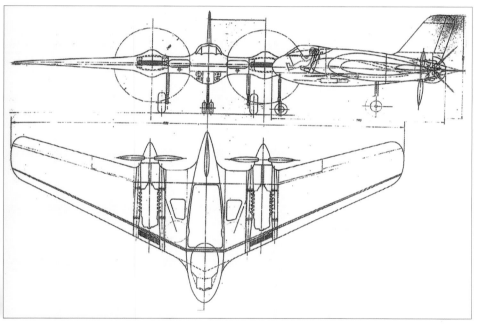

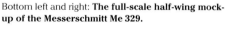

Messerschmitt Me 329.

Bottom left and right: **The full-scale half-wing mock-up of the Messerschmitt Me 329.**

Messerschmitt Me 609

1942

Messerschmitt Me 609 – data

Powerplants	2 x 2,000hp Junkers-Jumo 213E engines	
Dimensions		
Span	16.00m	52ft 6in
Length	9.52m	31ft 2¾in
Height	3.24m	10ft 7½in
Wing area	26.75m²	287.93ft²
Performance		
Max speed	760km/h at 8,500m	472mph at 27,890ft
Armament	2 x MK108 and 2 x MK103	
Bombload	2 x SC 250 or 1 x SC 500	

As early as 1940, the Messerschmitt design office concerned itself with the idea of fusing two fighters into a single airframe to provide a considerable increase in range and payload. The work was based on an RLM edict to simplify the number of combat aircraft to a few basic models. Known as the Bf109Z Zwilling (twin) and Me 609, the layouts were primarily intended as Zerstörer and fast bombers. Following a detailed examination, the most suitable solution capable of early production indicated the use of the 1,475hp DB 605A-powered Me109G. A considerable performance increase, however, was to be expected if 2,000hp Jumo 213 engines were installed in the Me109G fuselages, requiring only a few alterations, which enabled the single-seat twin Me 109 to incorporate complete major components of the standard aircraft. The modifications were limited mainly to the need for a completely new constant-chord wing centre section and tailplane that simplified manufacture. Besides relocation of the undercarriage attachment points and the use of larger wheels, the ailerons and outboard wing leading-edge slots were lengthened and auxiliary fuel tanks were installed in the starboard fuselage. The Bf109Z-1 and Z-3 Zerstörer variants were to be equipped with 30mm cannon[9] which could be replaced by heavier calibre weapons.

The deliberations and investigations led at the end of 1942 to the amalgamation of two Me109F airframes into the Bf109Z V1 prototype, equipped with two 1,350hp DB 601E motors which brought take-off weight to 5,890kg (12,985 lb). In the course of develop-ment work on the twin aircraft, the Me 309 was also considered and resulted in the Me 609 in Zerstörer and fast-bomber versions. The alterations and modifications corresponded in magnitude to that required for the Bf109Z-2 fast bomber. The more powerful DB 603G engine intended for the Me 609 led to improvements in flight performance.

[9] As a comparison, the Bf109Z-1 and Z-3 Zerstörer and Bf109Z-2 Schnellbomber each had the same dimensions but differed in weights and offensive loads as follows: span 13.27m (43ft 6½in), length 8.92m (29ft 3¼in), height 2.7m (8ft 10in) and wing area 23.7m² (255.1ft²). Loaded weight was c.7,300kg (16,094 lb). Armament consisted of two to four MK108s and bombloads of two SC 250s or one SC 500 could be carried. As a fast bomber, armament was two MK108s and bombload one SC1000 bomb. Maximum speeds at sea-level were 590km/h (366mph) with 1 tonne bombload, and 602km/h (374mph) without bomb. At 8km (26,250ft),corresponding top speeds were 734km/h (456mph) and 743km/h (462mph) respectively. Sea-level rates of climb, with and without 1 tonne bombload were 19.8m/sec (3,900ft/min) and 25.8m/sec (5,080ft/min), and at 7km (22,960ft) were 10.8m/sec (2,130ft/sec) and 15.5m/sec (3,050ft/min) respectively. Range at maximum continuous power was 1,700km at 673km/h (1,057 miles at 418mph) and 2,000km at 570km/h (1,242 miles at 354mph) max economical cruising power. Take-off run was 350m (1,148ft) and landing speed 160km/h (99mph).

Whereas the Bf109Z variants had a conventional tailwheel undercarriage, the Me 609 was equipped with nosewheels which retracted to lie flat beneath each engine. Nosewheel track was 5m (16ft 5in) and mainwheel track 3.1m (10ft 2in).

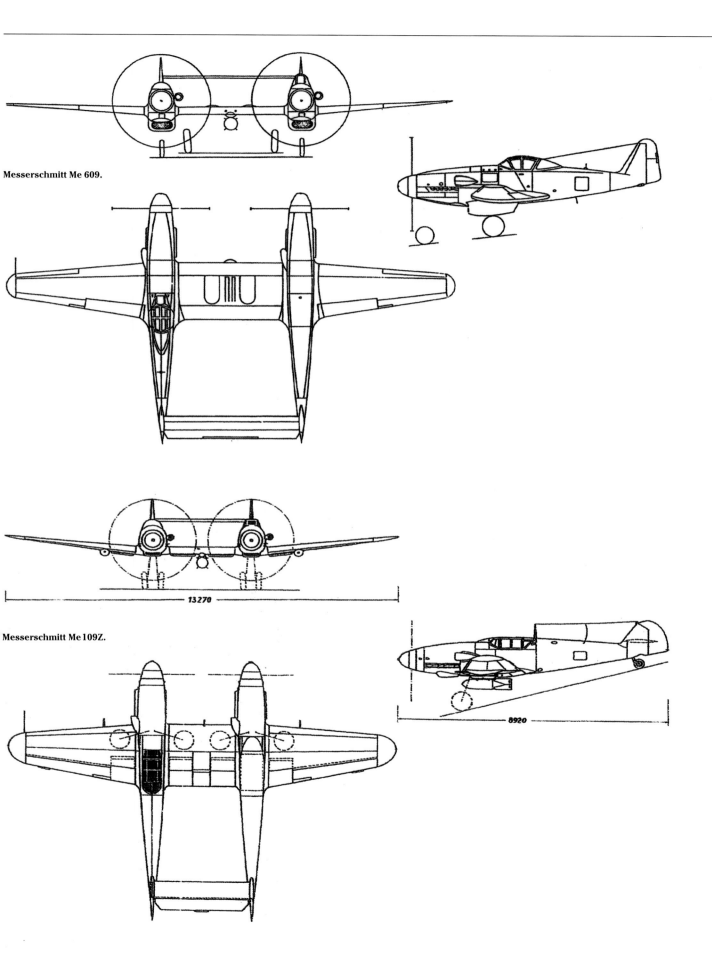

Messerschmitt Me 609.

Messerschmitt Me 109Z.

13.270

8920

Messerschmitt P 1079/18 'Schwalbe'

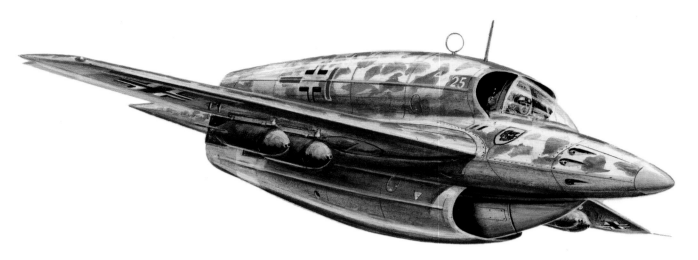

The P1079/18 'Schwalbe' (Swallow) project of 1942 was designed in the Messerschmitt Abteilung L (for Lippisch) under the leadership of Dipl-Ing Rudolf Seitz. This single-seat tailless aircraft with a wing leading edge sweep of 37° and fitted with fixed outboard slots, ailerons and elevators, was powered by two superimposed turbojets, the upper and lower air intakes bifurcated by the cockpit and nosewheel enclosures. As shown in the three-view drawing, the wide-track mainwheels retracted diagonally forwards into the wing roots, the aircraft having a total of six protected fuel tanks. The significance of the tail brakes is not known. Besides its use as a fast bomber and ground attack aircraft, it was also to function as a Zerstörer. Although this seemingly advanced project was stopped by the RLM in 1942, the P1079/17 which later received the RLM designation Me 328, was given the green light to proceed.

Messerschmitt P 1079/18 'Schwalbe' – data

Powerplants	2 x 900kg (1,984 lb) thrust Jumo 109-004 turbojets[10]	
Dimensions		
Span	9.05m	29ft 8¼in
Length	8.90m	29ft 2¼in
Height	2.75m	9ft 0¼in
Wing area	20.00m²	215.27ft²
Loaded weight	4,030kg	8,885 lb
Performance		
Max speed	950km/h	590mph

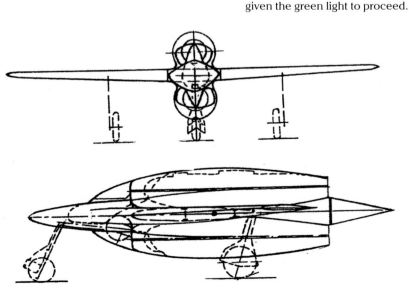

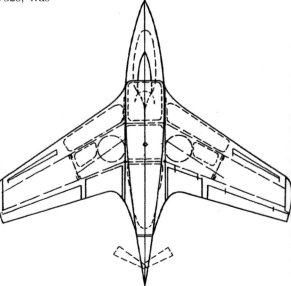

Messerschmitt Zerstörer Project

1941/42

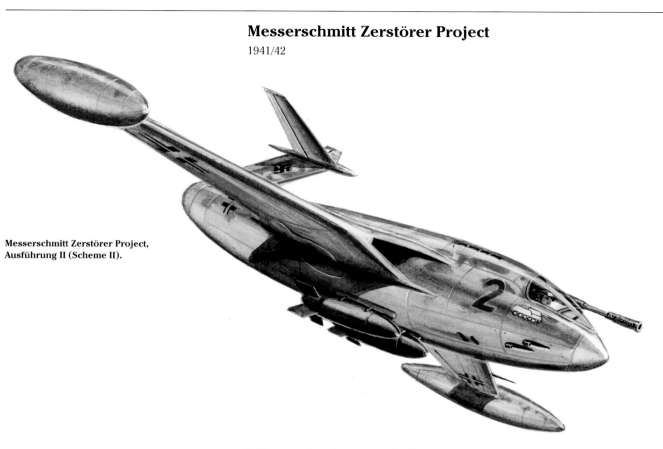

Messerschmitt Zerstörer Project, Ausführung II (Scheme II).

This consisted of a design study to make use of the expected turbojets that were under development. The oft-voiced assumption in the aeronautical press that the project consists of a 1945 development is refuted by the accompanying works sectional drawing in which a 'radial turbine engine' is shown in outline. This leads to the conclusion that Messerschmitt in 1942 still possessed no documentation on turbojet development or installation plans.

Gas turbine development in Germany was concerned from the very beginning with the axial-flow type, save for the experimental radial-flow turbojets developed by Dr-Ing von Ohain. The Messerschmitt Zerstörer project appears, accordingly, to consist of a study from the period 1941/42.

The high T-tailplane leads to the assumption of a later design period, since Dipl-Ing Hans Multhopp of Focke-Wulf employed the T-tailplane for the first time in 1945 in the design of the Tank Ta183. Messerschmitt, Heinkel and the other aircraft manufacturers followed suit hesitatingly, as too little was known of the related flying characteristics of this type of tail surface. When the Allies occupied Germany and split the war booty amongst themselves, they quickly recognised its advantages and adopted this design feature.[10]

[10] The three-view drawings of the Messerschmitt 'Schwalbe' (described here as the P1079/18) and the two Zerstörer designs of unknown designation were first seen by this Translator in the magazine *Luftfahrt International* 18 (Nov-Dec 1976) in an article entitled 'Little-known German aircraft projects of World War 2' pp.2841-2856.

The article described and illustrated a total of eight turbojet-powered projects stemming from a former Messerschmitt employee that were supplied to the publisher, who mentioned that it was not clear whether the projects bearing wild-life names originated during or after the Second World War. It included three-view drawings of the Me P.1106, P.1107, P.1108 as well as the 'Schwalbe' (Swallow) Zerstörer, the single-seat single-jet 'Wespe' (Wasp) and 'Libelle' (Dragonfly) light fighters, the six-jet 'Wildgans' (Wild Goose) bomber-transporter, and the T-tail unnamed Zerstörer Scheme I, together with some dimensions, weights and performance figures. Three-view drawings of the 'Schwalbe' and Zerstörer Schemes I and II are also illustrated by author Manfred Griehl in *Jet Planes of the Third Reich – The Secret Projects*, Vol.1, Monogram Aviation Publications, Sturbridge, Mass.,1998, pp.128, 191 and 193.

Unusual for Messerschmitt project drawings is that none of the dotted-outline turbojets in each of the drawings matches with the contours of any turbojets that were under development by BMW, Daimler-Benz, Heinkel-Hirth, Junkers and Porsche (which would be logical to expect), nor do the thrust figures quoted for them correspond to the known turbojet variants, the thrust for the 'Schwalbe' and unnamed Zerstörer being listed in the article as 1,195kg (2,634 lb) each, as opposed to the 900kg (1,984 lb) cited by the author.

Messerschmitt Zerstörer Project, Ausfühung I (Scheme I).

Messerschmitt Zerstörer (Scheme I) – data

Dimensions

Span	11.80m	38ft 8½in
Fuselage length	10.30m	33ft 9½in
Length, overall	12.20m	40ft 0¼in
Wing area	28.00m²	301.38ft²

Weights

Loaded weight	7,000kg	15,432 lb

Whereas the Jumo 004C, abandoned in January 1945, had a dry thrust of 1,015kg (2,238 lb) and 1,200kg (2,646 lb) with reheat, the turbojet in these projects had a larger maximum cross section of 90cm (35½in) and length without intake duct, of 4.2m (13ft 9¼in) – as measured from the drawings. Furthermore, as the total 1,200kg (2,646 lb) equipped load of the 'Schwalbe' (ie pilot, fuel, armament and bombload) probably consisted mainly of fuel, the wide-track mainwheels and ventral turbojet effectively precluded the installation of any external bombs and weapon loads for the Zerstörer role.

The author's remark that the 'radial-flow' type of turbojet was only featured in the early experimental units developed by Dr-Ing Hans-Joachim Pabst von Ohain (at Heinkel), indicates he is unaware that turbojets with single-and two-stage radial-flow compressors were also the subject of design studies between 1936 and 1942 by the DVL, BMW/Bramo, Focke-Wulf and Porsche due to their higher compressor pressure ratio, lower weight and simpler construction. The advantage of a smaller overall diameter provided by the axial-flow type for installation in high-speed aircraft led to the RLM abandoning the radial-flow type in favour of axials in early 1942.

The author's assumption that the Messerschmitt firm in 1942 still had no concrete turbojet design data or installation plans is contradicted by the considerable amount of wind-tunnel work conducted with models of the P.1065 (which became the Me 262 in 1940), the P.1070 twin-jet, and the P.1073B single-jet fighter (one BMW P.3304) of August 1940. Even the Lippisch-designed P.01-100 series of fighter projects at Abteilung L at Messerschmitt, Augsburg, from 1939 onward incorporated the clearly recognisable BMW F.9225, P.3302, P.3304 and Junkers T1 axial-flow turbojets shown in outline. In fact, as early as the autumn of 1938, Messerschmitt designers had calculated that to bestow a single-seat jet fighter with a speed of 850km/h at 1,500m (528mph at 4,920ft), a turbojet thrust of 600kg (1,323 lb) and a diameter of 60cm (23½in) was needed. It was precisely these figures that were aimed at when Dipl-Ing Helmut Schelp, in charge of jet-propulsion development at the RLM, placed the initial turbojet development contracts

with BMW (for the P.3302 and P.3304) and Junkers (for the T1) in 1939. In early aircraft projects with turbojets buried inside the fuselage, the complete turbojet nacelle was shown due to the RLM instruction not to interfere with the carefully-designed air intake lip and intake channel ahead of the compressor.

With regard to the T-tail arrangement, it should be mentioned that the high-positioned tailplane in the Arado E 654 project of 1941 was already tested with the British Westland P.9 Welkin prototype first flown on 11th October 1941. The first jet fighter to be flown – albeit with a 'low' T-tail (where the upper fin and rudder were removed), was the Gloster Meteor F.1 prototype, serial EE227/G, flown at the RAE Farnborough in 1944. Knowledge of the 'high' T-tail configuration employed in the Ta 183 of 1944 became known at the end of the war through the capture of German documents.

Finally, a further factor indicating that the 'Schwalbe' and unnamed Zerstörer designs were of later vintage than placed by the author is that the eight Messerschmitt projects mentioned in the magazine article *all* had a nosewheel tricycle undercarriage – a design feature that was only introduced into the tail-sitter Me 262 with the V5 prototype, following take-off difficulties experienced with earlier machines. As the Me 262 V5 only made its first flight on 6th June 1943 with fixed nose and mainwheels, followed by the Me 262 V6 where all three wheels were fully retractable and first flown on 17th October 1943, it is difficult to imagine that a retractable nosewheel undercarriage would have been included on other jet-propelled project designs some two years earlier, where the firm had not yet gathered any take-off and landing experience with this type of undercarriage on high-speed aircraft. Although nosewheel tricycle undercarriages were featured on the Lippisch P11, Me 265 and Me 329, these remained unbuilt projects. Not to be ignored is that the dotted-outline unknown turbojet in the 'Schwalbe' is identical to others in the later projects which did not exist until 1944-45. The rearward-reclining seat and flush canopy blending into the fuselage nose contours were not features of high-speed fighter-type designs of the 1941/42 period, having been introduced by Messerschmitt on the P1112 in 1945.

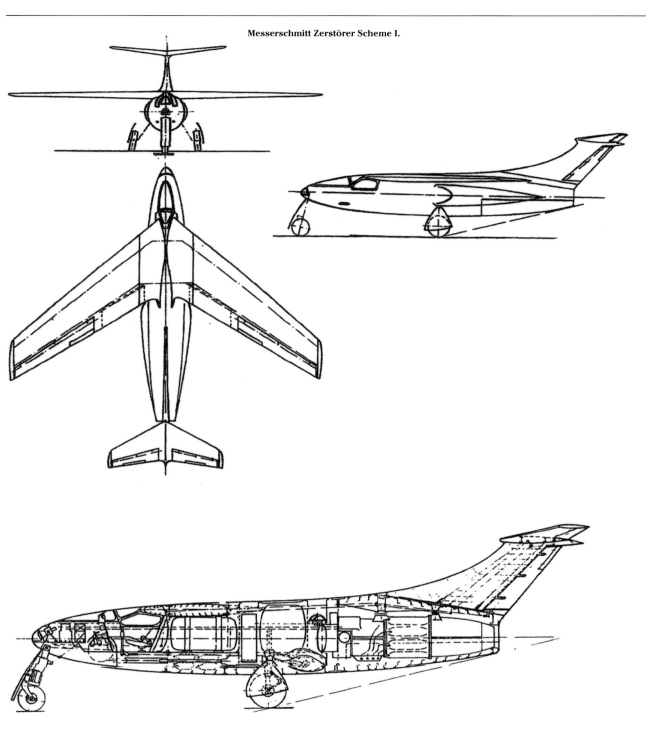

Messerschmitt Zerstörer Scheme I.

Messerschmitt Zerstörer Scheme I sectional side-view.

Chapter Six

Fast Bomber Projects 1944-1945

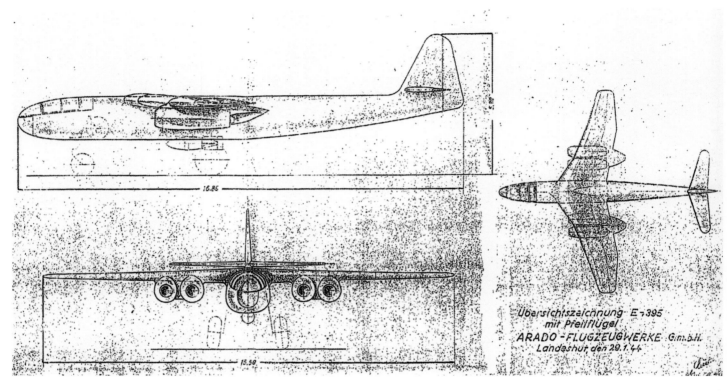

Übersichtszeichnung E-395 mit Pfeilflügel ARADO-FLUGZEUGWERKE G.m.b.H. Landshut den 29.1.'44

Arado E 395.01

1944

Under the designation E 395.01, a group of engineers headed by Dipl-Ing Lucht developed out of the Ar 234C a Zerstörer and fast bomber with exchangeable multi-sweep (crescent) wings as a counterpart to the Heinkel He 343. As can be seen from the accompanying drawings, the fuselage and wing dimensions were considerably larger than the Ar 234C-3 and the E 395.02 project. The compound sweep wing was developed according to the plans of Dipl-Ing R Kosin and Dipl-Ing Lehmann and pursued at the Avia firm in Prague. Details concerning the armament are uncertain, as development of the crescent wing had not yet been concluded when work on the project terminated.

Arado E 395.01 drawing with crescent wing of 15.5m (50ft 10¼in) span.

Arado E 395.01 data sheet of 15th May 1944 showing crescent and unswept wing variants. Interesting details in the text include the following: Powerplants: four HeS 011A or Jumo 012; crew of two in pressurised cabin; unswept wing is exchangeable with crescent wing.
Fuel quantity: 4,500kg (9,921 lb) in two protected fuselage tanks; 5,500kg (11,023 lb) at overload condition. In the Aufklärer (reconnaissance) mode, a total of 7,000kg (15,432 lb) of fuel can be carried using jettisonable auxiliary tanks.
Armament: two MG 151s fixed forward-firing and two MG 151s fixed rearward-firing, plus one MG 131

downward-firing via periscopic sight.
Bombload: 1,500kg (3,307 lb) normal, and 3,000kg (6,614 lb) at overload with reduced fuel. Carriage of a 'Fritz X' guided bomb possible.
Target aiming devices: A Lotfe 7D and BZA with fore and aft vision.
Radio installations: FuG X P, Fu Bl II, FuG 16 ZY, and FuG 25.

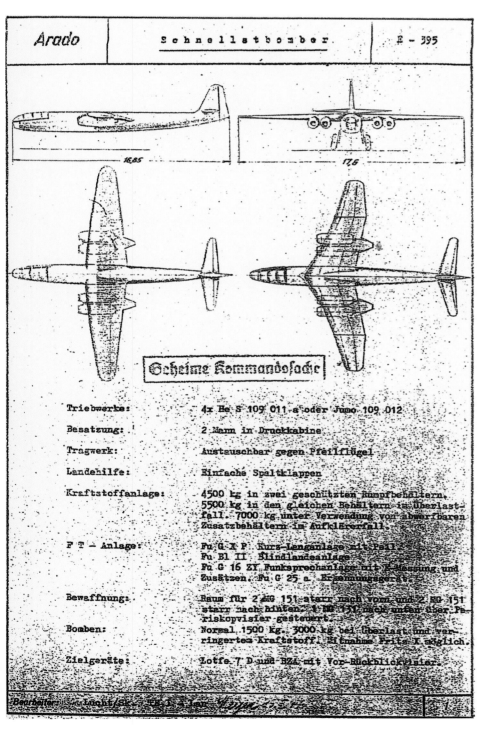

Arado E 395.01 – data

Powerplants	4 x 1,300kg (2,866 lb) thrust Heinkel-Hirth HeS 011 or two Junkers-Jumo 012 turbojets.	
Dimensions		
Span	17.60m	57ft 9in
Length	16.85m	55ft 3¼in
Height	5.30m	17ft 4¾in
Wing area	40.00m²	430.55ft²
Weights		
Empty weight	9,450kg	20,833 lb
Loaded weight	15,800kg	34,833 lb
Performance		
Max speed	887km/h at 6,000m	551mph at 19,685ft
Range	1,500km	932 miles
Service ceiling	14,500m	47,570ft
Armament	2 x MG 151/20 forward-firing 2 x MG 151/20 rearward-firing	
Bombload	1,500kg (3,307 lb) normal 3,000kg (6,614 lb) at overload	

Arado E 395.02
1944

The focal point for the E 395.02 was the single-seat Ar 234F fighter and fast bomber which was never finalised as a project but served as a basis for the single-seat E 395.02 fighter and Zerstörer. In addition to the revised cockpit arrangement, the fuselage was lengthened by 60cm (23½in) to an overall length of 15m (49ft 2½in). All other components such as the wings, fuselage, empennage and nosewheel were taken over from the Ar 234B. Powerplants were to have been two HeS 011 turbojets, the armament consisting of multiple 30mm, 55mm MK 112, 55mm MK 412 cannon and rocket-propelled weapons.

Whilst nothing is known of further progress with the project until the end of the war, published information suggests that an Ar 234F was scheduled for completion in January 1945. The three-view drawings illustrated enable a few details to be ascertained.

Arado E 395.02 – data

Powerplants	2 x 1,300kg (2,866 lb) thrust Heinkel-Hirth HeS011A turbojets	
Dimensions		
Span	14.40m	47ft 3in
Length	15.00m	49ft 2½in
Height	5.10m	16ft 8¾in
Wing area	27.70m²	298.15ft²
Weights		
Loaded weight	16,000kg	35,274 lb
Performance		
Max speed	887-900km/h	551-559mph
Range	1,500km	932 miles
Service ceiling	14,500m	47,570ft

Arado E 395.02.

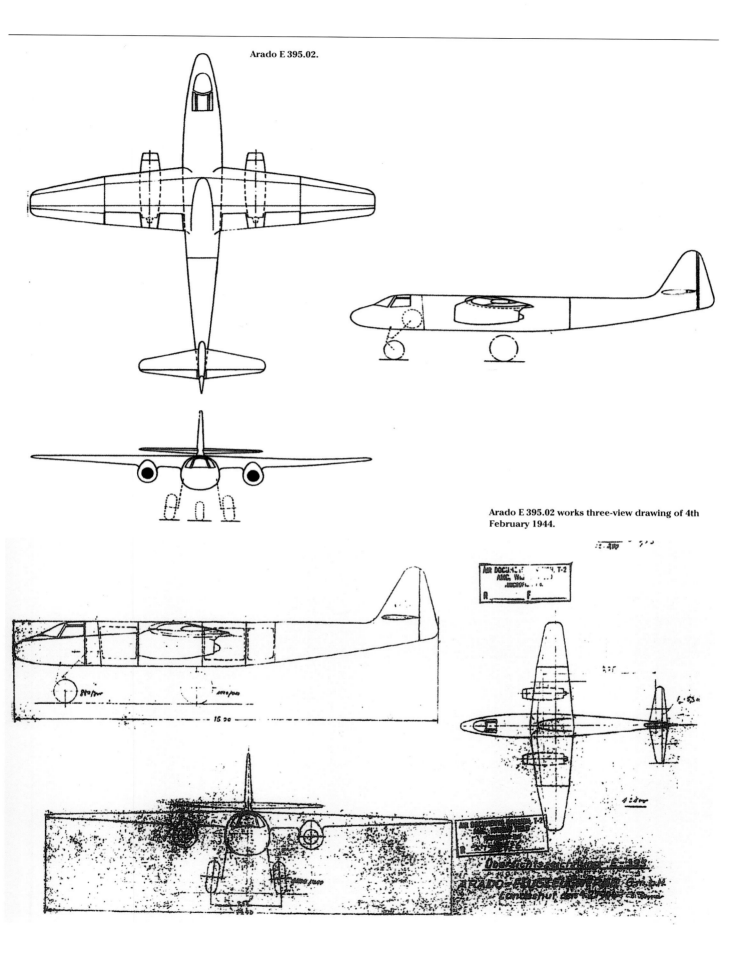

Arado E 395.02 works three-view drawing of 4th February 1944.

Arado E 583.05 'Project I'

1945

The E 583.05, better known as the Arado 'Project I', was a completely new design for a tailless Night- and Bad-weather Fighter and Zerstörer that was submitted to the EHK or Entwicklungs-Hauptkommission (Development Main Committee) on 15th March 1945. Designed to be of the simplest form for manufacturing purposes, the cantilever low wing with a leading edge sweep of 40° had a slightly-swept trailing edge, the crew of three housed in the pressurised cockpit. The twin fins and rudders were located at about one-third the semi-span, the ailerons acting as elevators. The two HeS 011A turbojets were placed in a paired nacelle beneath the rear fuselage astride the inboard trailing-edge split flaps, the mainwheels retracting into the wing roots on either side of the main fuselage fuel tank with the nosewheel retracting rearwards beneath the forward-firing armament and behind the nose radar dish. As the project would have required a long development and testing period, work on it was stopped in April 1945 in favour of the E 583.06 'Project II'.

Arado E 583.05 – data[1]

Powerplants	2 x 1,300kg (2,866 lb) thrust Heinkel-Hirth HeS 011A turbojets	
Dimensions		
Span	18.30m	60ft 0½in
Length	12.95m	42ft 5¾in
Height	5.45m	17ft 10½in
Wing area	66.00m²	710.40ft²
Armament	4 x MK 108 forward-firing, 2 x MK 108 rearward-firing 2 x MK 108 oblique upward-firing	

[1] Author's figures. Designed to meet a requirement of 27th January 1945 calling for a long-range, three-hour endurance night and bad-weather fighter, the early project submissions by Arado, Blohm & Voss, Focke-Wulf and Junkers resulted in designs each with a take-off weight of 19-20 tonnes – much too heavy to be easily put into production. Following new requirements laid down for armament, range and endurance, reducing the total weight to a more reasonable 12-13 tonnes, revised submissions were presented at the EHK conference at Bad Eilsen on 21st/22nd March 1945 by Arado, Blohm & Voss and Gotha. Like the other two firms' tailless submissions, the Arado 'Project I' was to carry no less than ten 30mm cannon – six in the fuselage nose, two in the tail, and two firing at 70° obliquely upwards. The weapons consisted of six nose MG 213C/30 (200rpg), two oblique MK 108 (100rpg) and 2 tail MG 213C/30 (200rpg), with provision for two SC 500 bombs. The normal J2 fuel load of 4,800 litres (1,055 gals) was housed in eight tanks – three in each wing and two in the fuselage, with 7,500 litres (1,650 gals) at overload. Data quoted for this variant were:

Dimensions		
Span	18.40m	60ft 4½in
Length	12.95m	42ft 5¼in
Height	3.80m	12ft 5½in
Wing area, early	75.00m²	807.27ft²
Wing area, later	66.00m²	710.40ft²

Weights		
Empty weight	7,700kg	16,975 lb
Fuel, normal	4,000kg	8,818 lb
at overload	6,250kg	13,779 lb
Loaded weight, normal	12,600kg	27,777 lb
at overload, with tail turret	13,200kg	29,100 lb

Performance		
Max speed at sea level	710km/h	441mph
at 9,000m (29,530ft)	810km/h	503mph
Endurance, 100% thrust	1.7 hrs at 10,000m	32,810ft
60% thrust	2.6 hrs at 10,000m	32,800ft
Service ceiling	12,600m	41.340ft

Arado E 583.05 'Project I' night and bad-weather fighter with fuel load 4,300kg (9,480 lb) in text of the three-view works drawing of 15th March 1945.

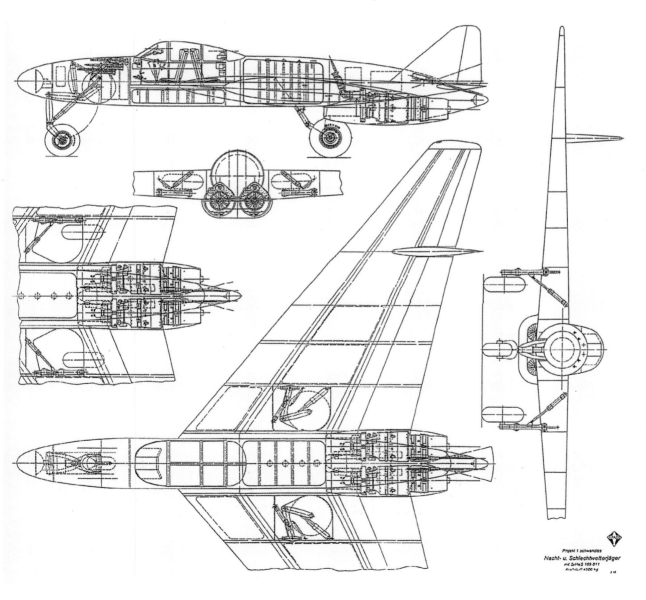

Arado E 583.06 'Project II'

1945

As a comparative design to the tailless E 583.05, the E 583.06 'Project II' had a more orthodox configuration, featuring a shoulder-mounted wing swept 35° at ¼ chord with its two Heinkel-Hirth HeS 011 turbojets in under-wing nacelles. The aim was to replace the Ar 234 in series production without major manufacturing effort, its increased speed and range allowing considerably more opera-tional possibilities. The 2.5m (8ft 2½in) nar-row-track tricycle undercarriage members all retracted into the fuselage, the powerful armament of four MK108s in the nose behind the radar dish and two MK108 oblique upward-firing cannon allowed the night fighter and high-altitude Zerstörer to fulfil its functions. As auxiliary armament, provision was made for a 50mm MK 214 Bordkanone as a Rüstsatz, and in the fighter-bomber role, it could carry a 1,000kg (2,205 lb) bombload. Like the earlier project, the crew of three were provided with ejection seats in the pres-surised cabin. As a special Rüstsatz, a brake parachute could be housed in the fuselage tail.

Arado E 583.06 'Project II' – data

Powerplants	2 x 1,300kg (2,866 lb) thrust Heinkel-Hirth HeS 011 turbojets	
Dimensions		
Span	14.95m	49ft 0½in
Length	17.30m	56ft 9in
Height	5.30m	17ft 4¾in
Wing area	42.80m²	460.68ft²
Weights		
Loaded weight	13,200kg	29,101 lb
Performance		
Max speed	750km/h+	466mph+
Armament	4 x MK108 in nose, 2 x MK108 oblique upward	
Bombload	1,000kg (2,205 lb)	

Dimensions		
Span	15.00m	49ft 2½in
Length	17.30m	56ft 9in
Height	5.30m	17ft 4¾in
Wing area	50.00m²	538.18ft²
Weights		
Empty weight	7,950kg	7,527 lb
Fuel weight, normal	4,000kg	8,818 lb
Loaded weight, normal	12,800kg	28,219 lb
at overload, with tail turret	13,200kg	29,101 lb
Performance		
Max speed at sea level	705km/h	438mph
at 7,000m (at 29,965ft)	775km/h	481mph
Endurance, 100% thrust	1.7 hrs at 10,000m	32,810ft
60% thrust	2.2 hrs at 10,000m	32,810ft
Service ceiling	11,400m	37,400ft

Author's data. Submitted to the EHK as an alternative to the 'Project I', this March 1945 proposal, of composite construction, had the same normal fuel load, armament, radio and radar as its predecessor, except that three SC 500 bombs could be carried. In a further variant, the swept tail surfaces were replaced by a swept butterfly or V-tail. Data for the 'Project II' were:

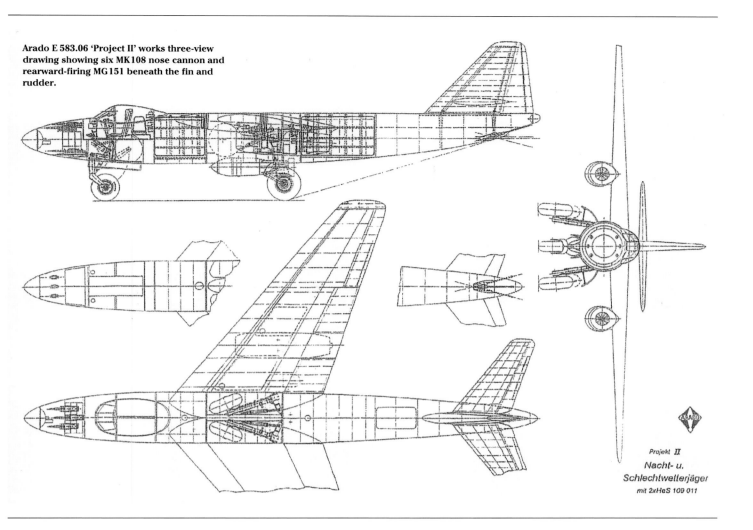

Arado E 583.06 'Project II' works three-view drawing showing six MK 108 nose cannon and rearward-firing MG 151 beneath the fin and rudder.

Projekt II
Nacht- u.
Schlechtwetterjäger
mit 2xHeS 109 011

Blohm & Voss P 203.01

1944

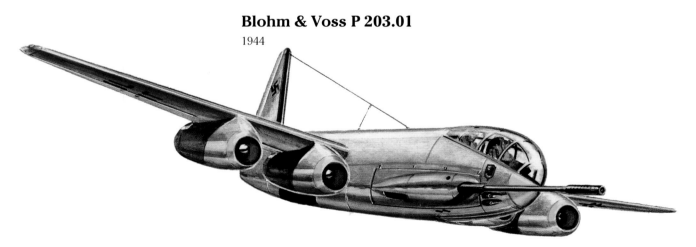

Somewhat primitive in overall appearance, the P 203.01 high-altitude Zerstörer of June 1944 resembled aircraft dating from the 1930s. Its simple 'lines' however, embodied a new development in the positioning of the powerplants that were revolutionary at the time.

With the P 203, Dr-Ing Richard Vogt sought to combine the reciprocating engine and tur-

bojet into one cohesive unit which, with reduced frontal area, promised considerable advantages: take-off from grass without RATO units, greater acceleration when used as a Zerstörer, increased range as a bomber, and increased duration as a night fighter and Zerstörer. A noteworthy feature was the double rectangular wing which Dr Vogt termed a 'step wing', where the inner portion was of constant chord and 14% thickness, the outboard rectangular sections of reduced depth having a thickness/chord ratio of 12%. In addition to simplifying manufacture, the increased-length ailerons (which could be used as landing flaps) and inboard flaps were interchangeable, as were the elevators. Dr Vogt had planned the entire load-bearing structure to be of monocoque construction with sheet metal covering and to utilise the wings to house the mainwheels and all the fuel so that fuselage space was used exclusively to accommodate the crew of two, armament, bombs, and equipment.

Main powerplants of the three P 203 variants were to have been two BMW 801J radials whose performance could be boosted by 35% over the BMW 801D with 150kg (331 lb) of methanol-water MW 50 injection for a total period of 30 minutes. Take-off weights and performance varied according to the type of turbojet installed, the latter housed in a short nacelle either beneath the BMW 801Js or in an extended nacelle to the rear and fed by a ventral air intake. The proposal also envisaged the installation of a heavy MK 114 Bordkanone.

Blohm & Voss P 203.01 three-view drawing.

Blohm & Voss P 203.01 – data

Powerplants 2 x 2,300hp BMW 801J radials with MW 50 and 2 x 1,300kg (2,866 lb) thrust Heinkel-Hirth HeS 011 turbojets.

Dimensions

Span	20.00m	65ft 7½in
Length	16.60m	54ft 5½in
Height	5.40m	17ft 8½in
Wing area	65.00m²	699.64ft²

Weights

Empty weight	12,380kg	27,293 lb
Fuel weight	4,000kg	8,818 lb
Loaded weight, with bomb	18,400kg	40,565 lb

Performance

Max speed, at sea level	670km/h	416mph
at 14 tonnes	920km/h at 11,900m	572mph at 39,040ft
Climb rate at Sea level	1,400m/min	4,593ft/min
at 13,000m (42,650ft)	480m/min	1,575ft/min
Range, at 16.3 tonnes		
at sea level	2,350km at 665km/h	1,460 miles at 413mph
at 10,000m (32,810ft)	2,650km at 870km/h	1,647 miles at 541mph
Service ceiling, at 14 tonnes	14,300m	46,915ft

Armament 2 x MG131 (400rpg) 2 x MG151 (200rpg) 2 x MK103 (100rpg) 1 x FHL131Z (400rpg)

Bombload 1,000kg (2,205 lb)

For the P 203.02 (2 x Jumo 004), equipped weight was 12,340kg (27,204 lb), take-off weight 18,290kg (40,322 lb) and maximum speed 815km/h at 11km (506mph at 36,090ft. For the P 203.03 (2 x BMW 003), equipped weight was 12,050kg (26,565 lb), take-off weight 18,000kg (39,683 lb) and maximum speed 710km/h at 11km (441mph at 36,090ft), both speeds at a mean flying weight of 14 tonnes (30,864 lb). The proposed BMW 801J, fitted with a Hirth turbo-supercharger, was only built in a small experimental series.

Dornier P 238/1

1944

As a further improvement on the Do 335, the Dornier Project Office developed a heavy high-altitude Zerstörer with a fuselage of revised design, capable of accommodating a larger quantity of fuel. According to information unconfirmed to date, the P 238/1 also bore the designation Ju 435[2] as the Junkers-Werke had been selected by the RLM to continue further design work on the project. Since Dornier was fully occupied with the P 237/3 high-altitude fighter project – a Do 335 powered by a combination of reciprocating and turbojet, the TLR (Technical Air Armament Board) in a decision of 28th August 1944 stopped further work on it. Because of delays in delivery of jet engines by the manufacturing firms, the RLM favoured the use of high-performance piston engines for high-speed fighter and Zerstörers for combating Allied bomber formations whose campaign of destruction of German cities in 1944 was near its zenith.

The P 238/1 with its new aerodynamic form and its two 24-cylinder Junkers-Jumo 222A-2 high-altitude radials driving tractor and pusher propellers, seemed ideal for this purpose. With a calculated top speed of over 800km/h (497mph), it had a ceiling of 14,000m (45,930ft) – an altitude which no airscrew-driven enemy fighter had thus far attained. Development proceeded apace up to the mock-up construction stage at the end of 1944 under the highest priority, but the approaching end of the war forced work on the project to be terminated.

[2] A similar two-seat radar-equipped night-fighter project designated Do 435 was also developed from the basic Do 335 and powered by two Jumo 222 radials. The main revision here consisted of a wider fuselage to accommodate the crew of two side-by-side in a pressurised cabin and long-span wooden outer wing panels. Loaded weight was almost 11,800kg (26,014 lb). Doubts have also been expressed in the past on the existence of the 'slim fuselage' design shown in the three-view drawing that was labelled as the Do 435.

Dornier P 238/1 – data

Powerplants	2 x 2,000hp Junkers-Jumo 222A-2 radial engines	
Dimensions		
Span	13.80m	45ft 3in
Length	17.10m	56ft 1¼in
Height	5.30m	17ft 4¾in
Wing area	38.50m²	414.40ft²
Weights		
Loaded weight	12,500kg	27,558 lb
Performance		
Max speed	800km/h	497mph
Service ceiling	14,000m	45,930ft

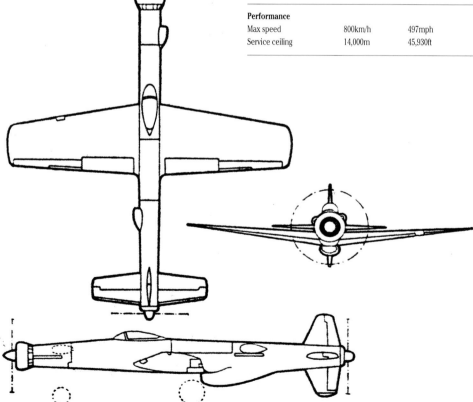

Dornier P 247/1

1944 to 1945

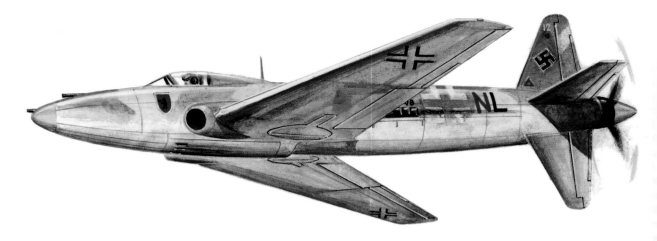

Dornier P 247/1 – data

Powerplant	1 x 2,600hp Junkers-Jumo 213T in-line engine with MW 50 driving 3.2m (10ft 6in) diameter propellers	
Dimensions		
Span	12.50m	40ft 0in
Length	12.06m	39ft 6¾in
Height	4.30m	14ft 1¼in
Wing area	26.00m²	279.85ft²
Performance		
Max speed at sea level	835km/h	519mph
Armament	3 x MK108 in fuselage nose	

The author's project description conforms to what is usually described as the P 247/6-01 – Translator.

Dornier P 247/1.

At the end of 1944 the Dornier Project Office submitted to the EHK Aircraft Committee in the RLM, their P 247/1 proposal for a single-seat Jagdbomber (fighter-bomber) and Zerstörer based on the Do 335. Powered by a single Jumo 213T motor developing 2,600hp at take-off power with MW 50 injection and driving pusher propellers behind the cruciform tail surfaces via an extension shaft, it offered the pilot a good forward view, especially for aiming the nose armament. By incorporating a low wing with a 30° swept leading edge, the estimated maximum speed of over 875km/h (544mph) lay close to the 900km/h (559mph) limit – a speed that had not until then been attained by any propeller-driven series-produced aircraft. The absence of the engine in the fuselage nose enabled three MK108 cannon to be housed there. The total 1,300 litres (286 gals) of fuel was held in a mix of protected and unprotected fuel tanks in the wings and fuselage. Circular air intake ducts for the engine radiator were located in the wing roots, the 3.9m (12ft 9½in) wide-track undercarriage wheels having low-pressure tyres like the Do 335.

Because of the critical war situation, this interesting project also did not progress beyond the drawing board. With it, a development ended in 1945 in which the reciprocating engine had reached the maximum achievable. It was due to these superb advances in German aircraft design in all realms of high-speed aerodynamics that paved the way for the entire world of aviation.

Dornier P 252/1

1945

Submitted with its descriptive brochure to the RLM on 27th January 1945, the P 252/1 Zerstörer project was a consequential further development of the Do 335 series, and with its central powerplant location driving contra-rotating pusher propellers, relinquished the fuselage nose airscrews. The use of contraprops possessed a whole number of advantages, namely: (1) highest airscrew efficiency, (2) undisturbed airflow over the airframe and lack of turbulence from a propeller airstream over the tail surfaces, and (3) elimination of a front annular radiator, replaced by suitably arranged cooling inlets in the wing roots, by which measure a high degree of aerodynamic streamlining was attained that enabled the propeller-driven aircraft to reach a maximum speed of over 900km/h (559mph) – a speed that was confirmed through precise calculations and measurements. In addition to that, the overall aerodynamic configuration of the P 252/1 with its wing leading edge sweep of 25° and the use of newly-developed contra-rotating airscrews with swept-back tips provided ideal conditions to achieve this speed in practice.

The tandem arrangement of the powerplants fore and aft of the wing centre section resulted in the following space advantages such as (4) most favourable view for the crew of three, housed in an armoured pressurised cabin and equipped with ejection seats, (5) accommodation for a large quantity of fuel with a new type of tank protection, and (6) exchangeable nose component for the installation of heavy armament and space for radar with parabolic dish antenna for night-fighting.

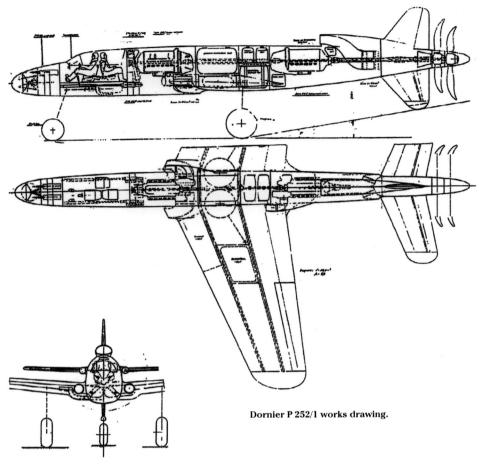

Dornier P 252/1 works drawing.

Dornier P 252/1 works drawing of 8th February 1945 showing alternate arrangements for the night-fighter version with nose 'Bremen' radar and two MK 108 and two MG 213C/30 cannon.

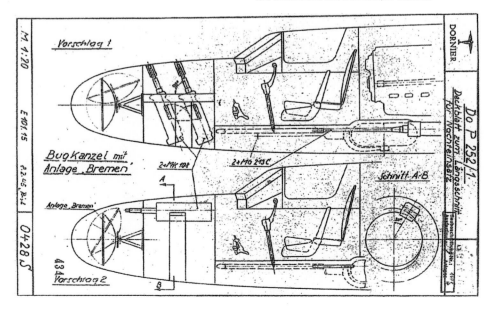

Dornier P 252/1 – data

Powerplants   2 x 2,000hp Daimler-Benz DB 603LA with MW 50 at take-off, or 2 x 2,300hp Junkers-Jumo 213J motors with MW 50 at take-off, each driving 3.20m (10ft 6in) diameter three-bladed VDM variable-pitch contraprops which could be blown off in an emergency.

Dimensions

Span	16.40m	53ft 9¾in
Length	15.20m	49ft 10½in
Height	4.95m	16ft 3in
Wing area	43.00m²	462.84ft²

Weights

Equipped weight	8,290kg	18,276 lb
Loaded weight	10,500kg	23,148 lb

Performance

Max speed	900km/h at 11,300m	559mph at 37,070ft
Service ceiling	13,400m	43,960ft

Armament[3]

Zerstörer	3 x MK 108 in fuselage nose, plus
	2 x MG 213C 30 at cockpit sides
Night-fighter	2 x MK 108 in fuselage nose, plus
	2 x MG 213C/30 at cockpit sides

[3] At least three main variants of the P 252 had been worked upon. The P 252/1 and P 252/2-01 were of the same general layout except for the swept wings.

The P 252/1 of February 1945 was a two-seat fighter-bomber powered by two 2,240hp Jumo 213J or two 2,100hp DB 603LA engines with MW 50 boost. Span was 16.4m (53ft 9¾in), wing area 43.2m² (464.99ft²), fuel capacity 1,900 litres (418 gals), MW 50 capacity 220 litres (48.4 gals), and maximum flying weight 10,730kg (23,655 lb). Underwing loads were two 250kg (551 lb) or two 500kg (1,102 lb) bombs. Maximum speed was 650km/h (404mph) at sea-level and 755km/h at 6km (470mph at 19,685ft), initial rate of climb 14m/sec (2,756ft/min), service ceiling 12,500m (41,010ft) and take-off distance 610m (2,000ft).

The P252/2-01 two-seat Zerstörer had a slightly shorter fuselage than the three-seat night-fighter. It differed from the others in having a sharply-swept wing with a 35° leading edge sweep, but of reduced root and tip chord. Between the two engines – the forward one at mid-fuselage above the wing centre section – was a 2,100 litres (462 gals) fuel tank. As with the other variants, the propeller shaft from the forward engine passed co-axially through the rear engine shaft to drive the contraprops with swept-back tips. The large single radiator for both engines was located behind the cockpit, with cooling air supplied from circular air scoops in each wing root and exhausting from a duct beneath the fuselage/wing junction. In this variant, the second crew member faced rearwards. Nose armament consisted of two MK 108s, two MG 213C/30s, and two MK 108s firing obliquely upwards. Span was 18.4m (60ft 4½in) and length 15m (49ft 2½in).

The P 252/3-01 had the same forward-firing armament as the P 252/2-01 but had a 25° swept wing of smaller span, the forward engine located directly behind the cockpit. Despite official disinterest after the end of February 1945 in newer piston-engined types beyond the Do 335, work on the P 252/3-01 appears to have continued. The installation of heavier calibre weapons such as two MK 214As was also envisaged for this variant.

Dornier P 254/1
1945

The P254/1 night-fighter and Zerstörer project of January 1945 was based on the aerodynamically favourable form of the Do 335. The necessity for further development with the turbojet enabled the rear engine, extension shaft, radiator and propellers to be removed. As opposed to the pure jet aircraft, the mixed-propulsion system offered particular advantages since on the approach flight and during the loiter phase, the jet engine could be shut down so as to achieve a considerable increase in range. Having carried out its mission, the reduced fuel consumption enabled the aircraft to reach its distant home airfield. By use of both propulsion units, the P254/1 could attain a speed of over 900km/h (559mph) – far higher than the Do 335's maximum speed.

In this two-seater design, a considerable amount of the Do 335 airframe components were utilised: the laminar-flow wing, undercarriage and forward fuselage, and the bomb bay allowing space for an auxiliary fuel tank. Powerplants were either a DB 603LA or Jumo 213J and an HeS 011 turbojet in the fuselage tail fed by lateral air intakes. As there was no need for a ventral fin and rudder, the upper components were accordingly increased in area.

The P 254/1 night-fighter (DB 603LA + TL) had a wing of larger span and area and a slightly higher loaded weight than the P254/2. (Jumo 213J + TL). In both variants, the second crew member's position was in front of the turbojet. Both were provided with ejection seats with stabilising chutes, the vertical fin and rudder being jettisonable in an emergency.

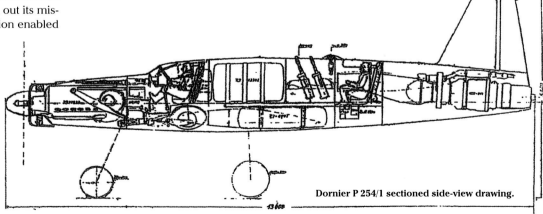

Dornier P 254/1 sectioned side-view drawing.

Dornier P 254/1 – data

Powerplants	1 x 2,300hp Daimler-Benz DB 603LA and 1 x 1,300kg (2,866 lb) thrust Heinkel-Hirth HeS 011 turbojet	
Dimensions		
Span	15.45m	50ft 8¼in
Length	13.40m	43ft 11½in
Height	5.64m	18ft 6in
Wing area	41.00m²	441.31ft²
Weights		
Empty weight	7,725kg	17,031 lb
Fuel capacity	2,950 litres	649 gals
Loaded weight	10,640kg	23,457 lb
Performance		
Max speed	865km/h at 11,000m	538mph at 36,090ft
Initial climb rate	14.0m/sec	2,756ft/min
Armament	2 x MG151/20 in nose, 1 x MK108 in nose, 2 x MK108 oblique upward	

Dornier P 254/2 – data

Powerplants	1 x 2,240hp Jumo 213J and 1 x 1,300kg (2,866 lb) thrust HeS 011 turbojet	
Dimensions		
Span	13.80m	45ft 3¾in
Length	13.40m	43ft 11½in
Height	5.64m	18ft 6in
Wing area	38.50m²	414.40ft²
Weights		
Empty weight	7,585kg	16,722 lb
Loaded weight	10,500kg	23,148 lb
Performance		
Max speed	822km/h at 7,500m	511mph at 24,600ft
Initial climb rate	11.3m/sec	2,224ft/min
Armament	2 x MG151/20 in nose, 1 x MK108 in nose, 2 x MK108 oblique upward	
Bombload	500kg (1,102 lb)	

The P 254/1-02 has been described by others as having a span of 13.8m and the P 254/1-03 as the one with span 15.45m, bubble cockpit for the rear gunner, and equipped with FuG 218 and FuG 350 'Naxos'.

Dornier P 256/1
1945

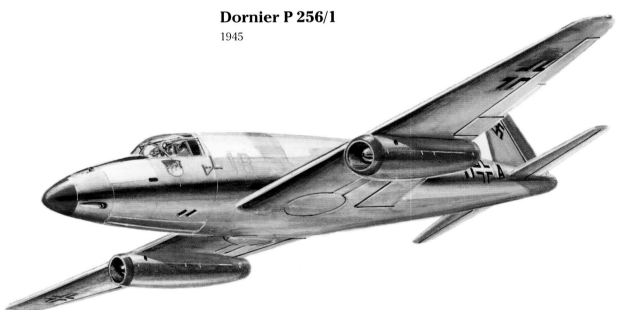

One of the latest developments at Dornier was the P256/1 night-fighter and high-altitude Zerstörer of March 1945, derived from the Do 335. Powerplants were two HeS 011 turbojets – the most modern and most powerful German turbojets at the end of the war. With a static thrust of 1,300kg (2,866 lb), the unit still delivered an impressive 500kg (1,102 lb) thrust at 10,000m (32,810ft) altitude. With a take-off weight of 12,000kg (26,455 lb), the P 256/1 had a penetration depth of 1,000km (621 miles) and a maximum speed of 824km/h (512mph). Fuselage nose arma-ment consisted of four MK108s, with provision for two MK108s firing obliquely upward. The pilot and radar operator sat side-by-side in the forward cockpit, the rearward-facing gunner operating the two MK108s located aft of the 3 fuselage fuel tanks of 3,900 litres (855 gals) capacity. To shorten the take-off run, RATO could be used, and for the night-fighting role, an FuG 240 'Berlin' radar with rotating parabolic dish antenna was housed in the plastic nose cone. The end of the war fore-closed work on the project.

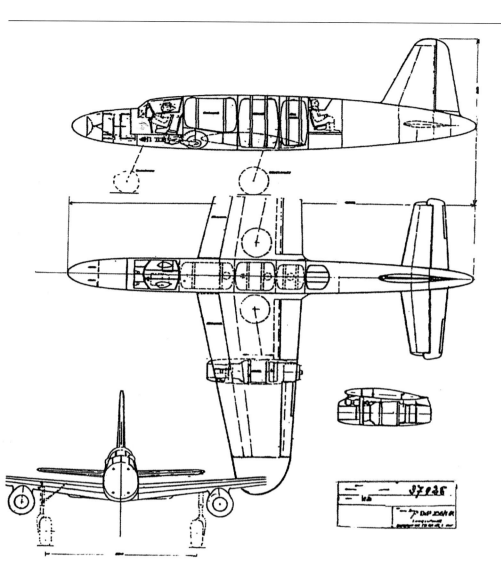

Dornier P 256/1 – data

Powerplants	2 x 1,300kg (2,866 lb) thrust Heinkel-Hirth HeS 011 turbojets and RATO units for take-off	

Dimensions		
Span	15.45m	50ft 8¼in
Length	13.60m	44ft 7½in
Height	5.50m	18ft 0½in
Wing area	41.00m²	441.31ft²

Weights		
Fuel capacity	4,650 litres	1,023 gals
Loaded weight	12,250kg	27,006 lb

Performance		
Max speed	830km/h at 6,000m	516mph at 19,685ft
Initial climb rate	13.6m/sec	2,677ft/min
Range	1,550km	963 miles
Endurance	1.75 hrs at 10,000m	32,810ft
Service ceiling	12,500m	41,010ft
Landing speed	170km/h	106mph

Armament	4 x MK108 in nose, 2 x MK108 oblique upward	

Focke-Wulf 0310239-01 and -10

1944

Focke-Wulf 0310239-01 Fast Bomber (swept wing).

Focke-Wulf 0310239-01 Fast Bomber (swept wing) – drawing dated 12th April 1944.

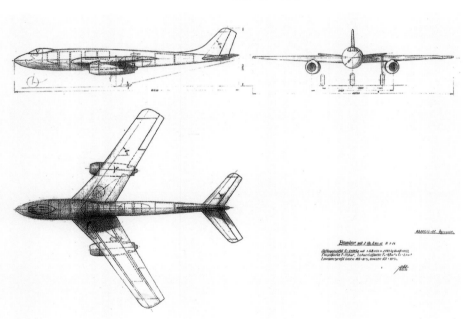

P11-121 (Delta VI) project of Dr Alexander Lippisch, the single-seat fast bomber was to attain a maximum speed of 1,000km/h (621mph) carrying a bombload of 1,000kg (2,205lb) over a distance of 1,000km (621 miles). Proposed powerplants were two HeS 011 turbojets. Since it was not possible to establish by preliminary calculations whether the task required was achievable with an aircraft of orthodox configuration or with a tail-

Under the designations 0310239-01 and 0310239-10, the Focke-Wulf Project Office in April 1944 under the leadership of Dipl-Ing H von Halem and Dipl-Ing D Küchemann (who made important post-war contributions to the development of the Anglo-French 'Concorde' airliner) drew up the designs for a Schnell-bomber and Kampfzerstörer. Similar to the

less layout, two design studies were initiated – the conventional 0310239-01 and the tailless delta-type 0310239-10. As a basic rule, high-subsonic aircraft are subject to Mach effects. The proposals therefore incorporated sweep-back on the wings and tail surfaces. Through selection of a thin profile of 11% thickness/chord ratio inboard and 10% at the wing-tips, and by reduction of the fuselage cross-section in the neighbourhood of the wing roots, the critical Mach value of 0.9 could be attained. An examination of various turbojet arrangements, however, indicated that the most favourable solution lay in the under-slung wing position. A disadvantage of this was the increased drag, but the engineers were convinced that this location was better than turbojet installation in the fuselage, for in terms of flying characteristics, this promised a higher degree of safe design planning when considering speed changes at high subsonic values which appeared to be more easily controllable. Equally advantageous was engine maintenance and rapid exchange.

The tailless 0310239-10 layout, as a result of its lower wing loading, promised improved climb performance and shorter take-off and landing runs, but no higher level speeds than the orthodox layout. Had it been possible to construct and operate the tailless variant and obtain a laminar-flow effect at the wing centre section despite the fuselage projection ahead of it, a higher normal speed would have resulted. Whereas the tailless proposal had a wing leading-edge sweep of 45°, the conventional type was given 42° inboard, and 35° of sweepback outboard of the turbojets and empennage surfaces. In order to reduce drag to a minimum on the tailless layout, von Halem and Küchemann placed the fuselage nose with its pressurised single-seat cabin and forward-retracting nosewheel ahead of the wing root as they were of the view that the wing, up to its main spar at about 40% chord, could be kept aerodynamically smooth. Both the mainwheels and the turbojets were placed behind the main spar, the fuel housed in tanks ahead of it outboard of the air intake ducts. Since with the exception of the crew compartment and nosewheel, all installations and loads were housed in the wings, this had an effect on the overall dimensions. The wing loading was lower but the profile thickness and hence drag was greater than desirable, but was compensated by the fact that the drag of the load-bearing components was lower than with a normal layout. On the tailless variant, the vertical fins and rudders were in the form of downturned surfaces at the wingtips whose camber was a continuation of the flow pattern at that point and had a favourable influence in aileron and rudder effectiveness in turning flight. The ailerons also served simultaneously as elevators, and in addition, small jet deflection flaps were located at the trailing-edge above the engine exhaust outlets, the landing flaps designed as split surfaces.

Until completion, work on both design schemes would certainly have taken several more months, if not years, had not the end of the war resulted in an early end to the project.

Focke-Wulf 0310239-01 – data (April 1944)

Powerplants	2 x 1,500kg (3,307 lb) thrust Heinkel-Hirth HeS 011 turbojets	
Dimensions		
Span, swept wing	12.65m	41ft 6in
Length	14.20m	46ft 7in
Height	3.75m	12ft 3½in
Wing area	27.00m²	290.62ft²
Weights		
Empty weight	4,225kg	9,314 lb
Fuel weight	2,775kg	6,118 lb
Loaded weight, with bomb	8,100kg	17,857 lb
Wing loading at take-off	300kg/m²	61.44 lb/ft²
Performance		
Max speed, without bomb	1,015km/h at 10,000m	631mph at 32,810ft
Initial climb rate	21.2m/sec	4,173ft/min
Range at 13,600m (44,620ft)	2,500km	1,553 miles
Service ceiling	13,500m	44,290ft
Bombload	1 x SB1000kg (2,205 lb)	

Focke-Wulf 0310239-10 – data

Powerplants	2 x 1,500kg (3,307 lb) thrust Heinkel-Hirth HeS 011 turbojets	
Dimensions		
Span, tailless	14.00m	45ft 11¼in
Length*	5.80m	19ft 0¼in
Height	2.75m	9ft 0¼in
Wing area	55.00m²	592.00ft²
Undercarriage track	5.70m	18ft 8.4in
Weights		
Empty weight	4,200kg	9,259 lb
Loaded weigh,t with bomb	8,100kg	17,857 lb
Wing loading, at take-off	147.27kg/m²	30.16 lb/ft²
Performance		
Max speed	1,060km/h	659mph
Range	2,500km	1,553 miles
Service ceiling	14,000m	45,930ft
Bombload	1 x SB1000kg (2,205 lb)	

* The oft-quoted figure. The three-view drawing shows a length of almost double – at 10.50m (34ft 5½in) – Translator.

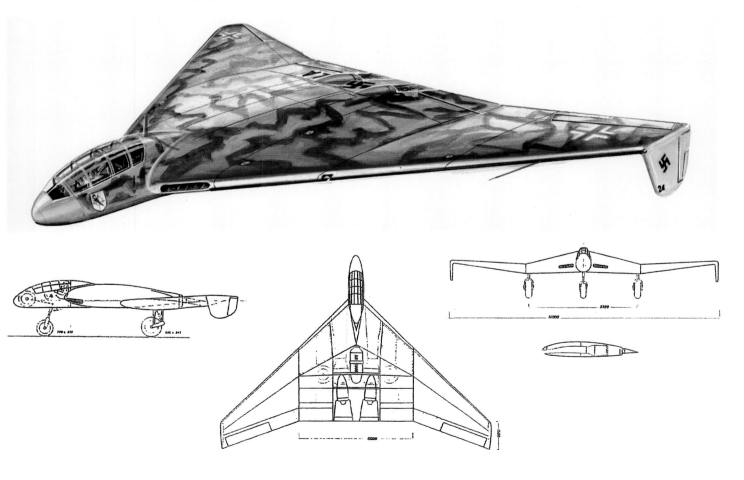

Focke-Wulf 0310251 –
A Night and Bad-weather Fighter/Zerstörer

1944

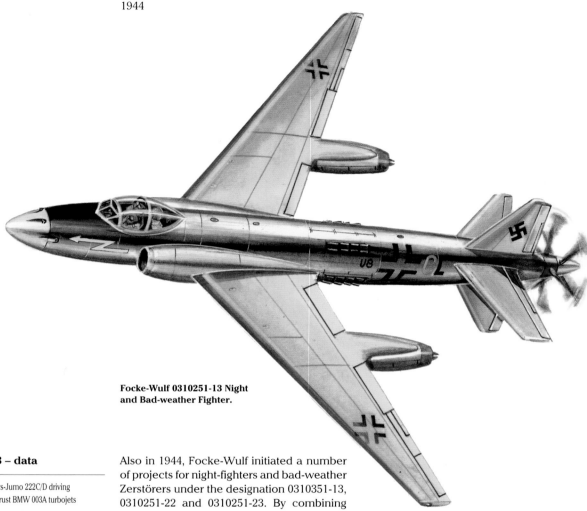

Focke-Wulf 0310251-13 Night and Bad-weather Fighter.

Focke-Wulf 0310251-13 – data

Powerplants	1 x 3,000hp Junkers-Jumo 222C/D driving
	contraprops and 2 x 800kg (1,764 lb) thrust BMW 003A turbojets

Dimensions

Span	21.00m	68ft 10¾in
Length	16.55m	54ft 3½in
Height	4.65m	15ft 3in
Wing area	55.00m²	592.00ft²

Weights

Empty weight	9,300kg	20,503 lb
Loaded weight (1)	12,000kg	26,455 lb
with fuel weight	2,000kg	4,409 lb
Loaded weight (2)	14,000kg	30,864 lb
with fuel weight	3,400kg	7,496 lb

Performance

Max speed	880km/h at 9,500m	547mph at 31,170ft
Initial climb rate	16.5m/sec	3,248ft/min
Service ceiling	14,000m	45,930ft

Armament	4 x MK 108 in nose

Also in 1944, Focke-Wulf initiated a number of projects for night-fighters and bad-weather Zerstörers under the designation 0310351-13, 0310251-22 and 0310251-23. By combining the reciprocating engine with the turbojet and swept flying surfaces, the aim was to achieve an increase in performance. The basic design consisted of a long fuselage of circular cross-section and a mid-set wing swept at 28° on the leading edge and cruciform empennage, the single high-powered piston engine in the fuselage driving pusher airscrews via a long extension shaft. As in the previous project described, the two turbojets were housed in underslung nacelles, but consisted of BMW 003As outboard of the main undercarriage wheels which retracted into the fuselage. The main fuselage engine, on the other hand, differed on all three variants. Among the most powerful under development in 1944 were the DB603N, Jumo 222C/D, and the Argus As 413. Because of their differing weights, the wingspan (of aspect ratio 8.0), wing area and loaded weights were altered so that wing loading was roughly the same in each case, the airframe construction allowing for the installation of the fuselage

motor with only minor alterations. Cooling air for the oil coolers and radiator was supplied via ducts of circular cross-section ahead of the wing roots, air being bled off from the starboard duct for the high-altitude two-stage turbo-supercharger attached to the engine. By favourable location of all the necessary installations in the fuselage, its surface area was held to a minimum, whereby the nose radar, fixed weapons, crew of three as well as the four fuel tanks, were adequately spread along its length. Despite various types of weapons proposed, the aircraft's speed over the enemy could be maintained, and in the event of being surprised when the turbojets were not in use, the all-round vision FuG 244 'Bremen' nose search radar with parabolic dish antenna provided early warning. Although a considerable amount of work was conducted on the projects between September 1944 and January 1945, a production contract did not result

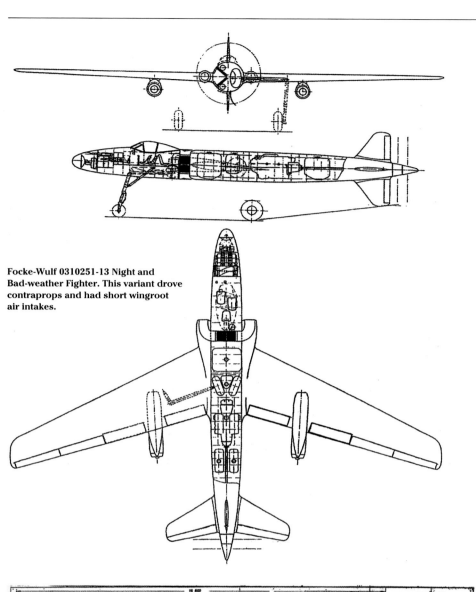

Focke-Wulf 0310251-13 Night and Bad-weather Fighter. This variant drove contraprops and had short wingroot air intakes.

Focke-Wulf 0310251-22 – data

Powerplants	1 x 2,830hp Daimler-Benz DB 603N and	
	2 x 800kg (1,764 lb) thrust BMW 003A turbojets	

Dimensions		
Span	20.40m	66ft 11¼in
Length	16.50m	54ft 1½in
Height	4.65m	15ft 3in
Wing area	52.00m²	559.71ft²

Weights		
Empty weight	8,800kg	19,400 lb
Loaded weight (1)	11,500kg	25,353 lb
with fuel weight	2,000kg	4,409 lb
Loaded weight (2)	13,000kg	28,660 lb
with fuel weight	3,000kg	6,614 lb

Performance		
Max speed	850km/h at 10,500m	528mph at 34,450ft
Initial climb rate	15.6m/sec	3,071ft/min
Service ceiling	13,400m	43,960ft

Focke-Wulf 0310251-23 – data

Powerplants	1 x 4,000hp Argus As 413 and	
	2 x 800kg (1,764 lb) thrust BMW 003A turbojets.	

Dimensions		
Span	22.80m	74ft 9½in
Length	16.75m	54ft 11½in
Height	5.10m	16ft 8¾in
Wing area	65.00m²	699.64ft²

Weights		
Empty weight	11,755kg	25,915 lb
Loaded weight (1)	15,000kg	33,069 lb
with fuel weight	2,500kg	5,516 lb
Loaded weight (2)	19,000kg	41,887 lb
with fuel weight	5,300kg	11,684 lb
Max speed	850km/h at 10,500m	528mph at 37,730ft
Service ceiling	13,000m	42,650ft

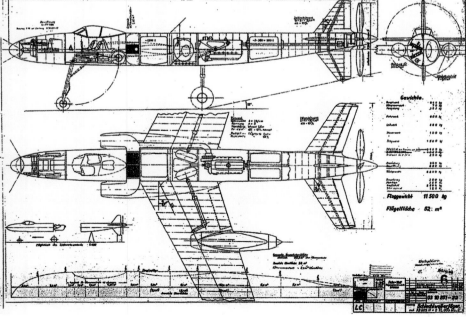

Focke-Wulf 0310251-22 works drawing. The DB 603N drives a single four-bladed propeller and has longer wingroot air intake ducts than the 0310251-13.

Gotha P 60C – A Night and Bad-weather Fighter

1945

Dr-Ing Rudolf Göthert (1912-1973), Chief Aerodynamicist at the Gotha firm, together with Dipl-Ing G Hühnerjäger, designed the Gotha P 60 projects.

As they could not obtain air superiority in terms of speed using propeller-driven aircraft, the TLR authorities were forced to concentrate towards the end of the war on jet-propelled day and night fighters. Various firms responded to the RLM specification of 27th February 1945, submitting their night-fighter proposals to the EHK in March 1945. These conformed to the requirements for the smallest dimensions, low flying weight and low wing loading, and powered by two HeS011 turbojets in or in the neighbourhood of the fuselage. This Gotha development was preceded by the following event:

Already in 1944, the Horten Brothers had earned the special attention of the EHK which had ordered the construction of an experimental Horten H IX prototype single-seat all-wing fighter-bomber. In the summer of 1944, the Gothaer Waggonfabrik Aircraft Construction Department had received an order to prepare for the series manufacture of 20 examples of the H IX under the designation 8-229. The H IX prototypes built by Gotha, and the projected further development into a two-seat night-fighter spurred the development engineers to initiate their own project, very similar to the H IX but with distinctive design features. Designated Gotha P 60, under the leadership of Dipl-Ing G Hühnerjäger and Chief Aerodynamicist Dr-Ing Rudolf Göthert, three versions of a night fighter were drawn up which fully met the 27th March 1945 specifications.

Their first proposal, the Gotha P 60A of January 1945, consisted of a 50° swept tailless fighter in which the crew of two lay in prone positions on either side of the flush-glazed fuselage nose. Powerplants were two BMW 003A turbojets superimposed above and below the rear fuselage. In this design, which had long-span leading-edge flaps on the outer wings to improve the stalling characteristics,

a pair of small narrow-chord vertical surfaces for directional control were located near each wingtip, and were pivoted so that when not in use, could be withdrawn completely within the wing. With a loaded weight of 7,500kg (16,535 lb) the fighter could attain a speed of over 950km/h (590mph).[4]

The Gotha P 60B, intended as a simplified further development with a conventional crew arrangement in the lengthened fuselage nose which housed an FuG 240 radar with parabolic dish antenna, had normal vertical fins and rudders at the wing trailing edges near the tips. Powerplants were two HeS 011 turbojets; weight rose to 10,000kg (22,046 lb) and estimated maximum speed 1,000km/h (621mph).[5]

The third variant, the Gotha P 60C, was to have utilised available construction materials and was to have had an FuG 244 night search radar with porcupine-type 'Morgenstern' (Morning Star) external antennae in the wooden pointed nose cone. The crew was increased to three, and intended powerplants were the available BMW 003A turbojets.[6] In all other respects, it was identical to the P 60B. From project documents that have survived, instead of the 'Morgenstern' antennae, provision was made for a parabolic radar dish antenna in a rounded nose. Whereas the pilot was seated beneath a raised cockpit canopy, the other two crew members had reclining seats slightly behind him in each wing root, forward vision being provided by long armoured-glass panels in the wing leading edges. Armament on all three versions consisted of four MK108 cannon, two each on either side and ahead of the mainwheel bays, but provision was made for two further oblique-firing MK108s behind the pilot. Construction of the P 60C ordered by the RLM was only hindered by the end of the war.

Gotha P 60C – data

Powerplants	2 x 1,300kg (2,866 lb) thrust Heinkel-Hirth HeS 011A turbojets	
Dimensions		
Span	13.50m	44ft 3½in
Length, fuselage	9.00m	29ft 6¼in
with rounded nose	10.90m	35ft 9in
with pointed nose	11.40m	37ft 4¾in
Height	3.50m	11ft 5¾in
Wing area	54.70m²	588.77ft²
Undercarriage track	3.20m	10ft 6in
Weights		
Empty weight	5,346kg	11,786 lb
Fuel capacity	3,500 litres	770 gals
Loaded weight	10,500kg	23,148 lb
Armament	4 x MK108 (120rpg) in nose	
	2 x MK108 (100rpg) oblique upward	

[4] The two-seat Gotha P 60A was originally designed to supersede the Horten Ho 229. In addition to the retractable directional control surfaces near the wingtips, it had combined elevators and ailerons on the wing trailing edge. At high speeds, only the outer section was used, whilst at lower speeds, the servo-tab operated inner surfaces were also employed. Salient details were: span 12.4m (40ft 8¼in), wing area 46.8m² (503.74ft²), normal loaded weight 7,450kg (16,424 lb) with 2,200 litres (482 gals) fuel and 8,550kg (18,849 lb) at overload. Maximum speed was 960km/h at 7km (596mph at 22,965ft), initial climb rate 14m/sec (2,756ft/min), normal range and endurance at 12km (39,370ft) altitude was 1,600km (994 miles) and 77 mins respectively. Armament was four MK108 cannon, two on each side of the crew. Provision was also made for the installation of a 2,000kg (4,409 lb) thrust HWK rocket motor.

[5] The Gotha P 60B fighter, an enlarged version of the P 60A and powered by two HeS 011As, had the same dimensions as the P 60C, except that the fuel capacity was almost twice that of the P 60A. Fuel capacity was increased to 4,200 litres (920 gals) and loaded weight 10,000kg (22,046 lb). Maximum speed was 980km/h at 5km (608mph at 16,400ft), initial climb rate 19m/sec (3,740ft/min), and range and endurance at 12km (39,370ft) was 2,650km (1,645 miles) and 3 hours 6 mins respectively. Equipped with an the additional 2,000kg (4,409 lb) thrust HWK rocket motor and 3,100 litres (680 gals) of turbojet and 1,800kg (3,968 lb) of rocket fuel, loaded weight was 11,000kg (24,251 lb). With the rocket motor in operation, initial climb rate was 50m/sec (9,842ft/min), the aircraft reaching an altitude of 9km (29,520ft) in 2.6 mins. With rocket motor first switched on at 3,000m (9,840ft), total time to climb to 12km (39,370ft) was 5.8 mins. If use of the rocket motor was delayed to 6km (19,685ft) altitude, a ceiling of 14.8km (48,556ft) could be attained in 9.5 mins.

The Gotha P 60C was entered in the 1945 night-fighter competition to which seven other designs including two tailless types were entered that included Arado (Project I), Blohm & Voss (BV P 215), Dornier, and Focke-Wulf (Projects II and III).

[6] This statement does not agree with the author's data table and three-view drawings clearly showing two HeS 011 turbojets. The BMW 003A, which would have made the P 60C seriously underpowered, was widely reported as the powerplant in magazine articles of German origin in the 1950s and 1960s.

Below: **Gotha P 60C with parabolic dish nose radar – drawing dated 21st April 1945.**

Bottom: **Gotha P 60C with 'Morgenstern' antenna in pointed nose.**

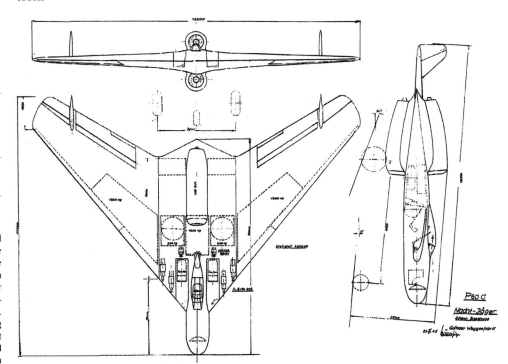

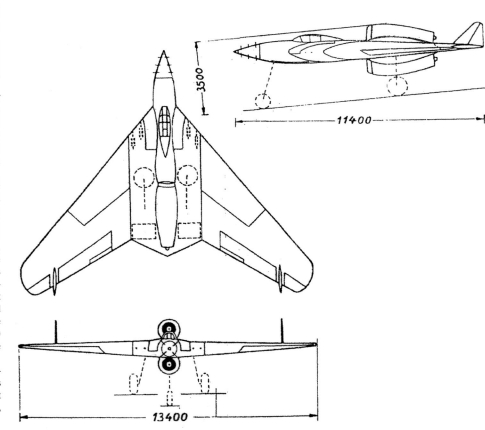

Heinkel P 1068.01-83

1943 to 1945

In January 1944, the Heinkel design office in Vienna submitted their P 1068.01-83 Zerstörer project to the RLM. It consisted of a strongly-armoured variant of the P 1068.01-76 Strahl-bomber (jet bomber)* with a two-man Zerstörer cockpit and additional weapon Rüstsatz in the bomb bay, capable of accommodating all current weapons up to 5cm calibre. Following completion of detail design, it was to become known as the He 343A-3. Except for a few components, the layout was identical to the He 343A-1 described in Volume 2. The He 343A-3 Zerstörer project, whose design was completed at the end of 1943, was to have entered series production under the highest priority category from January 1945. Despite receipt of a contract for the manufacture of three prototypes, without giving any reason the RLM ordered all design and construction work to be terminated on 23rd March 1945.

*See *Luftwaffe Secret Projects: Strategic Bombers 1935-1945*, p.81 ff.

Heinkel P 1068.01-83 – data

Powerplants   4 x 1,300kg (2,866 lb) thrust Heinkel-Hirth HeS 011A or 4 x 1,200kg (2,646 lb) reheat thrust Junkers-Jumo 004C turbojets

Dimensions			
Span		18.00m	59ft 0¾in
Length		16.50m	54ft 1½in
Height		5.35m	17ft 6½in
Wing area		42.25m²	454.76ft²

Weights			
Empty weight		10,770kg	23,744 lb
Fuel weight		4,820kg	10,626 lb
Loaded weight	004C	18,000kg	39,683 lb
	011A	19,550kg	43,100 lb

Performance			
Max speed,	004C	825km/h at 6,000m	513mph at 19,685ft
	011A	910km/h at 6,000m	565mph at 19,685ft
Service ceiling	011A	12,900m	42,320ft

Armament   4 x MK 103 (200rpg) in bomb-bay, plus 2 x MG 151/20 (200rpg) in rear fuselage, plus 1 x BK 5 (MK 214A) or 2 x MK 103 (100rpg), plus extra fuel*, plus 2 x MG 151/20 (200rpg)

* For the He 343A-1, A-2, A-3, normal fuel load was 4,820kg (10,626 lb) and 2,870kg (6,327 lb) in forward fuselage for the A-2. As a bomber, it could carry a total of 3,000kg (6,614 lb), made up of 4 x SC 500 or 2 x SC1000 or one Fritz X internally and 2 x SC 500 externally.

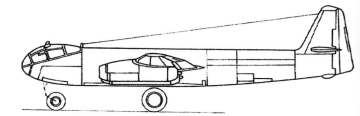

Heinkel P1079

1945

Among the latest Heinkel projects that Chief Designer Dipl-Ing Siegfried Günter and his colleagues Dipl-Ing Hohbach and Dipl-Ing Eichner were developing shortly before their capture by US Forces in Vienna-Schwechat on 27th April 1945 was the P1079 Zerstörer and night-fighter. The entire documentation was handed over to the so-called 'United States Air Forces Air Technical Intelligence Section', a US unit that sought after German scientists and their activities. From surviving documents, consisting exclusively of measurements and calculations dated 11th August 1945, which Dipl-Ing Siegfried Günter had to complete at the Landsberg/Lech internment centre for the Americans, the project was finally wound up.

The P1079 consisted of a two-seat Zerstörer and night-fighter powered by two HeS 011 turbojets and was expected to reach a speed of over 980km/h (609mph). By reducing total drag to a minimum, including the adoption of a swept butterfly or V-tail, measurement results indicated that the attainment of high Mach numbers was possible. The turbojets were housed in low-drag nacelles fed by intakes in the leading edges of

the 40° anhedral swept wing. The fuselage housed the two crew members seated back-to-back, the nose FuG 244 parabolic radar dish antenna and the four MK108 cannon on either side plus the two MG151/20 between the fuselage and engine nacelles. The 2.6m (8ft 6½in) narrow-track mainwheels retracted to rest vertically between the fore and aft fuel tanks, the nosewheel rotating to lie flat between the armament bays. In addition to the three protected fuselage fuel tanks, there was a further tank in each wing forward of the main spar outboard of the air intake channels. The wing main spar was of a special design at the wingroot/fuselage junction because of the need for the air intakes to pass through it.

Heinkel P1079 Zerstörer and Night-fighter.

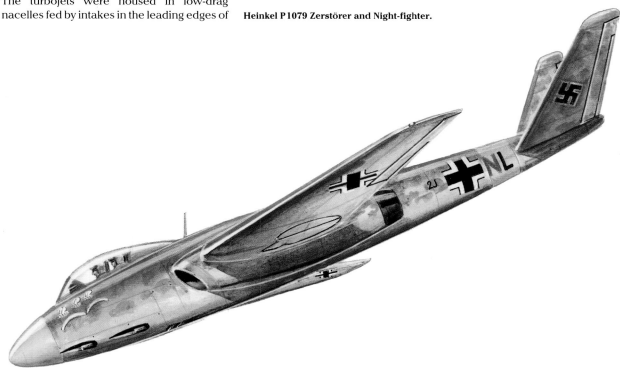

Heinkel P1079 – data[7]

Powerplants	2 x 1,300kg (2,866 lb) thrust Heinkel-Hirth HeS 011A turbojets	

Dimensions		
Span	13.00m	42ft 7¾in
Length	14.05m	46ft 1¼in
Height	4.60m	15ft 1in
Wing area	35.00m²	376.73ft²

Weights		
Wing loading	333kg/m²	68.2 lb/ft²

Performance		
Endurance	2.7 hours	

Armament	4 x MK108, plus 2 x MG151/20	

Several variants of the P1079 design existed: one with a butterfly tail as shown here, another with a broad-chord swept wing with downturned outer panels (as on the P1078B) with a single swept fin and rudder, and a pure flying-wing type without vertical fin, each powered by two HeS 011 turbojets buried in wingroot nacelles. One variant had a span of 12m (39ft 4¼in) and wing area 30m² (322.91ft²), whereas another had span 14m (46ft 3¼in) and wing area 40m² (430.55ft²). Design data for the latter variant were:

Powerplants	2 HeS 011 turbojets	2 Jumo 004 turbojets
Weights		
Fuel capacity	4,200 litres	3,600 litres
	920 gals	788 gals
Loaded weight	11,000kg	10,050kg
	24,251 lb	22,156 lb
Performance		
Max speed at sea level	825km/h	780km/h
	512mph	485mph
at 6,000m (19,685ft)	885km/h	790km/h
	550mph	490mph
Rate of climb at sea level	18.15m/sec	13.92m/sec
	3,570ft/min	2,740ft/min
at 6,000m (19,685ft)	11.20m/sec	6.00m/sec
	2,200ft/min	1,180ft/min
Max range at 11,000m	2,700km	2,700km
(36,090ft)	1,678 miles	1,678 miles
Endurance at 6,000m (19,685ft)		
normal	1.27 hours	1.12 hours
maximum	2.30 hours	–
Service ceiling	12,900m	10,000m
	42,320ft	32,810ft

The maximum range is with an additional 264 gals of fuel in 2 x 600 litre (132 gal) drop tanks.

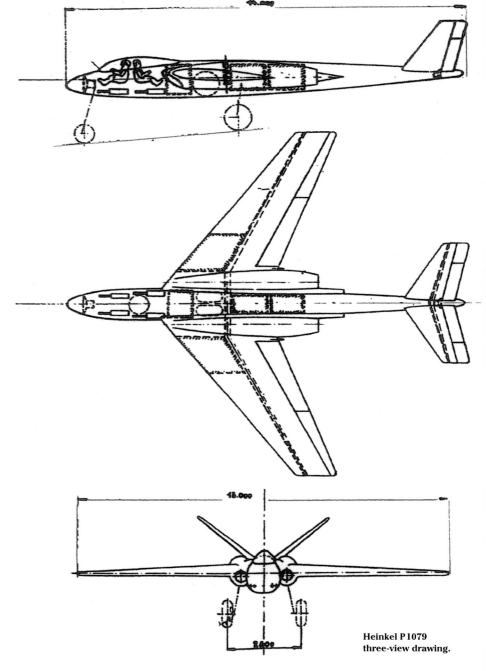

**Heinkel P1079
three-view drawing.**

Henschel P122

1944 to 1945

The Henschel P122 was one of the last heavy Zerstörer projects that had been submitted to the RLM at the beginning of 1945. Designed as a heavy high-altitude Zerstörer and Jagdbomber (fighter-bomber), the P122 was an all-metal tailless aircraft with a 30° swept high aspect ratio low wing housing the crew of two in a flush nose pressure cabin. Penetrating enemy airspace at an altitude of 17,000m (55,775ft), it was to attack enemy bombers during or shortly after take-off from their home airfields. With the aid of teleguided bombs and rockets, the two crew were to attack the airfields, thereafter escaping at a calculated speed of over 1,000km/h (621mph) on the return flight. The extreme altitude and high speed made a defensive armament superfluous since no enemy fighters could reach this altitude.[7]

[7] Described in a British Air Ministry report as a high-altitude *bomber*, the Henschel P.122 project obviously pre-dated that of the P.123 (Hs132) turbojet-powered ground attack aircraft which dated from the turn of 1943/44. How this large, fast-climbing unarmed high-speed bomber with its two

powerful BMW 018 turbojets designed to operate at altitudes up to 18,000m (59,055ft) where it produced some 15% of its sea-level static thrust could possibly have been usefully employed as a Jagdbomber (fighter-bomber) which signifies carrying offensive armament, or as a Zerstörer (twin-engined heavy fighter) is difficult to imagine, when one takes into account the necessity for a fully-pressurised cabin to maintain sustained level flight at extreme altitudes. Furthermore, as advanced radar equipment is not mentioned in the report for this project, it is difficult to imagine how the bomb aimer would have been able to make out pin-point targets such as aircraft taking off or ascending 17km (10.6 miles) beneath the aircraft. Even if targets such as ascending bombers and their airfields were to be attacked with remotely-guided bombs and rockets as the author states, the missiles themselves could not have been seen by the crew down to ground or target level, by which time the bomber itself would have advanced several kilometres along its flight-path.

One author has mentioned that the P.122 as a night-fighter (rather colossal for this role!), was to be fitted with six MK108 cannon each with 120 rounds/gun; four to have been fixed in the wings and two firing obliquely upward from the fuselage. The two small-scale original three-view drawings of the aircraft seen by this Translator show no sign of built-in offensive armament.

Henschel P 122 – data

Powerplants	2 x 3,500kg (7,716 lb) design static thrust BMW 018 turbojets*	

Dimensions		
Span	22.40m	73ft 6in
Length	12.40m	40ft 8¼in
Height	5.90m	19ft 4¼in
Wing area	c.70.00m²	753.46ft²

Weights		
Loaded weight	15,100kg	33,289 lb

Performance		
Max speed, at sea level	1,010km/h	627mph
at 10,000m (32,810ft)	935km/h	581mph
Initial climb rate	57m/sec	11,220ft/min
Range, at 17,000m (55,775ft)	2,000km	1,242 miles
at 10,000m (32,810ft)	1,110km	690 miles
Service ceiling	17,000m+	55,775ft+

* In his original data table, the author listed the BMW 018 engine thrust as only 1,250kg (2,756 lb) each – less than that of the HeS 011A, and the P122 loaded weight as only 5,100kg (11,243 lb), whereas each BMW 018 turbojet alone was expected to weigh 2,300kg (5,070 lb) – Translator.

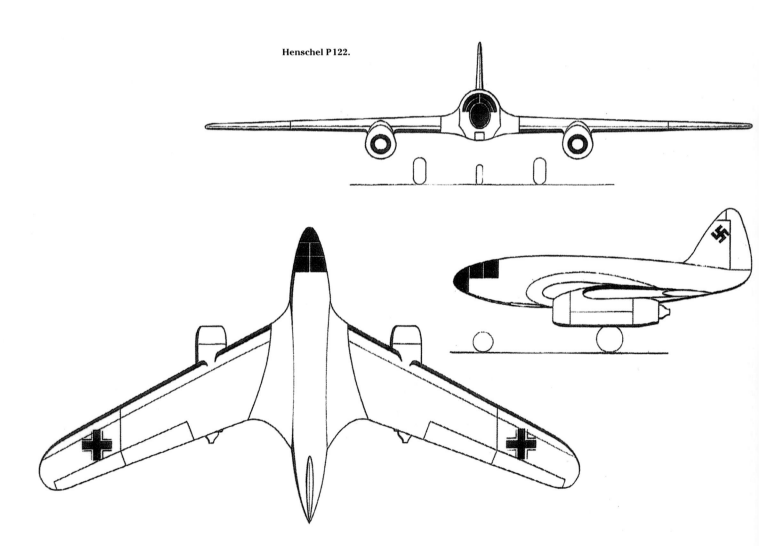

Henschel P 122.

Hütter Hu 211

1944

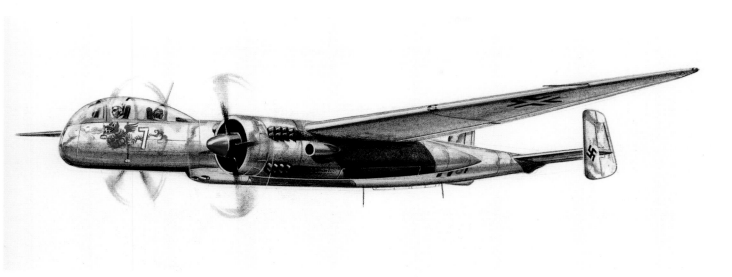

One of the most noteworthy of Wolfgang Hütter's projects was this long-range high-speed nocturnal reconnaissance aircraft with a range of 6,000km (3,728 miles). In all of Hütter's designs between 1942 and 1945, the use of major components of various aircraft and manufacturers is noticeable. Prefabricated parts such as the cockpit, fuselage, undercarriage and complete engine nacelles originated from known types such as the He 219, Ju 288 and Do 335. Only in respect of the wings and tail surfaces did Hütter go his own way. The influence of the glider pilot was unmistakable in all these designs, particularly the revision of the He 219E.

On the occasion of a meeting held on 25th May 1944 in Berchtesgaden with Reichsmarschall Hermann Göring and responsible RLM officials, Prof Dr-Ing Ernst Heinkel had suggested co-operation with Wolf Hirth and Wolfgang Hütter in connection with manufacture of the wings and tail surfaces out of wood in order to reduce the dural content of the He 219 from 3,900kg (8,600 lb) per aircraft to less than 1,000kg (2,200 lb). This meeting was decisive for the award of an RLM contract for a Fernerkunder (long-range reconnaissance aircraft) and Nachtzerstörer (heavy night-fighter). Bearing the highest priority classification, contract number GL/C-E2/A Nr. SS 5103/0010/44 was transferred to Wolfgang Hütter who used the opportunity to establish the Wolfgang Hütter Flugzeugbau GmbH in Kirchheim/Teck. The advanced state of the war compelled Hütter to design an aircraft similar to the He 219 in the shortest possible time, differing mainly in its high aspect ratio sailplane-type wing and butterfly tail.

The Hü 211 was to have been capable of reaching a speed of over 720km/h (447mph) at high altitude in order to combat the de Havilland Mosquito fighter. The designation number was used in agreement with Ernst Heinkel who had already been allocated it for the He 111Z twin-fuselage tow-craft. As there was nothing to improve in the He 219 fuselage and empennage, Hütter incorporated these structures into the design. The laminar-flow wings of aspect ratio 15:1 were completely new components of wooden monocoque construction, the wing planform resembling that of a high-performance sailplane. The centre section of the three-part

wings with electric de-icing in the leading edge, held a total of 3,600 litres (792 gals) of fuel plus two compartments each holding 180 litres (40 gals) of MW 50 engine injection additive for high altitudes. Three self-sealing tanks in the fuselage holding 5,100 litres (1,122 gals) brought the total fuel capacity to 8,700 litres (1,914 gals), a parachute landing brake being housed in the tail. As Wolfgang Hütter had mentioned in a letter dated 12th October 1944 to the AVA Göttingen, the first Hü 211 was expected to be ready to fly at the end of February 1945. Two prototypes were under construction in December 1944, but were destroyed in a bombing raid.

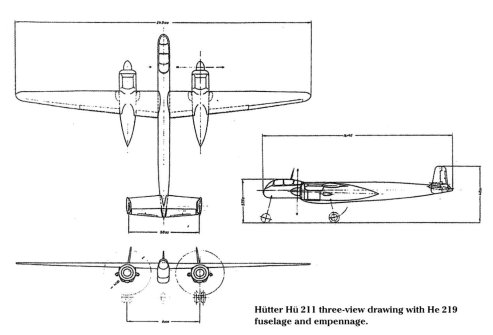

Hütter Hü 211 three-view drawing with He 219 fuselage and empennage.

Hütter Hü 211 – data

Powerplants	2 x 3,200hp Junkers-Jumo 222E/F radials	
Dimensions		
Span	24.50m	80ft 4½in
Length	16.50m	54ft 1½in
Height	4.80m	15ft 9in
Wing area	40.00m²	430.55ft²
Weights		
Empty weight	9,480kg	20,900 lb
Loaded weight	17,500kg	38,580 lb
Performance		
Max speed, at sea level	620km/h	385mph
at 7,300m (23,950ft)	720km/h	447mph
Range, econ cruise, at S/L	6,350km	3,496 miles
at 7,000m (22,965ft)	8,000km	4,971 miles
Service ceiling, at gross wt	11,100m at 16.5 t	36,420ft at 36,376 lb
	13,000m at 12.0 t	42,650ft at 26,455 lb
Time to 10,000m (32,810ft)	3.5 mins	
at gross weight	12 t	26,455 lb

Armament	2 x MG151/20 (200rpg) forward-firing, plus
	2 x MG151/20 (200rpg) rearward-firing

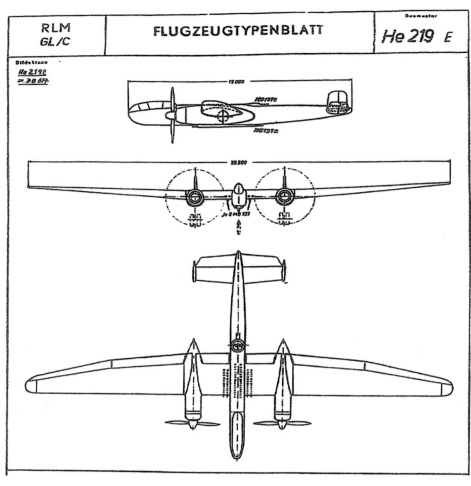

Heinkel He 219E three-view drawing of 7th July 1944 with Hütter wings.

Hütter Hü 211 model with butterfly or V-tail.

Junkers Ju 88C-6 –
From Dive-Bomber to Fast Night-fighter

Already at the beginning of the Second World War, the Luftwaffe's deficiency in night-fighters and Zerstörers became noticeable as the few available Bf110s and Do17Zs could not master the multifarious tasks to be accomplished. A suitably modified Ju 88 offered itself as a timely solution to the urgent need. In mid-1939, Junkers converted the Ju 88 V7 (GU+AE) into a Zerstörer and night-fighter. A fully-glazed cockpit hood replaced the aerodynamically streamlined metal nose, and had an armament of four MG17s and one MG FF cannon. Thus modified, the Ju 88 V7 served as the basis for all later Zerstörer and night-fighter variants. The first operational model, the Ju 88C-2 powered by two Jumo 211B engines, was delivered in July 1940 to the II. Gruppe of Nachtjagdgeschwader 1 (II. NJG 1), and as a Zerstörer, could carry a 500kg (1,102 lb) bombload. This was followed in 1941 by a limited series of Ju 88C-3 and C-4 night-fighters powered by air-cooled BMW 801MA radial engines.

At the end of 1941, the three-seat Ju 88C-6 which entered large-scale production served as the basic model for all the Ju 88 night-fighters. Designated Ju 88C-6a, this model was intended specifically for the night-fighter role. In addition to its standard armament plus two MG151/20s fitted as an oblique upward-firing Rüstsatz, it was equipped with FuG 212C-1 and FuG 220 SN-2 night search radar. The external radar antenna array, however, reduced the Ju 88C-6a maximum speed by some 40km/h (25mph). From May 1943, the Ju 88C-7 Zerstörer appeared, derived from the Ju 88A-4 but was built in only limited numbers. Further conversions followed, designated Ju 88C-7, C-7a, C-7b and C-7c, powered by the improved BMW 801MA or the more powerful Jumo 211J. These type designations such as Ju 88G-7a, G-7b and G-7c, were actually used by the Junkers-Werke as confirmed in original works documents and are therefore not 'post-war inventions' as some authors have stated, as they denoted the appropriate equipment fitted to each variant.

In the summer of 1943, a modified Ju 88C-5 served as the test prototype for the Ju 88G-1. Known as the Ju 88G-1a, it had enlarged vertical tail surfaces and an altered wing planform. Equipped with FuG 220 SN-2 night search radar with improved antenna array and four fixed MG151/20s in an underfuselage pannier, the Ju 88G-1a fulfilled the requirements for the night-fighter squadrons. It was followed by further variants, among them the Ju 88G-7b with FuG 228 'Lichtenstein SN-3' radar and 'Morgenstern' antennae. It attained a speed of 650km/h (404mph), and as a trustworthy night-fighter and nocturnal ground attack aircraft, was in service until May 1945.

Junkers Ju 88C-6. G W Heumann

Junkers Ju 88C-6 with FuG 220 SN-2 nose radar antenna array.

Junkers Ju 88C-6c – data

Powerplants	2 x 1,410hp Junkers-Jumo 211J-1 or J-2 or BMW 801engines	
Dimensions		
Span	20.00m	65ft 7½in
Length, with radar array	14.96m	49ft 1in
Wing area	54.50m²	586.62ft²
Weights		
Loaded weight	12,500kg	27,557 lb
Performance		
Max speed	500km/h at 5,000m	310mph at 16,400ft
Range, max	2,000km	1,242 miles
Service ceiling	9,900m	32,480ft

Junkers Ju 88C-6 with FuG 228 SN-3 central nose radar antenna.

Junkers Ju 88C-6 three-view drawing.

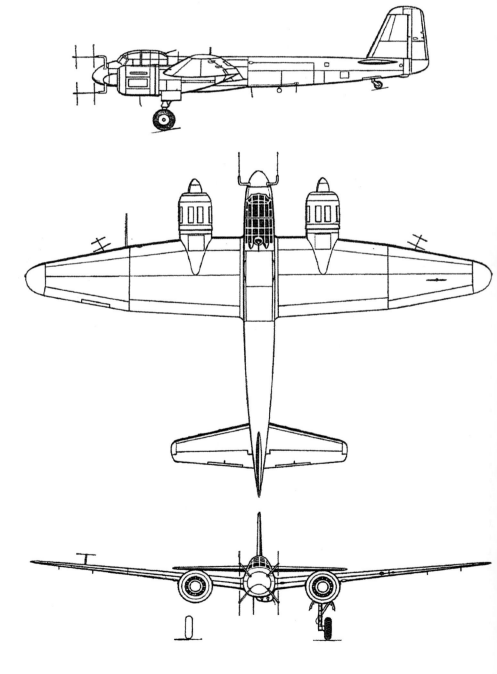

Lippisch P11-121 –
The World's First Turbojet-powered Zerstörer
1943 to 1945

This project, belonging to the P11-series, was also designated Delta VI and was one of the 'classical' designs which Dr Alexander Lippisch had developed during his time at the Messerschmitt AG. Following his departure from the company on 28th April 1943, Lippisch continued work on the P11-series from 1st May 1943 at the new Luftfahrtforschungs-anstalt Wien (Aeronautical Research Institute, Vienna). A full-size mock-up of the P11-121 was built and inspected by a delegation from the Technische Luftrüstung (Technical Air Armament Board), and it was only the absence of more powerful turbojets that were not available because of delivery delays that hindered the manufacture of a flyable version of the aircraft. Despite numerous aerodynamic and technical improvements, among them the provision of movable wingtips to increase lift and controllability, the P11-121 single-seater did not convince the TLR officials, and it was only after 1945 that this and other pioneering Lippisch developments gained acceptance in modern aircraft construction. The attitude of the Acceptance Committee documents the straight-laced mentality of RLM higher-ups who hindered the construction of one of the most modern aircraft.[8]

[8] In his book *Ein Dreieck Fliegt (A Delta Flies)*, pp.93-95, Lippisch illustrates a three-view drawing of the single-seat P11-121 Schnellbomber (fast bomber) dated 17th May 1943, similar in overall layout to the single-seat P11 V1 glider shown here, except that it is powered by two Jumo 004C turbojets (of 1,015kg (2,238 lb) static thrust without reheat installed in the c.1m (3ft 3¼in) deep and 6m (19ft 8¼in) long wing centre section, exhausting between a *single* fin and rudder with provision on either side of it for upper and lower jet exhaust deflection flaps – another advanced Lippisch innovation. In this design, the 500 x 180mm nosewheel retracted *forwards* ahead of the cockpit, the 840 x 300mm mainwheels folding upwards to rest vertically within the wings. Outboard of the undercarriage bays, the 35° swept wings each housed a 1,200kg (2,646 lb) fuel tank. A 1,000kg (2,205 lb) bomb was housed inside a deep, streamlined ventral fairing beneath the fuselage, the aircraft featuring wide-span constant-chord landing flaps ahead of the trailing-edge ailerons and elevators. The P11-121 of May 1943 had a span 10.6m (34ft 9¼in), length 6.8m (22ft 3⅜in), height 2.7m (8ft 10¼in), track 2.44m (8ft) and wing area 50m² (538.18ft²). With a fuel weight of 2,400kg (5,291 lb) and the above bombload, loaded weight was c.7,260kg (16,005 lb).

The project was continued until December 1943 as a twin-engined P11 (Delta VI) Zerstörer up to the mock-up stage, and prototype construction of the P11 V1 was begun. A whole series of wind-tunnel measurements were conducted on the project at the AVA Göttingen. For the prototype, the bombload fairing was removed, and at the centre between the *twin* fins and rudders 1.9m (6ft 2¾in) apart, the two turbojets were replaced by two solid-fuel RATO units each 2.2m (7ft 2⅜in) in length. In the P11 V1, the 560 x 200mm nosewheel retracted *rearwards*. Fuselage depth from canopy hood to bomb enclosure was 1.75m (5ft 9in). The P11-twin mainwheels of 710 x 185mm retracted forwards to lie vertically within the wings. The P11 V1 (Delta VI V1), fabricated of wood, was to have been flown as a glider, but during the evacuation ahead of the advancing Russian forces, had to be abandoned by the roadside. The Americans supposedly found it in a shed or garage of the Grand Hotel in Strobl am Wolfgangsee. Data for the P11 V1 glider are as shown in the three-view drawing. In the powered P11 V2, each wing housed a 1,800 litres (396 gal) fuel tank. Salient data quoted for this variant were: powerplants two Jumo 004C turbojets, span 10.8m (35ft 5¼in), length 7m (22ft 11¼in), wing area 50m² (538.18ft²), fuel capacity 3,600 litres (792 gals) and loaded weight c.7,300kg (16,094 lb). Estimated maximum speed was 1,040km/h at 6km (646mph at 19,685ft), cruising speed 850km/h (528mph), range 3,000km (1,864 miles) and fixed armament two MK103 or one large-calibre cannon.

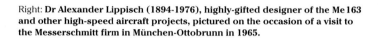

Above left and right: **Lippisch P11-121 model photos with wingtips upswept (above) and flat (right).**

Right: **Dr Alexander Lippisch (1894-1976), highly-gifted designer of the Me 163 and other high-speed aircraft projects, pictured on the occasion of a visit to the Messerschmitt firm in München-Ottobrunn in 1965.**

Lippisch P11-121 – data

Powerplants	2 x 1,300kg (2,866 lb) thrust Heinkel-Hirth HeS 011 turbojets	
Dimensions		
Span	10.80m	35ft 5¼in
Length	7.485m	24ft 6¾in
Height	2.76m	9ft 0⅜in

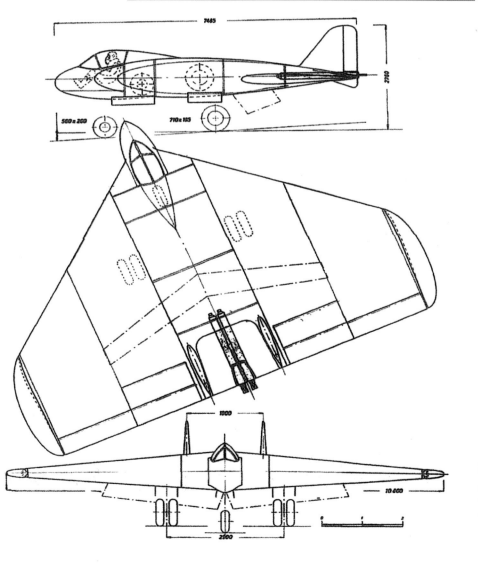

Lippisch P11-121 V1 three-view of the glider
prototype with two solid-fuel RATO units.

Messerschmitt P1099

1944

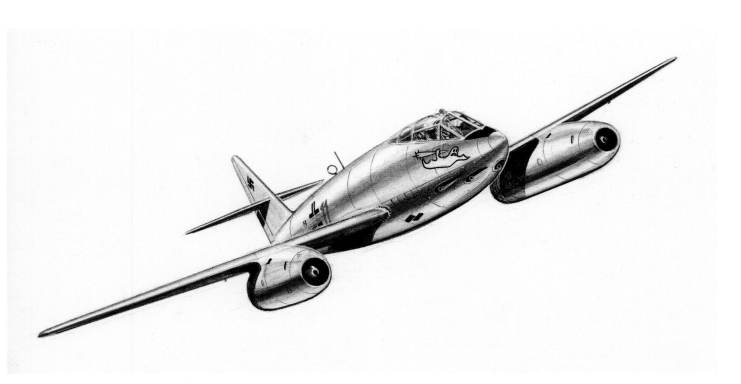

During the year 1943 the Messerschmitt Project Office studied a number of proposals for modifying the basic Me 262 airframe for the carriage of additional loads such as cameras and bombs, whereby the minimum of modifications would be made to the production airframe. Initial proposals of July 1943 resulted in the well-known Interzeptor (interceptor), Aufklärer (reconnaissance), and Schnellbomber (fast bomber) variants. In order to reduce the drag of externally suspended bombs, later proposals of September 1943 envisaged an enlarged fin and rudder coupled with a deepened fuselage to house additional fuel tanks and bombload internally. These interim solutions, however, were to give way to newer proposals incorporating an entirely new circular cross section fuselage of 1.76m (5ft 9¼in) diameter to better accommodate the increased fuel tankage and heavier armament in the P1099 and P1100 projects.

In February 1944, the P1099 design was submitted to the RLM as a two-seat bad-weather fighter and Zerstörer that, apart from the new fuselage and enlarged fin and rudder, retained the wing and tailplane of the basic airframe but with the turbojets moved further aft on the wings to take into account the altered centre of gravity position and twin mainwheels to cater for the increased loaded weight. Powerplants were to have been two Jumo 004C turbojets, replaceable with two HeS 011s when these became available. The two crew members were seated side-by-side in a pressurised cabin near the extreme nose. In addition to normal and heavy large-calibre forward-firing armament in the weapons bay, a noticeable feature was the provision of lateral remote-controlled rearward-firing FPL 151 barbettes on the rear fuselage. In one proposal, the weapons bay housed an MK108 with 80 rounds and a Rheinmetall-Borsig 55mm MK112 with 40 rounds. This weapon, with an installation weight of 320kg (705 lb), had a rate of fire of 300 rds/min which could be experimentally increased to 450 rds/min. Alternatively, the weapons bay could house two MK103s and a single Mauser 50mm MK 214 with 40 rounds. The MK 214 weighed over 700kg (1,543 lb) and had a rate of fire of 150 rds/min.

The P1099 and P1100 projects were worked on simultaneously over a period of several months in 1944.

Messerschmitt P1099 – data

Powerplants 2 x 1,200kg (2,646 lb) reheat thrust Junkers-Jumo 004C or 2 x 1,300kg (2,866 lb) thrust Heinkel-Hirth HeS 011 turbojets

Dimensions

Span	12.61m	41ft 4½in
Length	12.00m	39ft 4½in
Height	4.40m	14ft 5¼in
Wing area	22.00m²	236.80ft²
Undercarriage track	3.20m	10ft 6in

Weights

Fuel weight	3,300kg	7,275 lb
Loaded weight	8,800kg	19,400 lb

Performance

Max speed, 100% thrust	800km/h at 9,000m	497mph at 29.530ft
120% thrust	940km/h at 9,000m	584mph at 29,530ft
Rate of climb, at sea level	17.4m/sec	3,425ft/min
at 9,000m (29,530ft)	2.5m/sec	492ft/min
Range, from take-off	1,310km at 7,000m	814 miles at 22,965ft
Service ceiling, 100% thrust	9,800m	32,150ft

All the above weights and performance figures (with the Jumo 004C) are from Translator. In addition to the above, optimum range (normal fuel) was 1,450km (901 miles), and with 1,250kg (2,756 lb) extra fuel, optimum range was 1,980km (1,230 miles). Endurance (100% thrust) and normal fuel was 0.86 hrs at sea-level and 1.8 hrs at 7km (22,965ft). With the additional fuel and 120% thrust, endurances were 1.8 hrs at sea-level and 2.52 hrs at 7km (22,965ft). Take-off run at 120% thrust and two 2,000kg (4,409 lb) thrust RATO units was 830m (2,723ft). Landing speed was 185km/h (115mph) at landing weight 6,100kg (13,580 lb) with 20% fuel and full offensive load.

As a fighter, armament could comprise four MK108s, or two MK103s, or two MK103s plus two MK108s. The heavier calibre weapons for the Zerstörer (bomber destroyer) role were as above. As a two-seat night fighter, armament was four MK108 and two MK108Z oblique upward-firing weapons in the fuselage.

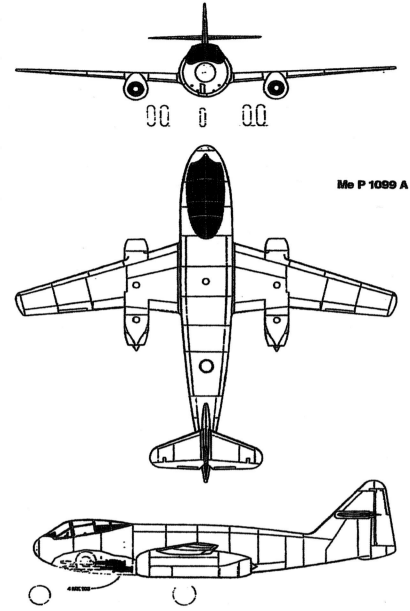

Me P 1099 A

Right: Messerschmitt P1099 three-view drawing with four MK108 cannon.

Below: Messerschmitt P1099 side-view with one MK214 and two MK103s in the weapons bay plus two FPL 151 rear barbettes.

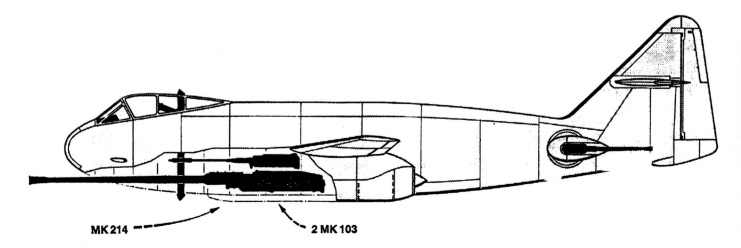

MK 214   2 MK 103

Messerschmitt P1100

1944

Similar in overall design to the P1099 above except that the turbojet nacelles retained their forward position, the P1100 two-seat unarmed Schnellbomber (fast bomber) and all-weather Zerstörer of January 1944 also featured a variety of armament combinations. In one bomber configuration the pilot sat in a raised cockpit offset to port, the second crew member seated completely within the fuselage on the starboard side, whilst in another, both crew members sat side-by-side in the widened cockpit. Beneath the five fuselage fuel tanks of 3,900 litres (858 gals), the aircraft could carry a total of 2,500kg (5,512 lb) of bombs in various combinations within the widened fuselage. Armament could consist of two forward and two rear-ward-firing MK108s, or one movable FHL 151 in the nose, two MK103Zs behind the cockpit and two FPL 151s in the fuselage barbettes. For combating closer-range targets such as

enemy bomber formations, the heavier calibre MK112 or MK214, each with 40 rounds, could be fitted. Powerplants were two Jumo 004C turbojets but could be replaced with the more powerful HeS 011 at a later date, as in the enlarged proposal of March 1944 which had a completely new wing swept 35° at ¼ chord, swept tailplane and engines moved to the wing roots flanking the mid-positioned wings, this proposal having single mainwheels of larger diameter.

Seen as a whole, the P1099 and P1100 projects represented the preliminary culmination of a series of developments to equip heavy fighters, fast bombers, and Zerstörers with a maximum of armament and bombs making use of several complete components of the Me 262, in particular the wings, turbojet nacelles and tail surface layout. Having progressed to the cockpit mock-up stage, the projects were terminated in early 1945.

Messerschmitt P1100 with fuselage weapons (March 1944).

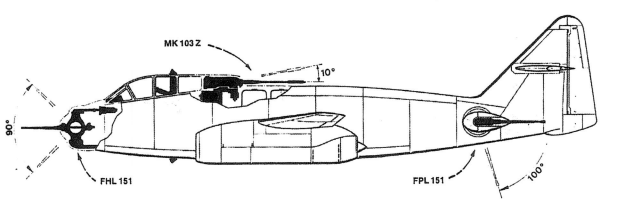

MK 103 Z

10°

90°

FHL 151

FPL 151

100°

Messerschmitt P1100 – data

Powerplants 2 x 1,200kg (2,646 lb) reheat thrust Junkers-Jumo
004C or 2 x 1,300kg (2,866 lb) thrust Heinkel-Hirth HeS 011 turbojets

Dimensions

Span	12.61m	41ft 4½in
Length	12.00m	39ft 4½in
Height	4.40m	14ft 5¼in
Wing area	22.00m²	236.80ft²
Undercarriage track	3.20m	10ft 6in

Weights

Loaded weight	10,262kg	22,624 lb

As a fast bomber, the P1100 could carry combinations of AB 250 and
AB 500 containers, SC 250, SC 500, SC1000, SB1000, BM1000 bombs or a
PC1400 bomb, divided between the forward and rear bomb bays. With a
normal bombload of 1,500kg (3,307 lb) and fuel load 3,320kg (7,319 lb),
take-off weight was 9,120kg (20,105 lb); with maximum bombload
2,500kg (5,512 lb) and 3070kg (6,768 lb) of fuel, take-off weight was
10,200kg (22,486 lb). At either load, maximum speed at 100% thrust was
780km/h (486mph) at sea-level and 820km/h at 9km (515mph at
29,530ft); at 120% thrust, corresponding speeds were 860km/h (536mph)
and 950km/h (590mph) respectively. Rate of climb was 16m/sec
(3,150ft/min) at sea-level and 6.8m/sec at 9km (1,340ft/min at 29,530ft)
and service ceiling 9,400m (30,840ft). At 100% thrust and with the lower
bombload, range was 680km (424 miles) at sea-level and 1,330km at
7km (826 miles at 22,965ft) and endurances 0.86 hrs and 1.85 hrs at
those altitudes. At 120% thrust and with two 2,000kg (4,409 lb) thrust
RATO units, take-off run was 1,200m (3,937ft), and landing speed
175km/h (109mph).

**Right: Messerschmitt P1100 fast bomber three-view
drawing (January 1944).**

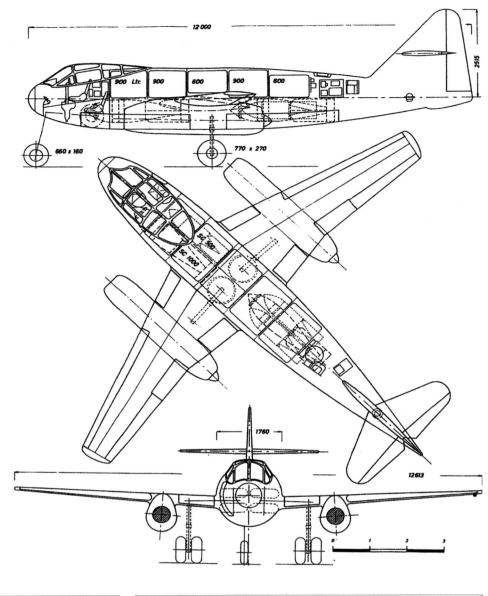

**Messerschmitt P1100 mock-up with pilot's cockpit
offset to port. This variant had an increased
wingspan and area.**

Messerschmitt P1101-92

1944

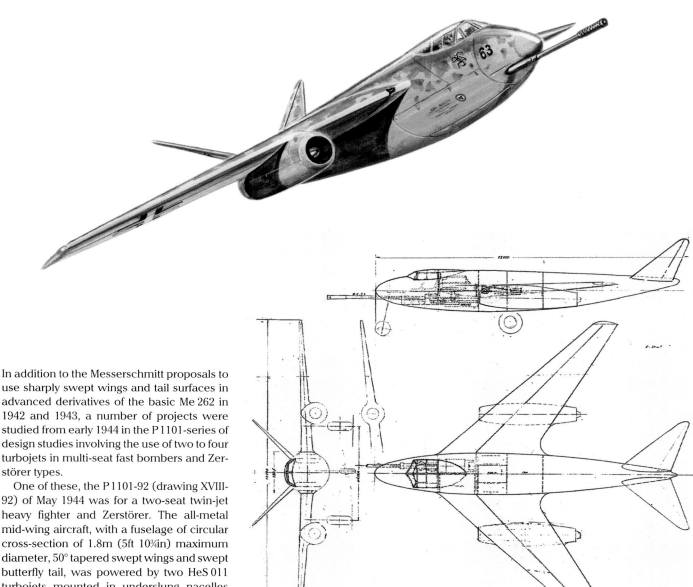

In addition to the Messerschmitt proposals to use sharply swept wings and tail surfaces in advanced derivatives of the basic Me 262 in 1942 and 1943, a number of projects were studied from early 1944 in the P1101-series of design studies involving the use of two to four turbojets in multi-seat fast bombers and Zerstörer types.

One of these, the P1101-92 (drawing XVIII-92) of May 1944 was for a two-seat twin-jet heavy fighter and Zerstörer. The all-metal mid-wing aircraft, with a fuselage of circular cross-section of 1.8m (5ft 10⅝in) maximum diameter, 50° tapered swept wings and swept butterfly tail, was powered by two HeS 011 turbojets mounted in underslung nacelles outboard of the mainwheels which retracted inwards to rest vertically in the fuselage, the nosewheel retracting rearwards to rest flat beneath the cockpit. An alternative proposal, as seen in the drawing, placed the turbojets symmetrically through the wing structure. A special feature of the design was the installation of a single BK7.5 cannon on the starboard side, its long barrel projecting well ahead of the nose. Lack of availability of the powerplants caused this and similar projects to be laid aside.

An exception to the jet-propelled series of studies was the tailless P1101/XVIII-97 of 22nd May 1944 that was powered by two Junkers-Jumo 222 engines driving pusher propellers.

Messerschmitt P1101-92 – data

Powerplants	2 x 1,300kg (2,866 lb) thrust Heinkel-Hirth HeS 011 turbojets	
Dimensions		
Span	13.28m	43ft 6¾in
Length, without BK7.5	13.10m	42ft 11¾in
Height	4.10m	13ft 5½in
Wing area	35.00m²	376.73ft²
Undercarriage track	4.06m	13ft 3¾in
Armament	1 x 7.5cm BK7.5 cannon	

Messerschmitt P1101-99

1944

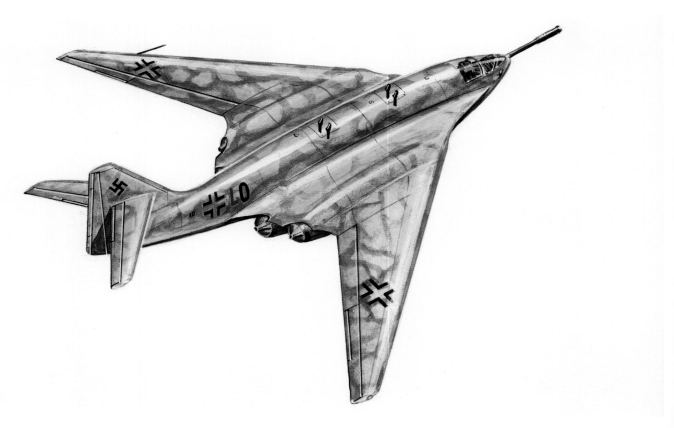

Another of the basic high-speed swept wing concepts under the P1101 label was the P1101 (drawing XVIII-99) of 6th June 1944 for a two-seat heavy Jagdzerstörer (pursuit-destroyer) whose mid-wing and tailplane had a leading-edge sweep of 45°. Other studies in the P1101 series involved aircraft with folding wings (swept 60° in the folded position) as well as bombers with variable-sweep wings* Project leadership lay in the hands of Dipl-Ing Hans Hornung who was also responsible for overall design. In this proposal, the four HeS 011 turbojets were housed in staggered wingroot pairs fed by combined elliptical air intakes in the wing leading edges, and in its layout, was seen as the optimum for an attack and Zerstörer aircraft. As in the earlier study described, armament comprised a 7.5cm Pak 40 Bordkanone housed this time on the fuselage port side beneath the crew compartment. Additionally, it had five MK112 cannon – one in the starboard wingroot between the fuselage and air intake duct and the other four as oblique upward-firing weapons in staggered pairs fore and aft of the mainwheel bays surrounded by the seven fuselage fuel tanks. Like successive project studies of this nature, it did not pass beyond the drawing board.

*See *Luftwaffe Secret Projects: Strategic Bombers 1939-1945*, pp.109-112.

Messerschmitt P1101-99 – data

Powerplants	2 x 1,300kg (2,866 lb) thrust Heinkel-Hirth HeS 011 turbojets	
Dimensions		
Span	15.20m	49ft 10¼in
Length	15.40m	50ft 6¼in
Height	4.90m	16ft 1in
Wing area	47.00m²	505.89ft²
Undercarriage track	5.96m	19ft 6½in
Empty weight	12,730kg	28,065 lb
Weights		
Fuel weight	5,700kg	12,566 lb
Loaded weight	18,630kg	41,072 lb
Performance		
Max speed	960km/h	597mph
Armament	1 x BK7.5 (12 rounds), total weight 830kg (1,830 lb) plus 5 x MK112 (40rpg), total weight 2,100kg (4,630 lb)	

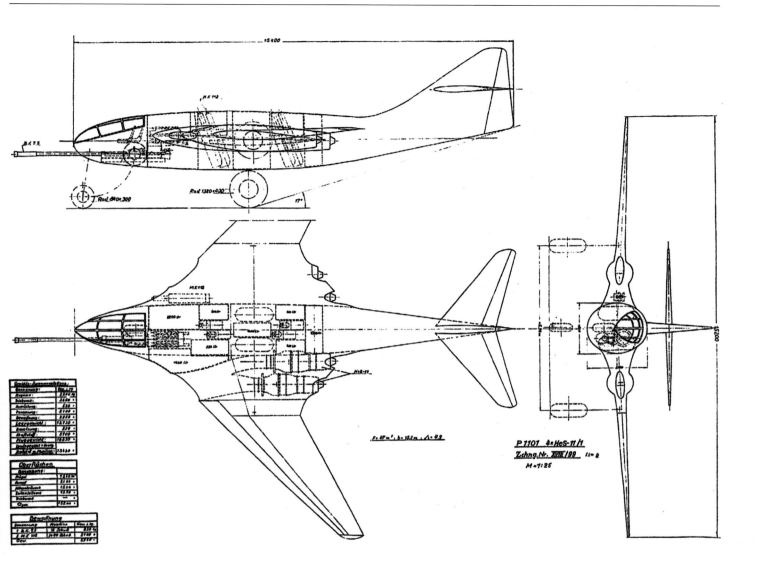

Messerschmitt P1101-99 three-view works drawing and data table.

Dipl-Ing Hans Hornung (1900-1978) was head of Preliminary Development at Messerschmitt. He was responsible for several high-speed projects and was particularly involved with the design of the P1101-99.

Messerschmitt P1101-104

1944

This heavy two-seat Zerstörer which followed at the end of June/early July 1944, in its layout and construction was a further development of the P1101-99. For the first time, however, a 'sickle' or crescent wing planform was employed, the four HeS 011 turbojets mounted in underwing paired nacelles projecting ahead of the wing, which had a leading-edge sweep of 50° inboard and 37° outboard of the turbojets. As in the earlier project, it had a swept butterfly tail. Fixed forward-firing armament consisted of a BK7.5 (Pak 40) cannon to starboard and three MK108 on the port side, between which the retracted nose-wheel rested vertically. A further three MK108 cannon were arranged to fire obliquely upwards at mid-fuselage surrounded by the fore and aft fuselage fuel tanks, with additional tankage in the wing box main spar structure. At the rear was a remote-controlled FDL 108Z tail barbette. Like the earlier projects, Dipl-Ing Hans Hornung had design leadership for the P1101/XVIII-104. Priority given to the Jägernotprogramm (Fighter Emergency Programme) of July 1944 caused this proposal to be confined to the drawing board.

The essentially identical P1101/XVIII-105 fast bomber with a reduced fuel capacity was able to carry an internal bombload of nine SC 500 or four SC 1000 bombs in its capacious 6.6m (21ft 7¾in) long bomb bay, the oblique upward firing MK108s having been deleted.

Messerschmitt P1101-104 – data

Powerplants	4 x 1,300kg (2,866 lb) thrust Heinkel-Hirth HeS 011 turbojets	
Dimensions		
Span	17.35m	56ft 11in
Length	18.10m	59ft 4½in
Height	4.10m	13ft 5½in
Weights		
Fuel capacity	7,530 litres	1,650 gals
Performance		
Max speed	860km/h	534mph
Armament	1 x BK7.5 (Pak 40) in fuselage, plus 3 x MK108 in fuselage, plus 3 x MK108 oblique upward-firing	

Dipl-Ing Woldemar Voigt (1907-1980) headed the Messerschmitt Project Office and had great influence on all the company aircraft developments and projects.

Messerschmitt P1101-104 works drawing.

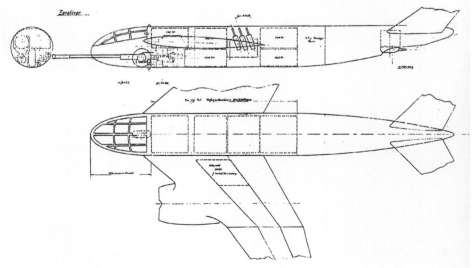

Messerschmitt P1101-28

1945

This project study of 11th April 1945 for a two-seat Schnellbomber (fast bomber) and Zerstörer constituted a further development of the P1099 and P1100 proposals of 1944. Whereas the basic fuselage, cockpit and tail surfaces were retained, the two HeS 011 turbojets were relocated in mid-positioned nacelles to which the new wings having a leading edge sweep of almost 40° were attached. An interesting feature of the design was that the mainwheels were to retract inwards to rest vertically in the fuselage between the fore and aft fuel tanks. Exactly how this was to be accomplished with the turbojets in the way is not clear from the documents. Although the final form of the fuselage nose portion had not been decided, the end of the war brought an early end to the project.

Messerschmitt P1101-28 – data

Powerplants	2 x 1,300kg (2,866 lb) thrust Heinkel-Hirth HeS 011 turbojets	

Performance		
Span	14.30m	46ft 11in
Length	12.35m	40ft 6¼in
Height	4.40m	14ft 5¼in
Undercarriage track	5.00m	16ft 4¾in

Performance		
Max speed	910km/h	565mph

Dipl-Ing Walter Rethel (1892-1977), head of design, was influential on the design of the Me 210, Me 410 and Me 323 as well as the projects described above. The photograph shows him (centre) with Prof Willy Messerschmitt (right) together with other engineers of the well-known firm.

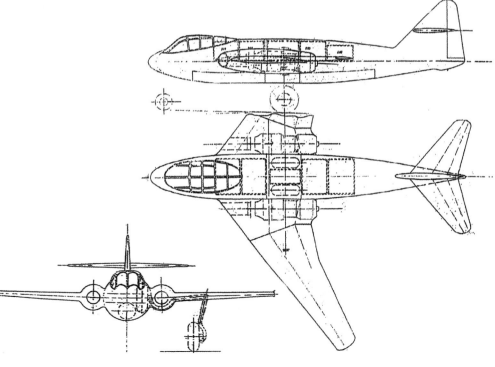

Messerschmitt P1112

1945

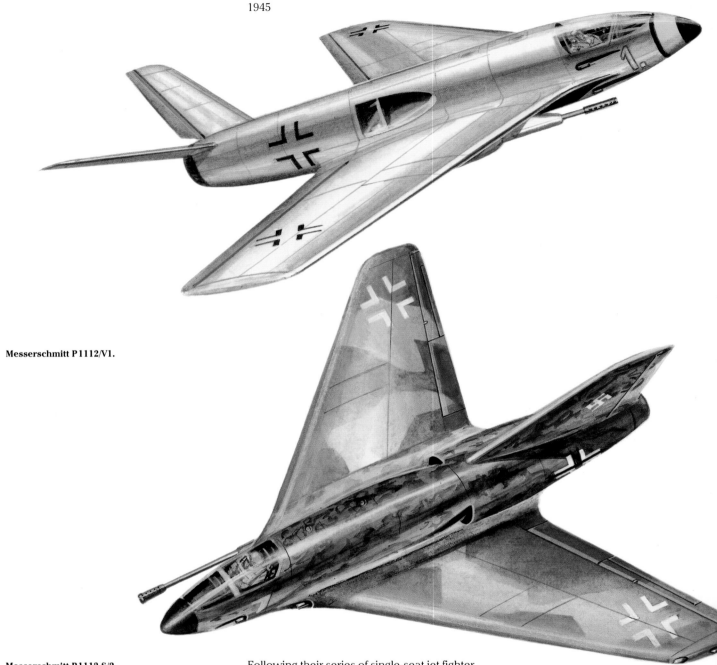

Messerschmitt P1112/V1.

Messerschmitt P1112 S/2.

Following their series of single-seat jet fighter designs begun in 1944 where the turbojet was completely enclosed in the fuselage as proposed in the various P1110 and P1111 layouts, an attempt was made to eliminate the faults in the P1111 tailless design in the P1112 on which work was begun on 25th February 1945. Although excellent aerodynamically, it was realised that the P1111 could still be further improved upon, principally by alterations to the wing, fuselage, fuel tank locations, and wingroot air intakes that were replaced in one design variant by fuselage scoops closer to the engine intake duct. During the period between 3rd and 30th March 1945, this project, whose design had advanced to the construction of a cockpit mock-up at Ober-ammergau, was laid out in three distinctive versions:

The P1112 S/2 of 3rd March 1945 had broader chord wings with a sweep of 50° at ¼ chord and wingroot air intakes for the HeS 011 turbojet. Span was 7.8m (25ft 7in), length 8.2m (26ft 10¾in), height 3.05m (10ft), wing area 22m² (236.8ft²), fuel weight 1,550kg (3,417 lb) and armament four MK108s – two in the wing roots and two in the fuselage. A further S/2 variant (drawing XVIII/167) of 27th March 1945 featured a lengthened fuselage with lateral air intakes and narrower-chord 45° swept wings of longer span.

The P 1112 S/1 of 27th March 1945 was a variant with identical narrower-chord wings having a leading-edge sweep of 45°, flush intakes in the fuselage sides close to the turbojet intake duct, and the same longer fuselage as the S/2 just mentioned. Span was 8.74m (28ft 8in), length 8.25m (27ft 0¾in), height 3.16m (10ft 4½in), wing area 22m² (236.8ft²), and armament four MK 108 cannon, all of which were in the fuselage.

The P 1112/V1 design study of 30th March 1945 had a revised wing featuring full-span leading-edge slots, fuselage flush lateral intakes, but as opposed to the two previous models which had a single swept fin and rudder, had a swept butterfly tail and lengthened fuselage. All three had a fuselage maximum diameter of 1.1m (3ft 7¼in).

In their basic configuration, each featured a flush cockpit blending into the fuselage nose contours, the 10cm (4in) thick armoured-glass windscreen providing the rearward-inclined pilot with a good forward and downward vision because of refraction through the glass and 60mm (2¼in) thick armoured side panels, the pressurised cabin divided into a number of compartments. As

visible in the drawing, the four MK 108s were grouped in pairs behind and on either side of the pilot. In all three schemes, the nosewheel retracted rearwards behind the cockpit and the wide-track mainwheels diagonally forwards into the thickened wingroot/fuselage junction. Performance calculations were not completed, but as with the P 1111, maximum speed was expected to be in the region of 1,000km/h at 7,000m (621mph at 22,965ft) with service ceiling 14,000m (45,915ft). According to Dipl-Ing Hans Hornung, the P 1112 was the last development of the Messerschmitt Project Office.*

* A post-war CIOS Report mentions the P 1113 high-speed jet bomber, but gave no details. Some authors have even referred to P 1114 and P 1115 single-seat turbojet-powered projects with variable-sweep wings, but this may be speculation related to patent applications. The often quoted P 1116 variant of the P 1106 has been subject to doubt as the two projects were almost identical in layout.

Messerschmitt P 1112/V1 – data

Powerplant
1 x 1,300kg (2,866 lb) thrust HeS 011 A-0 turbojet, to be replaced later by the 1,500kg (3,307 lb) thrust HeS 011B-0 when it became available.

Dimensions
Span	8.16m	26ft 9in
Length	9.24m	30ft 3¾in
Height	2.84m	9ft 3¾in
Wing area	22.00m²	236.80ft²
Undercarriage track	2.54m	7ft 6in

Armament (proposed)
4 x MK 108 in fuselage. 1 x MK 112 (17.3.1945) or 1 x MK 214 (16.3.1945)

Messerschmitt P 1112/V1 three-view drawing of 30th March 1945.

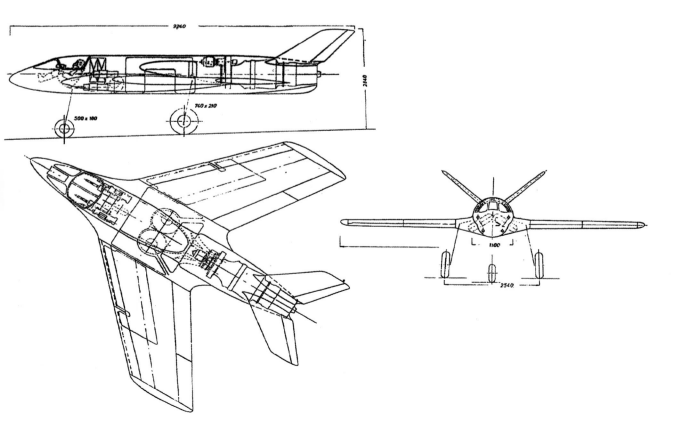

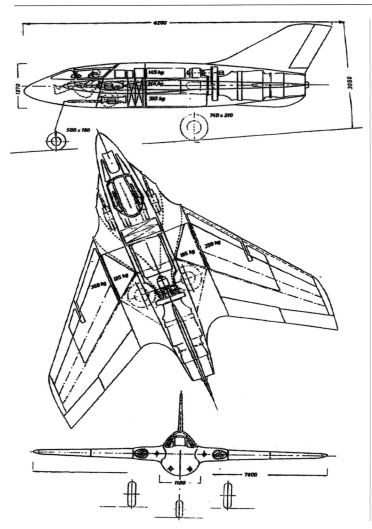

Messerschmitt P1112 S/2 three-view drawing of 3rd March 1945.

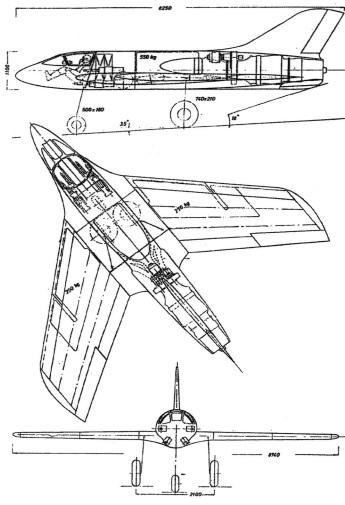

Messerschmitt P1112 S/1 three-view drawing of 27th March 1945.

Below left and right: **Messerschmitt P1112/V1 mock-up cockpit of March/April 1945.**

Chapter Seven

Special Purpose Aircraft 1941-1945

With the exception of 'Mistel' (Mistletoe) composites and air-launched towed aircraft of various types, this chapter describes the Sonderflugzeuge (special purpose aircraft) projects that did not fall into the categories previously described. These encompassed various manned Flak-Raketen (anti-aircraft rocket interceptors) and radar-equipped Frühwarnflugzeuge (early-warning aircraft), as well as conventional types fitted with auxiliary turbojets as a means of achieving temporary performance increases. Additionally, a brief history of pulsejet and rocket motor development is included that portrays the significance that was attached to these new forms of motive power incorporated in early design studies although little actual flying experience with them had been gathered up until then. With the exception of the Bachem 'Natter' (Viper) interceptor and Junkers Ju 248 (Me 263), few of these projects passed the prototype construction stage at the end of the war.

Turbojet Attachments to Orthodox Aircraft

The attachment of turbojets to conventional propeller-driven aircraft had the primary aim of increasing maximum speed. Thanks to the 'performance advantage', the Zerstörer (heavy fighters), Schlachtflugzeuge (ground attack aircraft) and Schnellbomber (fast bombers) upon completion of their tasks would be able to escape pursuit by enemy fighters, which was the main purpose.

At the end of 1941, the RLM suggested turbojet attachment as a special Rüstsatz (equipment set) in order to achieve a short-time tactical speed increase. According to the RLM guidelines, the auxiliary turbojet should be so installed in the fuselage that when put into operation, it could be extended into the airstream. Additional turbojets were experimentally considered for attachment beneath the fuselage by various firms for a variety of aircraft, without impeding the weapons systems. Moreover, the jet exhaust was also not to cause damage to the fuselage underside and tail surfaces. The RLM idea was that the trial suspension of these Sondertriebwerke (special engines) should, if possible, be at the bomb attachment points.

Arado Ar 240A-0 with a Jumo 004 special Rüstsatz (equipment set).

The primary intention was that ground attack and night-fighter aircraft should be equipped with these 'special' powerplants and resulted in very diverse calculations of maximum speed increases. The Arado Ar 240A-0 for instance, whose maximum speed with airscrew engines was 764km/h (475mph), with the addition of a Jumo 004 would have a speed increase of 144km/h (89mph). Trials with the Heinkel He 219A-010 with a BMW 003 turbojet gave a speed increase of only a moderate 60–70km/h (37-43mph). More promising were installation studies by the Messerschmitt firm, where a BK 5 cannon-equipped Me 410A-2/U4 with two turbojets would recoup the speed loss suffered by the heavy armament. The installation, however, was never carried out.

Among the first firms to actually carry out flight trials with a turbojet attachment was the Junkers Flugzeugwerke. As early as 1941, a Ju 88A-4 had been fitted with a Junkers turbojet with which test flights were conducted until 1944.[1]

[1] The very first turbojet flying test-beds were the Heinkel He 118 V2 (one HeS 3A) in 1939 and the He 111H-6 (one HeS 8A) in 1940. After the Ju 88A-4 came the Messerschmitt Bf 110 (one Jumo 004) flown in December 1941, the Me 262 V1 (two BMW P.3302) unsuccessful flight attempt on 26th March 1942, the Junkers Ju 88A-5 (one BMW 003A) in October 1943, and the Ju 88 V41 (DE+DK) with a Jumo 004B in July 1944. An He 219 was also earmarked to conduct flight trials with an early HeS 011 prototype but the war ended before preparations were completed.

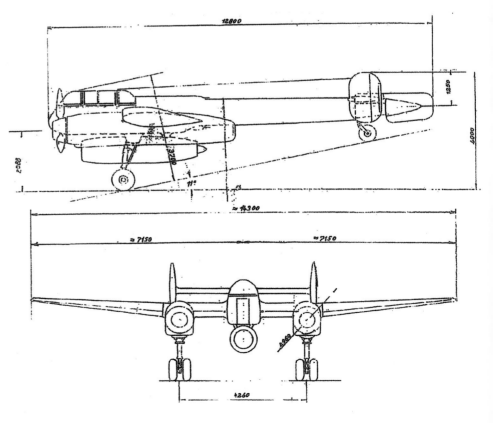

Messerschmitt Me 410A-2/U4 with nose BK 5 cannon and two Jumo 004 turbojets.

Arado Ar 240A-0 works drawing with dimensions and a Jumo 004 turbojet.

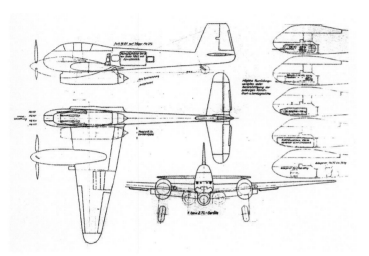

Top: **Messerschmitt Me 410B-2 with auxiliary Jumo 004 turbojet.**

Above: **Messerschmitt Me 410 works drawings with one or two turbojets and various armament, bombs, cameras and fuel loads within the maximum permissible take-off weight.**

Above right: **Heinkel He 219A-010 with auxiliary turbojet.**
G W Heumann

Centre and bottom: **Two views of the Junkers Ju 88A-4 (GH+FQ) flying test-bed with Jumo 004A V11, flown in Rechlin from early 1943 onwards.**

Pulsejet Development by Dipl-Ing Paul Schmidt

As early as 1928, the young Munich engineer Dipl-Ing Paul Schmidt, as a sideline to his consulting activities on fluid dynamics, became interested in developing an resonator type of propulsion unit for aircraft. Alongside earlier ideas and experiments by Georges Marconnet and others abroad from 1909 onwards on both resonating and non-resonating units, this type of propulsion initially termed an 'impulse duct', became generally known after the Second World War as a 'pulsejet'.

Between 1928 and 1932, Paul Schmidt was granted no less than four German and two British patents on the subject, and as he stated in a lecture almost three decades later, had no knowledge of previous developments by others before him. His initial practical work, begun in 1931 to develop a 'powerful tubular jet engine' commenced with tests on ignition systems and the effects of shock waves, using an intermittent fuel feed and a continuous air supply for the combustion process. In order to raise funds for further investigation, with the assistance of Prof Dr Georg Madelung in 1934, the aerodynamics were worked out for a pulsejet-powered interceptor fighter, a light bomber and a pilotless flying-bomb, the latter launched from a near-vertical inclined ramp. This had encouraged further research to be financed by both the RVM and the HWA, the RLM subsequently providing the funding from 1935, some four years before it actively supported turbojet development.

Without going into technical details, Paul Schmidt and his co-workers had progressed between 1938 and 1940 to conducting static tests of pulsejets with duct diameters of 12cm (4¾in), 20cm (8in) and 51cm (20in). The last, with a length of around 3.5m (11ft 5⅜in), gave a static thrust of 500kg (1,102 lb) at the beginning of 1940. By the end of that year his SR 500 duct, now developing 550kg (1,212 lb) thrust had been statically tested, where specific fuel consumption (sfc) was 2.8kg/kg thrust/hr. Due to the fact that with the Schmidt duct, air resistance increased with forward speed and efficiency decreased, hence requiring careful design to minimise, this may have been one reason why work was ordered by the RLM to be temporarily stopped in early 1941. Up to this time, Schmidt had only tested his ducts on stationary test-stands, the whole unit mounted vertically with the exhaust nozzle uppermost. This eliminated unburned residual fuel accumulating in the duct as happened later when tested in the horizontal position.

Paul Schmidt nevertheless continued to make improvements, and by conically enlarging the thrust tube diameter from 45cm (17¾in) at the front end to 56.5cm (22¼in) at the exhaust nozzle over a length of around 3.8m (12ft 5½in), in 1942 achieved a static thrust of 750kg (1,653 lb) and an sfc of 2.75, having heard that the pulsejet was foreseen as an auxiliary take-off aid. Although he made further improvements to the flap valves and other mechanisms, *none* of his ducts were ever flight-tested. At the close of 1942, Paul Schmidt was asked to send his 750kg thrust pulsejet for wind-tunnel testing to the LFA Braunschweig (Brunswick). The fuel pump provided at the LFA was too small, and even after replacement with a larger pump, only delivered two thirds of the normal fuel supply needed. Thrust measured was only 375kg (827 lb) at an airspeed of 350km/h (217mph). Because soft-springing was not used in the LFA tests, the duct ran rather noisily and the shocks threatened to destroy the building. Thereafter, no further wind-tunnel tests were conducted at the LFA with his pulsejets. Computations by Schmidt had shown that with this unit, the 750kg (1,653 lb) static thrust decreased to 630kg at 350km/h (1,389 lb at 217mph), rising again to 700kg at 750km/h (1,543 lb at 466mph) but at a constant slight increase in fuel consumption. For his pioneering work, overshadowed by the successful range of pulsejets developed by the Argus Motoren GmbH, the pulsejet was officially decreed from June 1944 to be designated as the Argus-Schmidtrohr.

Pulsejet Development by the Argus Motoren GmbH

In the autumn of 1938, Dipl-Ing Helmut Schelp of the RLM visited various firms of the German aero-engine industry to urge them to undertake work on continuous combustion jet propulsion units, ie in the main, radial- and axial-flow turbojets. Despite the previously hesitant attitude of the manufacturers, the BMW, Bramo and Junkers firms agreed to commence work in earnest, and during the course of 1939, were awarded the first development contracts. After Dipl-Ing Paul Schmidt had rejected the RLM suggestion in mid-1939 to transfer his pulsejet work to Argus, the latter firm was asked in November 1939 to undertake work on an intermittent combustion propulsive duct or pulsejet. At that time, they were neither advised of Paul Schmidt's on-going work nor of other developments abroad during the previous three decades.

Unlike Paul Schmidt, the group of engineers at Argus, led by Dr-Ing Fritz Gosslau and Dr-Ing Günther Diedrich (who later left the firm) decided upon the use of continual fuel injection via vaporising nozzles and self-ignition of residual gases. Following construction, static tests and road tests on a special vehicle of a number of small experimental pulsejets, it was only in February 1940 that the RLM informed Argus of the work of Paul Schmidt. At the meeting which took place the following month in Munich between the two parties when Schmidt demonstrated his SR 500 pulsejet, Argus realised that with his development of the spring flap valve, Schmidt had made more progress than their own, and used a modification of it in their subsequent designs. Further development proceeded rapidly to the point where the first Argus Versuchs-Schubrohr (experimental thrust tube), the VSR 9a of 120kg (265 lb) thrust and tube diameter 30cm (11¾in), was test-flown beneath a Gotha Go145 two-seat biplane at Diepensee on 30th April 1941. The Argus pulsejet was subsequently flown on a variety of other test-bed aircraft, among them the Bf109, Bf110, DFS 228, DFS 230, Do17, Fw190, Go 242, He111, He 280, Ju88, and Me 328. It was also planned to fit the Argus pulsejet on the Do 217, He162, He 219, Ju188, Me163A, Me 262 and Me 321, but did not reach the flight-test stage. Several of the latest projects by Argus, Blohm & Voss, Gotha, Heinkel, Henschel and Junkers during the period 1944-45 also envisaged the use of the As 014 and As 044 pulsejet. Its only operational use, however, was in the well-known V-1 flying bomb of which some 32,000 examples had been built by the end of the war.

Top: **The first test flight with a 120kg (265 lb) thrust Argus VSR 9a pulsejet took place beneath the Go145 (D-IIWS) biplane piloted by Flugkapitän Staege at the E-Stelle Diepensee on 30th April 1941. Clearly visible is the ventral fuselage protective sheet-metal covering due to the heat radiation and hot exhaust gases when the pulsejet was in operation. At this stage, it did not feature the well-rounded aerodynamic intake fairing ahead of the box-shaped inlet flap-valve grid. At take-off and landing, the propulsive tube lay close to the fuselage underside and was lowered in flight when put into operation.**

Centre: **Experimental installation of an Argus 109-014 pulsejet on the rear fuselage of a Messerschmitt Bf110 (GI+AZ) flying test-bed. Used in the Fieseler Fi103 (V-1) flying bomb of 1944, the 3.6m (11ft 9⅜in) long pulsejet developed a static thrust of 350kg (772 lb) for a weight of 138kg (304 lb).**

Right: **Prof Dr-Ing Fritz Gosslau (1898-1965), Director of Engine Development at the Argus Motorenwerke GmbH, Berlin, headed development of the 109-014 and 109-044 pulsejets.**

Messerschmitt P1079/1 and P1079/10c

1941

At an early date the Messerschmitt Project Office became interested in pulsejets developed by Dipl-Ing Paul Schmidt and in 1941, incorporated his large SR 500 pulsejet in the design of the P1079 series of projects. From the numerous design configurations under this project number, two of the most conspicuous selected for description here are the P1079/1 of April 1941 and the P1079/10c of May 1941. In the first, the SR 500 pulsejet was enclosed completely within the fuselage, whilst in the second, the thrust tube projected into the freestream beyond the aircraft structure.

Because of an RLM-imposed information ban on the engine manufacturers, the aircraft firms were only scantily informed of the thermal loads and performance of the jet tubes – an observation particularly noticeable in the P1079 series of projects. Thus the Schmidt duct, with its wall temperature of 500°C and thrust of just 500kg (1,102 lb) could hardly have been capable of transporting heavy external loads, let alone make the aircraft airborne. This was only one of several unsolved problems which the 'Arbeitsgemeinschaft Strahltriebwerke' (Jet Propulsion Working Committee) established under RLM pressure in December 1942, tackled and released the necessary information to the aircraft firms. As a result of Paul Schmidt's complaints concerning the Argus firm's use of his flap-valve idea, the Argus-Schmidt designation was officially decreed, but only in 1944. From then on, the intermittent combustion jet tube, regardless of origin, was designated the Argus-Schmidtrohr.

Messerschmitt P1079/1. The two underwing X4 air-to-air rockets shown by the artist did not exist at project date as development only commenced in 1942 and was first test-flown in 1944.

Messerschmitt P1079/1 – data as of 15.4.1941

Powerplant	1 x 500kg (1,102 lb) static thrust Schmidt SR 500 pulsejet	
Dimensions		
Span	6.32m	20ft 8¾in
Fuselage diameter	1.00m	3ft 3¼in
Length, fuselage	6.40m	21ft 0in
Height, skid retracted	1.87m	6ft 1¼in
Tailspan	1.94m	6ft 4¼in
Weights		
Fuel weight	905kg	1,995 lb
Performance		
Max speed	760km/h	472mph
Bombload	1 x SD1000, SD1400 or SD1700B	

As visible in the drawing, this single-seat mid-wing aircraft featured 40° of sweepback on the wing and empennage surfaces. The fuselage nose housed 120kg (265 lb) of fuel, the remainder held in an annular tank surrounding the pulsejet fed by a ventral air intake ahead of the landing skid.

Messerschmitt P 1079/10c – data as of 20.5.1941

Powerplant	1 x 500kg (1,102 lb) static thrust Schmidt SR 500 pulsejet	
Dimensions		
Span	5.00m	16ft 4¾in
Fuselage max width	0.90m	2ft 11½in
Length, fuselage	5.10m	16ft 8¾in
overall	7.20m	23ft 7½in
Height, skid extended	1.80m	5ft 11in
Tailspan	2.00m	6ft 6¾in
Weights		
Fuel capacity	800 litres	176 gals
Performance		
Max speed	750km/h	466mph
Bombload	1 x SC 1700B or SD 1700B (maximum)	

In one design study, this shoulder-wing single-seater, likewise with 40° swept wing and tail surfaces, carried the external bomb beneath the fuselage which had tandem landing skids, the pulsejet this time fed by a dorsal air intake. In another variant, the bomb was enclosed in an aerodynamic fairing.

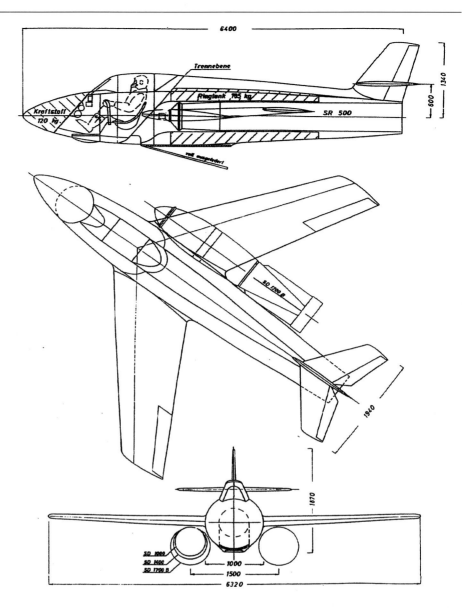

Messerschmitt P 1079/1 three-view works drawing.

Messerschmitt P 1079/10c.

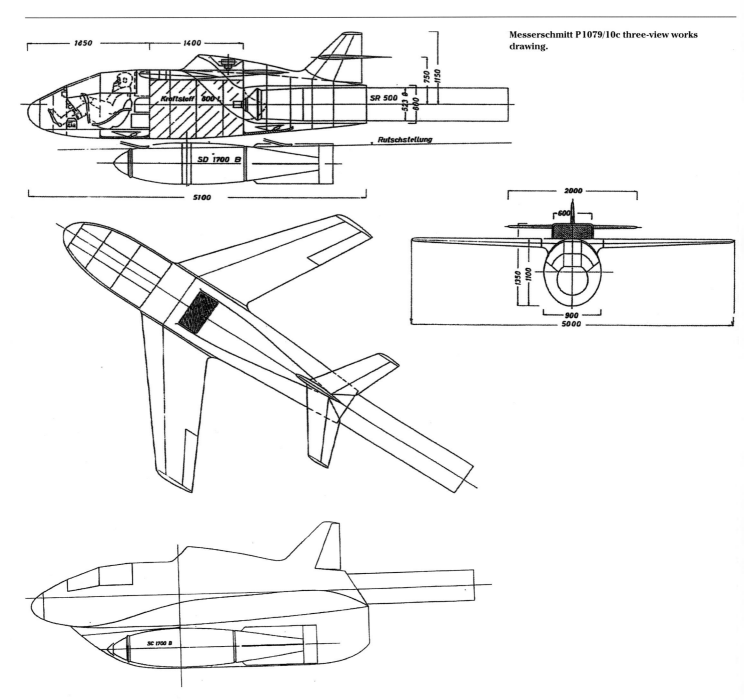

Messerschmitt P 1079/10c three-view works drawing.

Messerschmitt P 1079/10c variant with streamlined bomb enclosure.

The Paul Schmidt SR 500 pulsejet.

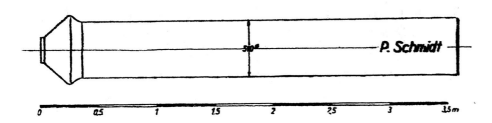

The Combined Pulsejet-ramjet – A Further Argus-Schmidt Development

As early as 1943, Dipl-Ing Paul Schmidt had voiced his views in favour of a combined pulsejet-ramjet unit. This led in January 1945 to Dr-Ing Fritz Gosslau of Argus and Paul Schmidt initiating tests with combined pulse-jet-ramjet units embodying an As 014 or As 044 pulsejet partially enclosed by a surrounding ramjet as shown in the illustration, the ramjet brought into operation through a set of fuel injection nozzles at the pulsejet's exhaust orifice. The expectation was that this type of propulsion unit would enable super-sonic speeds to be attained. Although the RLM initially showed a keen interest in the series of satisfactory tests that had been conducted, in April 1945 it ordered an immediate halt to further work without giving any reason.

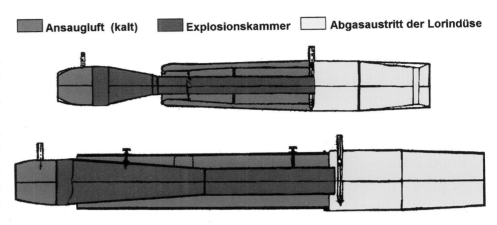

| Ansaugluft (kalt) | Explosionskammer | Abgasaustritt der Lorindüse |

Schematic arrangement of an Argus As 014 or As 044 pulsejet combined with an auxiliary ramjet to increase thrust at high speeds.

The Walter 109-509 Liquid-Propellant Rocket Motor

The 109-509 rocket motor, developed by the Hellmuth Walterwerke (HWK) of Kiel, was one of the most successful rocket propulsion units used in high-speed aircraft. Its predecessor was the HWK RII-203 'cold system' rocket motor of 750kg (1,543 lb) thrust, installed in the Me 163A V4 (KE+SW) prototype with which test pilot Heini Dittmar reached a speed of 1,003km/h (623.25mph) or Mach 0.85 on 2nd October 1941. This was the first time ever that the 'magic' 1,000km/h (621mph) mark had been exceeded, but for reasons of wartime secrecy, was not publicly announced.

This led the RLM, up until that time hesitant, to promote rocket motor development and establish a programme for its use in a rocket-propelled interceptor. Prof Dr-Ing Hellmuth Walter, dissatisfied with the low thrust developed by the 'cold system' rocket motor, then developed the 'hot system' rocket motor with an initial thrust of 1,500kg (3,307 lb) in the 109-509A-0, later increased in stages to 2,000kg (4,409 lb) thrust. The rocket motor components, fuel pumps and all accessories were grouped together within a box-shaped rectangular framework from which projected the bracing supports and fuel feed ducts for the combustion chamber and exhaust nozzle at the rear of the aircraft fuselage. In its sev-

Walter 109-509A-1 rocket motor in the Me 163B.

eral variants, the HWK 509 motor was planned to be installed in numerous aircraft projects. Those actually flown with this rocket unit were the Ba 349, Me 163B, Me 163C, and Me 262C-1a. A disadvantage of the unit was that its thrust could not be economically throttled for cruising flight, so that for later models such as the HWK 509A-2, 509B and 509C, a smaller 300kg (661 lb) thrust auxiliary cruising chamber was added that could be operated alone when the main chamber was switched off to enable the aircraft to consume less fuel in cruising flight at moderate speeds. It was also possible to operate both chambers together or alternately. The improved HWK 509C was equipped with an auxiliary cleaning system and a chemical starter.

HWK 109-509A-1 – data

Length, overall	2.5m	8ft 2½in
Width, overall	86.5cm	34in
Height, overall	81.5cm	32in
Installation weight	168kg	370 lb
Maximum thrust	1,600kg	3,527 lb
Minimum thrust	100kg	220 lb
Cruising chamber	none	

HWK 109-509A-2 and B – data

Total thrust, both chambers	2,000kg	4,409 lb
Max thrust, main chamber	1,700kg	3,748 lb
Min thrust, main chamber	200kg	441 lb
Cruising chamber thrust	300kg	661 lb
Installation weight	177kg	390 lb
Exhaust velocity	1,950m/sec	6,398ft/sec
Fuel consumption	7.5kg/sec	16.53 lb/sec

HWK 109-509C – data

Total thrust, both chambers	2,400kg	5,291 lb
Max thrust, main chamber	2,000kg	4,409 lb
Min thrust, main chamber	400kg	882 lb
Cruising chamber thrust	400kg	882 lb
Installation weight	188kg	414 lb
Exhaust velocity	1,975m/sec	6,980ft/sec
Fuel consumption	9.3kg/sec	20.50 lb/sec

Other versions developed were the 109-509D, S1, S2, and 109-559.

Top right: **Walter 109-509 simplified sectional arrangement drawing.**

Centre right: **Walter 109-509A-2 rocket motor with auxiliary cruising chamber.**

Bottom right: **Walter 109-509A-2 schematic arrangement. A = main chamber, B = cruising chamber.**

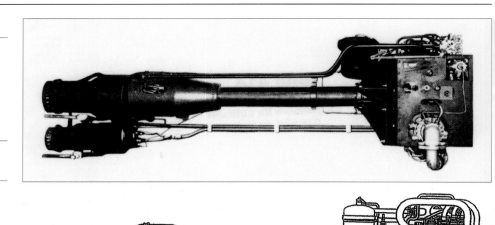

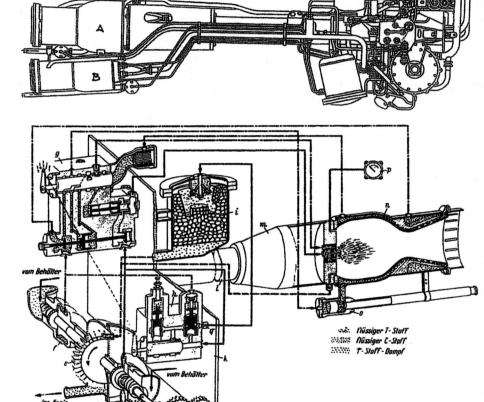

Left: **Professor Dr-Ing Hellmuth Walter (1900-80) was a genius in the realm of rocket propulsion technology. His particular merit was the design and manufacture of the liquid-propellant rocket motor, developed in a very short time into a usable propulsion unit for aircraft. He also became widely known for his revolutionary development of the T-Stoff (hydrogen peroxide) turbine drive for submarines.* Certainly no other German engineer and inventor was so versatile in the range and realisation of extraordinary ideas and developments in the 20th century.†**

*For more details see Eberhard Möller: *Marine Geheimprojekte. Hellmuth Walter und seine Entwicklungen (Navy Secret Projects – Hellmuth Walter and his Developments)*, Motorbuch Verlag, Stuttgart, 2000 – Author.

† Other comprehensive accounts are to be found in: Hellmuth Walter: *Development of Hydrogen Peroxide Rockets in Germany, AGARDograph 20*, Verlag E Appelhans & Co, Brunswick, 1957, pp.263-280 (on rocket motors for aircraft and missiles).
Emil Kruska & Eberhard Rössler: *Walter U-Boote (Walter Submarines)*, J F Lehmanns Verlag, Munich, 1969 (Walter hydrogen peroxide-driven submarines, torpedoes, and some post-war developments).
Eberhard Rössler: *The U-Boat*, Cassell & Co, London, 2001 (The Walter Process, pp.168-182).

Arado Ar 234R Höhenaufklärer

1944

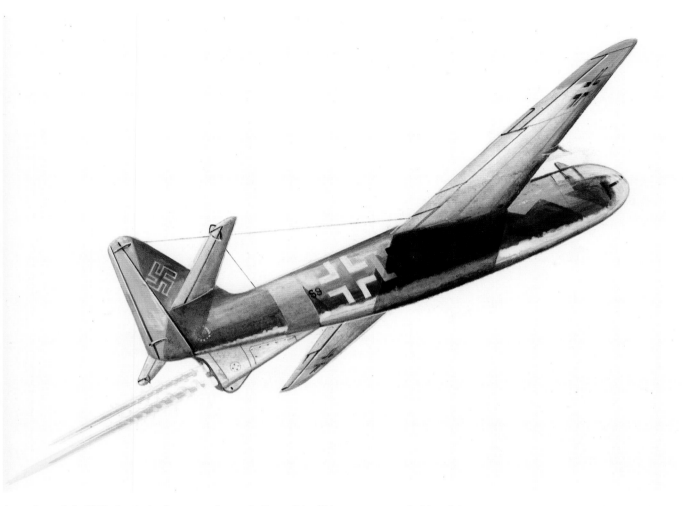

As early as July 1943, Arado had proposed a short-range rocket-powered Ar 234B capable of reaching 17,000m (55,775ft) which led between February and May 1944 to Dipl-Ing Wilhelm van Nes supervising development of the single-seat rocket-powered high-altitude Ar 234C reconnaissance aircraft, intended to fly over the airfields in England from which the Allied bomber formations operated and report their approach path to the anti-aircraft and fighter command centres. Two high-resolution cameras were to make it possible to cover airfields and radar positions even in conditions of poor visibility.

Designated Ar 234R, in this Höhenaufklarer (high-altitude reconnaissance aircraft) of which two main variants had been considered, the two Jumo 004 turbojets were replaced in one design study by two under-wing-mounted Walter HWK 509A rocket motors. Take-off was to have been under its own power, ascending to 16,500m (54,130ft) where it would photograph its objectives in a

shallow glide. This was superseded in a later Ar 234R study by the more favourable location of a twin-chamber rocket motor installed beneath the empennage. The upper main chamber, used for climbing, delivered 1,500kg (3,307 lb) thrust, the lower one of 400kg (882 lb) used for cruising flight. As all of the 5,500kg (11,023 lb) of rocket fuel for the 9,100kg (20,068 lb) weight aircraft would have been used up during the climb, it was soon abandoned in favour of using the HWK 509C with a 2,000kg (4,409 lb) main chamber and 400kg (882 lb) thrust cruising chamber. Towed to 8,000m (26,250ft) altitude by a Heinkel He177 over a distance of just over 200km (124 miles), the Ar 234R would then be released and accelerate in ascending flight under full power to its operating altitude of around 18,000m (59,050ft). With a maximum endurance of about 21 minutes during which the rocket motor would be switched on at intervals, the Ar 234R would then return in a 250km (155 miles) glide to its home air-

Arado Ar 234R high-altitude reconnaissance aircraft.

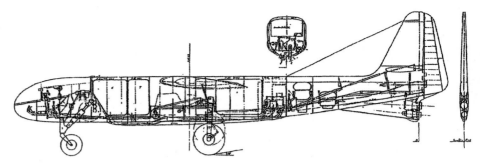

field. For the long glide from its peak altitude, the Ar 234R was to have had a laminar-flow wing whose maximum thickness lay between 50% and 60% chord, enabling a gliding angle of 1:14 and a reduction in boundary-layer turbulence and drag compared with a normal profile. The Ar 234R models each featured a pressurised cabin adopted from the Ar 234C series.[2]

[2] The Ar 234R high-altitude capabilities were exceeded by the rocket-powered DFS 228 high-altitude reconnaissance aircraft which, following a tow to altitude, could theoretically reach 25,000m (82,000ft) and with intermittent powered and glide periods, could cover a distance of 750km (466 miles) after release from the tow-craft.

Ar 234R – data (from data sheet)

Wing area		27.00m²	290.62ft²
Equipped weight	(1 & 2)	3,500kg	7,716 lb
Crew	(1 & 2)	100kg	220 lb
Fuel weight	(1)	3,500kg	7,716 lb
	(2)	4,000kg	8,818 lb
Loaded weight	(1)	7,100kg	15,653 lb
	(2)	7,600kg	16,755 lb
Wing loading	(1)	263.0kg/m²	53.86 lb/ft²
	(2)	281.5kg/m²	57.63 lb/ft²

Top right: **Arado Ar 234R works drawing of 22nd March 1944 with HWK 509C in rear fuselage.**

Dipl-Ing Wilhelm van Nes (1900-1979) headed the Arado Development Department where the Ar 234R also originated.

Right: **Arado Ar 234R data sheet of 20th February 1944. Drawing key: (1) climbing chamber, (2) cruising chamber, (3) turbine pump unit, (4) auxiliary T-Stoff tank, (5) rear T-Stoff tank, (6) forward T-Stoff tank, (7) C-Stoff tank, (8) armour protection, (9) pressurised cabin.**

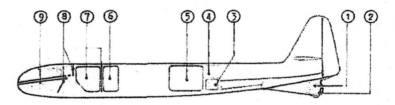

Arado	Nebenverwendungszweck	Ar 234

d) Kurzstreckenhöhenaufklärer

1.) Steigofen
2.) Marschofen
3.) Turbinen-Pumpenaggregat
4.) Zusatzbehälter für T-Stoff
5.) T-Stoff-Behälter (hinten

6.) T-Stoffbehälter (vorn)
7.) C-Stoffbehälter
8.) Panzerung
9.) Druckkabine

Rüstgewicht	3500 kg	
Betriebsstoff	3500 kg	(4000 kg)
Besatzung	100 kg	
Startgewicht	7100 kg	(7600 kg)
Steigschub	2000 + 400 = 2400 kg	
Marschschub	400	kg
Flächenbelastung G/F	263 ÷ 133,5	kg/m²
Schubbelastung G/Go	0,338	kg/kg
Zuladungsverhältnis B/Go	0,493	kg/kg

1.) Reichweite: Zur Erzielung brauchbarer Reichweiten muß das Flugzeug auf Höhe geschleppt werden. Vorgesehen ist als Schleppflugzeug die He 177 mit DB 610. Dabei sind beim Start die Flächenbelastungen beider Flugzeuge etwa gleich. Als Schlepphöhe können etwa 8000 m erreicht werden. Unter dieser Voraussetzung werden die Leistungen gerechnet.

2.) Triebwerk: Als Triebwerk wird das Walter-Schubgerät mit 2000 kg Steigschub und 400 kg Marschschub verwendet. Steig- und Marschofen werden unter das Rumpfheck gebaut, sodaß am Flugzeug selbst nur geringe Bauänderungen vorgenommen werden müssen. Die Leitungen von den Behältern zur Pumpenanlage und weiter zu den Öfen werden außenbords verlegt und abgedeckt.

3.) Antriebe: Zum Antrieb von Atemluftpresser, Hydraulikpumpe und Generator wird entweder ein Seppeleraggregat, das unter dem Flügel an dem TL-Aufhängebeschlägen angebracht ist, oder eine kleine T-Stoff-Turbine mit Getriebe verwendet.

4.) Druckkabine: Als Druckkabine soll eine vorgezogene Kabine der Ar 234 C angebaut werden, was einige bauliche Änderungen im vorderen Rumpfspant erfordert.

Bearbeiter: Bl.1

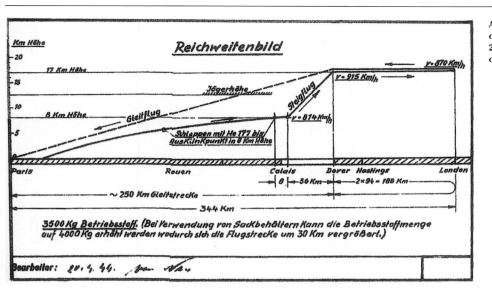

Arado Ar 234R flightpath, speeds and range sheet of 20th April 1944. A later performance table of 24th May 1944 indicated a maximum endurance of 140 mins.

Arado Ar 234B-1
Early-Warning Project

1944

Arado Ar 234B-1 fighter-control and early-warning aircraft with the FuG 244 'Berlin N' panoramic search radar.

The decisive factor for the development of the rotating antenna for the FuG 244 'Berlin' panoramic night-fighting search radar was a Rüstungsstab (Armament HQ Staff) notice dated 16th November 1944 which required development of low-drag search antennae. Based on antenna development of the FuG 350 'Naxos A' target-approach and search radar with a vertically rotating 'Stiehlstrahler', Arado engineers together with Telefunken technicians developed a discus-shaped antenna housing in which four ceramic 'Stiehlstrahler' intended for installation in the

FuG 224 'Berlin N' panoramic search radar rotated at 1,000rpm. Calculations made by Dipl-Ing Rudiger Kosin dated 22nd December 1944 showed a discus of 1.82m (6ft) diameter. The rotating search antenna equipment, itself of 1.5m (4ft 11in) diameter mounted on a pylon 21cm (8¼in) wide and 36cm (14¼in) high above the Ar 234B-1, could cover a range of 45km (28 miles) at speeds of 800km/h at 8,000m (497mph at 26,250ft) around the aircraft. The FuG 224 'Berlin N' worked at a

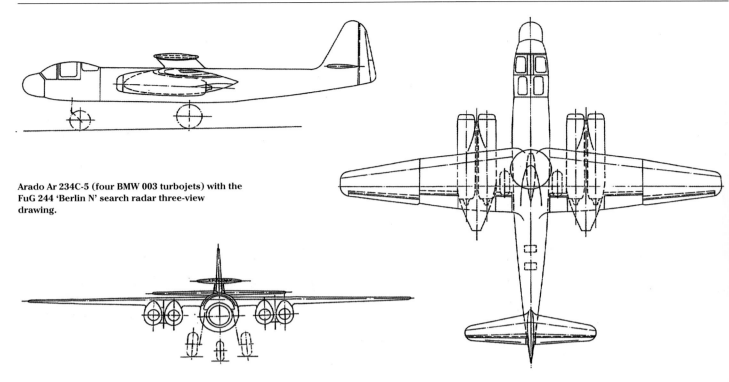

Arado Ar 234C-5 (four BMW 003 turbojets) with the
FuG 244 'Berlin N' search radar three-view
drawing.

frequency of 3,300MHz on the 9cm (3½in)
wavelength, with a peak impulse of 20kW
and was to have been mounted on an
Ar 234B-1 above the wingroot/fuselage junction. The discus-shaped device bore the code
name 'Obertasse' (upper cup) or 'Drehteller'
(rotating plate). The final shape of the discus
was decided upon in February 1945 based on
wind-tunnel measurements made at the AVA
Göttingen and the DVL Berlin-Adlershof. An
Arado proposal was for the 1.5m (4ft 11in)
diameter discus to be mounted above the
Ar 234C-5.

Also under development by Telefunken
was a panoramic search antenna of only
90cm (35½in) diameter for the FuG 244, but
no examples of the 'Obertasse' were completed at the end of the war. The antenna
housing shape and its installation form above
the fuselage became a war booty of the Allies,
who further developed this German technology used today in the AWACS strategic search
and observation radar installations in the
armed forces of both East and West.

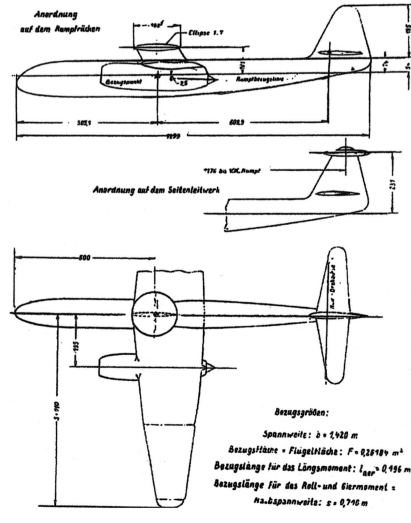

A 1/10th scale model of the Arado Ar 234B of 14.2m
(46ft 7in) wingspan showing two locations
investigated for the FuG 244 'Berlin N' panoramic
search radar – as an elliptical housing centred on a
dorsal pylon 5m (16ft 4¾in) from the fuselage nose
or else mounted near the top of the fin and rudder.

Bachem Ba 349 'Natter'

The Bachem 'Natter' in a ram attack on a Boeing B-17 bomber. The fuselage nose bores into the fin and rudder and tail gunner's position.

As a result of an RLM requirement in the spring of 1944 calling for a 'Verschleiß-flugzeug' (literally 'wear-and-tear' or short-life aircraft) fast fighter capable of the simplest manufacture, several firms, among them Heinkel, Junkers and Messerschmitt responded. Following inspection of the design studies in summer 1944, however, an absolute outsider made the running – Dipl-Ing Erich Bachem. He had not been provided with a copy of the RLM requirements, and his idea of a minimum-dimensioned vertical take-off rocket-propelled interceptor built entirely of wood had been rejected by the RLM. Only after he had made his proposal known to Reichsführer SS Heinrich Himmler who was known to be in favour of unusual ideas, did the RLM relent and accorded the Bachem BP 20 proposal the designation Ba 349 'Natter' (Adder or Viper). From then on, a development ensued which in its speed was certainly impressive: from the initial hand-made sketches in July until the first

unpowered towed flight on 3rd November 1944, less than four months had passed! Even the completion of this book from manuscript to printing took multiples of this time.

That merely three months after receipt of a construction order, the first experimental prototypes were able to leave the manufacturing and assembly lines in Waldsee/Württemberg for their flight trials, was due not only to the energy displayed by Erich Bachem and his colleagues, but above all, to the SS-Hauptamt (Main Office) and its Chief, Obergruppen-führer and General der Waffen-SS Gottlob Berger, who pushed forward the work on the 'Natter' with the highest priority rating.

A near-vertical take-off and rocket projectile armament were imprints of the term 'manned anti-aircraft rocket' or 'manned projectile', although this designation was not strictly accurate. Employment of the Bachem 'Natter' as a target-defence interceptor against the streams of Allied bombers had been initially envisaged by the planners as follows:

A Bachem 'Natter' on target-approach, firing its salvo of 24 R4M 'Orkan' air-to-air rocket projectiles.

Upon approach of the bomber formation, the 'Natter' would take-off from its steeply-inclined ramp. After reaching an altitude of up to 12,000m (39,370ft), the attack would follow in a dive. The pilot would then fire his salvo of 24 R4M 'Orkan' (Hurricane or Typhoon) rockets. Initially 33 RZ 65 projectiles had been planned to be released from the nose honey-comb container, and then utilise the kinetic energy of the 'manned projectile' for a ram attack. By blowing off the nose and crew compartment, the pilot was to save himself at the last moment. The brake parachute streamed from the tail end of the 'Natter' would cause such a powerful deceleration that the pilot would be thrown forward and out, the remainder of the aircraft with its precious rocket Walter rocket motor being saved from impact with the bomber. This type of mission, however, was discarded for a variety of reasons. Instead, after firing its nose armament in a dive, the 'Natter' was to break away and land. Thanks to its superior speed, the danger from enemy fighters was minimal.

Although the Ba 349 did not come to be used operationally, several successful towed flights and ramp-launched take-offs were made. The sole manned flight on 1st March 1945 at the troop training ground at Heuberg in the Swäbische Alb, following a successful take-off, ended as a result of various unfortunate circumstances in tragedy in which the pilot Lothar Sieber lost his life. His name entered the annals of aeronautical history as the pioneer of the first manned rocket-powered take-off.*

*See Horst Lommel: *Die erste bemannte Raketenstart der Welt. Geheimaktion Natter (The world's first manned rocket take-off. Secret enterprise Natter)*, Motorbuch Verlag, Stuttgart, 1998.
Also by Horst Lommel: *Das bemannte Geschoss (The Manned Projectile) Ba 349 'Natter'*, VDM Heinz Nickel, Zweibrücken, 2000.

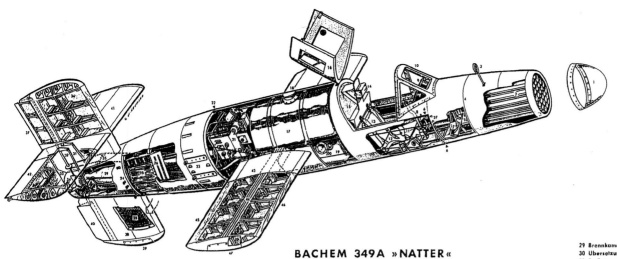

BACHEM 349A »NATTER«

1. abwerfbare Plexiglaskappe, gibt Raketenbatterie frei
2. Raketenlafette für 24 „Föhn"-Raketen
3. Kreisvisier
4. Lage der vorderen Panzerung
5. Sammler (für Instrumente und Raketenbetätigung)
6. Trennspant des abwerfbaren Rumpfbugs
7. Seitensteuerpedale
8. Steuerknüppel für kombiniertes Höhen-/Querruder
9. Instrumentenbrett
10. Windschutz, Stahlblech mit Panzerglas
11. Sitzwanne
12. Anschnallgurte
13. Rückenpolster
14. gepolsterte Kopfstütze
15. Rückenpanzerung
16. Cockpithaube
17. Behälter für T-Stoff
18. Einfüllstutzen
19. Behälter für C-Stoff
20. Einfüllstutzen
21. Aggregat 109–509A (Walter)
22. Rumpf-Sollbruchstelle
23. vordere Beschläge und Kontaktleisten für Zusatz-Startraketen
24. hintere Aufhängeösen für Startraketen
25. Fallschirm für Zelle und Triebwerk, gepackt
26. Fallschirmbehälter mit Ausstoßmechanik (Auswurfplatte)
27. Auslösekabel für Fallschirm, bei Abwurf des Rumpfbugs betätigt
28. Abdeckklappe in Abwurfstellung
29. Brennkammer 109–509A
30. Übersetzung für Höhen-/Querruder
31. Stoßstangen
32. Seitenruderwelle mit Verlängerung li Strahlruder
33. Stoßstange für Höhen-Strahlruder
34. Strahlruder
35. Kabelumlenkung für Seitenruderbedie
36. oberes Seitenleitwerk, Holzkonstrukti
37. Seitenruder
38. unteres Seitenleitwerk, Holzkonstrukti
39. Gleitbeschlag für Start
40. Seitenruder
41. Höhenleitwerk, Holzkonstruktion
42. differentialgesteuertes Höhen-/Querru
43. Flügel in Holzkonstruktion
44. vierfach verleimter Hauptholm
45. vorderer Hilfsholm als massive Flügel
46. hinterer Hilfsholm, massiv
47. Randbogen mit Gleitbeschlag für Sta

Bachem 'Natter' sectional drawing. The jettisonable nosecap was never fitted.

Bachem Ba 349A 'Natter' – data

Powerplant   1 x 1,700kg (3,748 lb) thrust HWK 509A-2 liquid-propellant rocket motor and 4 x 1,200kg (2,646 lb) thrust Schmidding SG 34 solid-propellant RATO units

Dimensions

Span	3.60m	11ft 9¾in
Length	5.72m	18ft 9¾in
Height	2.20m	7ft 4½in
Wing area	3.60m²	38.75ft²

Weights

T-Stoff weight	990kg	2,182 lb
C-Stoff weight	600kg	1,323 lb
Loaded weight	2,050kg	4,159 lb

Performance

VTO speed	880km/h	547mph
Max level speed	1,000km/h at 5,000m	621mph at 16,400ft
Initial climb rate	180m/sec	35,433ft/min
Time to 3,000m (9,840ft)	21 secs	
to 6,000m (19,685ft)	36 secs	
to 9,000m (29,530ft)	48 secs	
to 12,000m (39,370ft)	63 secs	
Range, at 3,000m (9,840ft)	45.5km	28 miles
at 6,000m (19,685ft)	97.5km	61 miles
Endurance at 3,000m (9,840ft)	4.60 mins	
at 6,000m (19,685ft)	5.15 mins	
at 9,000m (29,530ft)	4.85 mins	
at 12,000m (39,370ft)	3.90 mins	

Armament   24 x R4M or 33 x RZ73

Bachem Ba 349A three-view drawing.

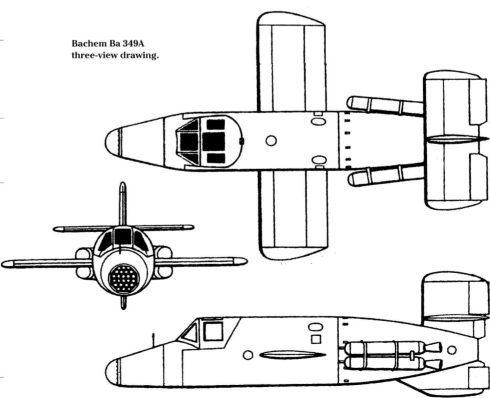

Dipl-.Ing. Erich Bachem pictured post-war beside a model of the 'Natter'.

Born in Mühlheim/Ruhr on 12th August 1960, Erich Bachem from 1925-1930 studied mechanical engineering at the Technische Hochschule Stuttgart, where he discovered his interest in aviation. Together with his like-minded colleagues and friends Wolf Hirth, Wolfgang Hütter and Armond Protzen, in 1925 he established the Akaflieg or Akademische Fliegergruppe (Academic Flying Group). Following his qualification as a Flugbaumeister (Master of Aircraft Engineering) in Berlin and some futuristic publications, in 1933 he became Technical Director at the Fieseler-Werke in Kassel, where he headed the Development Department in 1938. On 10th February 1942 he founded the Bachem Werk GmbH in Waldsee/Württemberg where the 'Natter' originated. Erich Bachem also conducted pioneering work in other fields of endeavour, such as with the 40kg (88 lb) weight 'Lerche' Lark – the forerunner of the ultra-light aircraft. From 1948 to 1952 he was engaged in the manufacture of violins and guitars in Argentina, thereafter becoming Technical Director at the Maschinenfabrik Schwarz & Dyckerhoff in Ruhrtal. In 1956 he returned to Waldsee where, based on his aircraft construction experience, he designed a series of robust caravans for the Karosseriewerk Hymer. Erich Bachem passed away on 25th March 1960 as a result of a cirrhosis.

The Bachem BP 20 'Natter' M23 prototype with which Lothar Sieber undertook the world's first manned rocket-powered vertical take-off on 1st March 1945.

Heinkel P 1077 'Julia'

1944

Since both anti-aircraft artillery as well as fighter aircraft had almost reached their limits against the masses of incoming bomber formations confronting them, the Project Office of the Heinkel Flugzeugwerke GmbH in Vienna in the early summer of 1944 evolved the 'Julia', which can be regarded as a cross between a manned flak-rocket and a fast miniature fighter. The idea for such an aircraft was not new, as the Bachem Werke GmbH in Waldsee/Württem-berg had already conducted similar investigations with the manned, expendable Ba 349 'Natter' at a time when the P 1077 'Julia I' was still in its infancy. The 'Natter' and 'Julia' were both primarily designed as protectors of industrial centres (eg oil refineries) which could best be

accomplished by using transportable ramps for steep-angle or vertical take-offs.

A common feature of the various 'Julia' design variants was its high shoulder wing, almost circular fuselage, twin fins and rudders and fuselage landing skid(s), the aircraft to be built largely of wood and partly of sheet-metal covering. In one variant, the pilot was accommodated in a prone position in a flush cockpit, an alternative arrangement providing for conventional seating beneath a raised canopy. As can be seen from the drawings, the C-Stoff tank was housed forward of the wingroot leading edge, the larger T-Stoff tank occupying the fuselage cross-section beneath the wing. As an alternative to the 'hot' HWK 509 rocket motor with main and cruising

Heinkel P 1077 'Julia II' with HWK 509A-2 rocket motor.

Heinkel P 1077 'Romeo II' with Argus As 044 pulsejet.

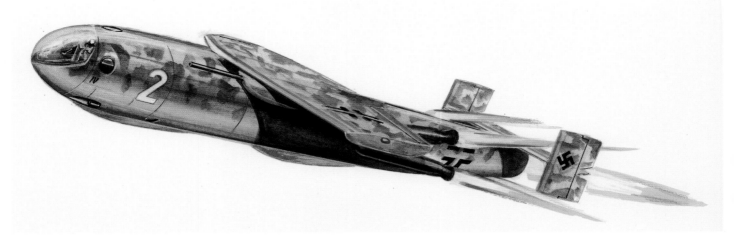

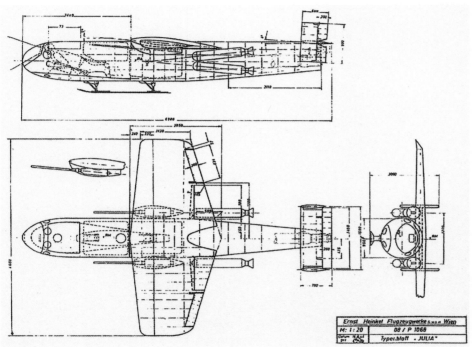

Heinkel P 1077 'Julia I' with prone pilot and four RATO units.

Heinkel 'Julia I' works drawing of 16th August 1944. The project, originally P 1068, was later redesignated P 1077.

chambers, the lower-thrust Argus 109-044 of 475kg (1,047 lb) static thrust was also considered. Although this caused a reduction in maximum speed attainable, the pulsejet's lower fuel consumption per unit thrust enabled a longer range and duration to be achieved for a similar quantity of fuel. Designated as the 'Julia II', this variant was capable of over 600km/h (373mph), and with its four solid-fuel RATO units and a jettisonable auxiliary undercarriage, could take off in a conventional manner. On 16th November 1944, Heinkel submitted the P 1077 project to the RLM which, after some hesitation, approved the construction of three prototypes at the beginning of 1945. Whereas mock-ups of the 'Julia' I and II had been completed, the three prototypes were still incomplete at the end of the war.[4]

[4] On 8th September 1944, an order was placed for the construction of 20 pre-production aircraft to be built in woodworking factories in the Vienna region and preparations were made by the parent firm to tool up for 300 aircraft per month. The P 1077 Baubeschreibung (construction description) submitted to the RLM on 15th October 1944, included provision for a prone pilot glider trainer armed with two wing-mounted MG 1 MG 151/20s, the aircraft to be towed to altitude. As planned, initial pre-production aircraft were to have the twin-chamber HWK 509 rocket motor and four fuselage-mounted RATO units. On subsequent models, armament consisted of two fuselage-mounted MK 108 cannon (70rpg), with a conventionally seated pilot.

From a Heinkel document of 15th October 1944, the Argus As 044-powered variant with both prone and normal cockpit layout, designated 'Romeo' I and II, was designed to take off with the aid of a catapult or two or four solid-fuel RATO units. To simplify large-scale production, an internal Heinkel-Werke meeting on 16th October 1944 decided to adopt a seated pilot position, rectangular wings, a single fin and rudder, increased fuselage girth to house larger fuel tanks, and a single extensible skid in place of the earlier ones in tandem. Other armament alternatives to the two MK 108s considered were the R4M, two MK 103s, the 7-barrel SG 117, the 21-barrel (3 x 7) SG 118 and the 49-barrel (7 x 7) SG 119 developed by Rheinmetall-Borsig to combat the bomber formations. In the Heinkel Report of 16th October 1944, the horizontal take-off favoured for the 'Julia' by the authorities was recommended to be replaced by a vertical or steep-angled ramp take-off and rapid climb to altitude. For more details see:

(1) E J Hoffschmidt: *German Aircraft Guns & Cannons*, WE Inc, Old Greenwich, Conn., 1969.
(2) Geheimprojekt 'Julia', *Luftfahrt International* Vol.1, No.1 (Jan-Feb 1974), pp.113-122.
(3) J Dressel & M Griehl: *Die deutschen Raketenflugzeuge 1935-1945*, Weltbild Verlag GmbH, Augsburg, 1994, pp.38-45.
(4) Manfried Griehl: *Jet Planes of the Third Reich – The Secret Projects, Vol.1*, Monogram Aviation Publications, Sturbridge, Mass., 1998, pp.149-153.

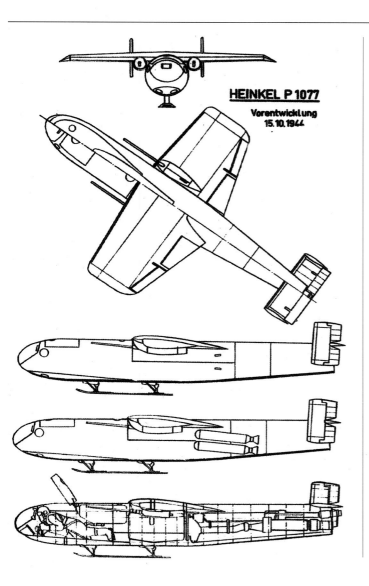

HEINKEL P1077
Vorentwicklung
15.10.1944

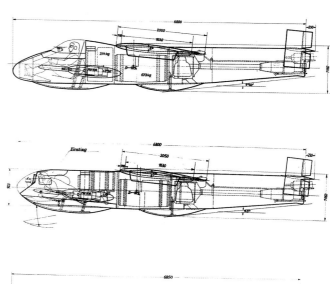

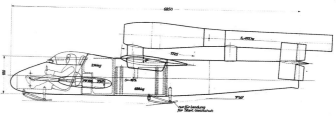

Left: **Heinkel P1077 prone pilot variant with and without RATO units.**

Above: **Heinkel 'Julia' II and I and 'Romeo' II side-view drawings of 15th October 1944 with (top) normal cockpit, (centre) prone cockpit, and (above) with Argus As 044 pulsejet. Wing root and tip chord on this variant was less than the rocket-powered models but fuel load was slightly increased.**

Heinkel P1077 'Julia I' – data as of 14.11.1944

Powerplant

One HWK 509A-2 with 1,700kg (3,748 lb) thrust main chamber and 300kg (661 lb) thrust cruising chamber, plus 4 x 1,200kg (2,646 lb) Schmidding SG 34 solid-fuel RATO units each of 10 seconds duration.

Dimensions

Span	4.60m	15ft 1in
Length, fuselage	6.80m	22ft 3¾in
Length, overall	7.03m	23ft 0¾in
Height	1.50m	4ft 11in
Wing area	7.20m²	77.50ft²

Weights

Equipped weight	945kg	2,083 lb
T-Stoff weight	650kg	1,433 lb
C-Stoff weight	200kg	441 lb
Loaded weight, w/o RATO	1,795kg	3,957 lb
with RATO	2,175kg	4,795 lb

Performance

Mean take-off acceleration	2.06 g	
Initial climb rate after 10 secs	202m/sec	39,763ft/min
Altitude after 10 secs	1,000m	3,280ft
Take-off run to 50m (164ft)	350m	1,148ft
Climbing time, VTO mode, at 40° climb angle		
to 5,000m (16,400ft)	31 secs	
to 10,000m (32,810ft)	52 secs	
to 15,000m (49,210ft)	72 secs	
Max speed	980km/h at 5,000m	609mph at 16,400ft
Range, normal chamber		
at 5,000m (16,400ft)	50km at 800km/h	31 miles at 497mph
at 5,000m (16,400ft)	46km at 900km/h	29 miles at 559mph
at 10,000m (32,810ft)	51km at 800km/h	32 miles at 497mph
at 15,000m (49,210ft)	26km at 900km/h	16 miles at 559mph
Range, cruising chamber		
at 5,000m (16,400ft)	73km at 800km/h	45 miles at 497mph
at 10,000m (32,810ft)	51km at 900km/h	32 miles at 559mph
at 15,000m (49,210ft)	49km at 800km/h	30 miles at 497mph
Endurance		
at 5,000m (16,400ft)	4.85 mins at 800km/h (497mph)	
at 10,000m (32,810ft)	5.00 mins at 900km/h (559mph)	
at 15,000m (49,210ft)	3.20 mins at 800km/h (497mph)	
Landing speed	160km/h	99.4mph

Junkers EF127 'Walli' – A Target-defence Interceptor

1944

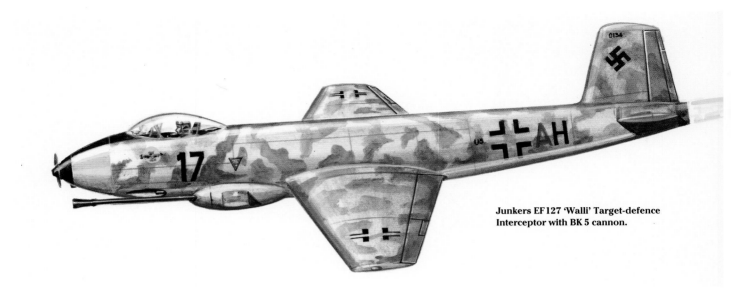

Junkers EF127 'Walli' Target-defence Interceptor with BK 5 cannon.

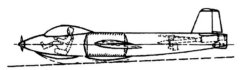

Junkers EF127 'Walli' three-view drawing.

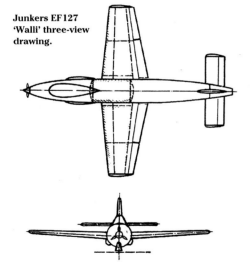

In response to an RLM specification for an Objektschutzjäger (target-defence interceptor), the Junkers Flugzeugwerke submitted its EF127 proposal of 16th December 1944, capable of attaining operational readiness in the shortest possible time. By using construction materials and components from the EF126 'Elli' already described, two design schemes had been laid out: the EF127 (I) with a retractable tricycle undercarriage and capable of unaided take-off and landing, and the EF127 (II) to have taken off from a mobile take-off ramp and landing on its extended fuselage skid.

In order to provide adequate space for the T-Stoff and C-Stoff tanks for the HWK 509A main rocket motor and auxiliary cruising chamber, fuselage length was increased over that of the pulsejet-powered EF126 and tailplane planform was also revised. Take-off assistance was provided by two Schmidding SG 34 RATO units, the aircraft climbing rapidly to its operational altitude of 10,000m (32,810ft), returning in gliding flight to its home airfield.

Junkers EF127 – data as of 16.12.1944

Powerplant   One Walter HWK 509C rocket motor with 2,000kg (4,409 lb) thrust main chamber and 400kg (882 lb) thrust cruising chamber plus 4 Schmidding SG 34 RATO units

Dimensions

Span	6.65m	21ft 9¾in
Length	8.00m	26ft 3in
Height	2.35m	7ft 8½in
Wing area	8.90m²	95.80ft²

Weights

Equipped weight	1,030kg	2,276 lb
T-Stoff weight	1,100kg	2,425 lb
C-Stoff weight	500kg	1,102 lb
Loaded weight	2,790kg	6,151 lb
with RATO units	2,960kg	6,526 lb

Performance

Max speed at sea level	1,015km/h	630mph
at 10,000m (36,090ft)	900km/h	559mph
Initial climb rate	133m/sec	26,180ft/min
Climbing time	75 secs to 10,000m	(32,810ft)
Range, after climb		
to 5,000m (16,400ft)	107km at 700km/h	66.5 miles at 435mph
Range at 10,000m (32,810ft)		
at cruising thrust	240km at 700km/h	149 miles at 435mph
Endurance		
at 5,000m (16,400ft)	9.6 mins at 700km/h (435mph)	
Take-off run, without RATO	340m	1,115ft
with RATO	167m (in 4 secs)	548ft
Landing run, on skids	180m	591ft
with brake chute	140m	459ft
Landing speed	132km/h	82mph

Armament (proposed)   2 x MK 108 or 2 x MG 151/20 or 2 x MG 213C

Junkers EF127 wind-tunnel half-model with revised wing planform.

Messerschmitt Me 163B in
Luftwaffe Service

The first live use of the Me 163B on 16th August 1944 near Leipzig. Leutnant Helmut Ryll (in white 11) of JG 400 is attacking US bombers of the 91st Bomber Group over Bad Lausick. Air Classics, May 1977

The effect on morale which the Me 163B caused to Allied flying crews is reflected in the combat reports that were released for publication after the end of the Second World War. At the time of its appearance, the Me 163B was a completely unknown aircraft to the British and the Americans. The latter were convinced at the time that their operational missions were numbered. In point of fact, the destruction caused by the Me 163B was insignificant as the rocket-powered interceptor, by reason of its propulsion unit, was not mature enough for frontline service.

With its eight minute flight duration, its normal radius of action was seldom more than 40km (25 miles) and in order to retain its operational capability, had to return to its home airfield for refuelling and maintenance. As a range-increasing measure, Messerschmitt had planned to install an improved Walter rocket motor, the HWK 509C-4, with a 1,700kg (3,748 lb) thrust main chamber and a 300kg (661 lb) thrust auxiliary cruising chamber. The two chambers could be operated either singly or together in the event of an emergency.

Messerschmitt Me 163B. G W Heumann

Messerschmitt Me 163B works three-view drawing

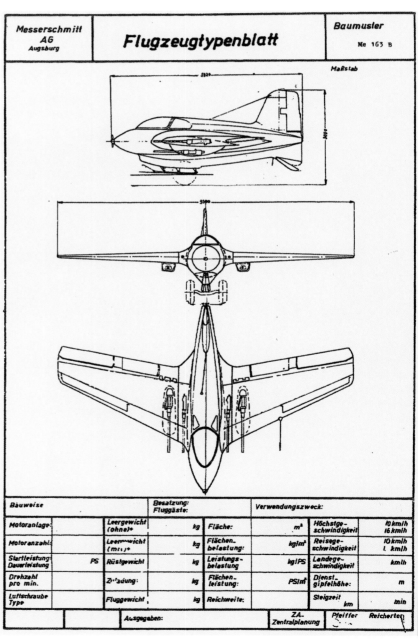

Messerschmitt Me 163B with BT 700 Bomb-Torpedo

Only recently accessible files of the AVA Göttingen contained photographs and a number of file notations that described measurements of special urgency made on the Me 163B model equipped with two underwing BT 1000 Bombentorpedos (bomb-torpedoes). From the documents of the Messerschmitt Project Office in Oberammergau and the AVA of 18th February 1944, it is evident that at the wish of Prof Willi Messerschmitt, wind-tunnel measurements of an Me 163B with underwing loads were to be urgently conducted, as the rocket-propelled aircraft was scheduled to be used at an early date against ship targets. The documents were annotated as being of the highest urgency. From various AVA reports, it is evident that Dipl-Ing Hubert of the Messerschmitt Project Office was placed in charge of this work. In a communication from the Junkers Flugzeugwerke dated 22nd September 1944 to the AVA, it is mentioned that measurements made in the Junkers wind-tunnel would be accepted, where information from the Junkerswerke concerned measurements made on a Messerschmitt fighter with suspended underwing loads. These measurements were conducted by Junkers on an Me 163B with two BT 700s – smaller than the BT 1000, and which enabled the Me 163B to take off on its normal wheeled dolly. As far as is known, take-off trials were made at Junkers in December 1944 with BT mock-ups, but no documentation on their progress is to hand. Since a great deal has already been published concerning the Me 163B, other than its use as an operational fighter described above, mention will only be made here of the BT 700 and the BT 1000.

These BT-designated air-launched missiles (where the figure following denotes the weight) used against ground and seaborne targets were developed by the former Forschungsanstalt (Research Institute) Graf Zeppelin or FGZ in Stuttgart-Ruit. The BT missiles were manufactured partly at the Tripel-Werke factories in Mohnsheim/Elsass. Their high-explosive contents were furnished by the Luft-Munitionsanstalt (Air Ammunition Institute) in Boodstedt, the propulsion unit consisting of a powerful Rheinmetall-Borsig solid-propellant rocket motor. Release of the BT missile was to take place by the so-called 'Schleuderwurf' (catapult-release) method with the aid of a Reflexvisier (reflex bomb-sight) or Revi specially developed for it. Depending on the attack altitude, the release distance was up to 3,000m (9,840ft).

BT 700 Bomb-Torpedo – data[5]

Dimensions		
Length	3.50m	11ft 5¾in
Diameter	42.6cm	16¾in
Tailfin span	1.10m	3ft 7¼in

Weights		
Weight, without rocket unit	780kg	1,720 lb
Explosive weight	330kg	728 lb

[5] Of interest is that an unpowered version of the BT 700 developed for use against naval vessels and merchant ships and released from the Fw 190, was intended not to achieve a direct hit but to first travel some 60-80m (200-260ft) underwater before impact. The triple tail fins stabilised the missile during descent and were designed to separate upon striking the water. Two types of fuses were fitted,

one of which was activated during the missile's underwater path and the other in the event of an unintentional direct impact above the water line. The BT 700 C2 had an overall length of 3.358m (11ft 0¼in), maximum body diameter 45.6cm (18in) and weight 700kg (1,543 lb), of which 375kg (827 lb) was the high-explosive filling.

Other than the similarly-shaped BT 200 C1 and C2 with 105kg (231 lb) explosive, the BT 400 C1 and C2 with 215kg (474 lb) explosive, the BT 700 C1 and C2 and BT 1000, the heaviest was the BT 1400 built of thin-walled steel sheet, its tailfins blown off on impact with the water by electro-pyrotechnic means instead of by shear screws as with the other Bomb-Torpedoes.

BT 1000 Bomb-Torpedo – data

Dimensions		
Length	4.24m	13ft 11in
Diameter	48cm	19in
Tailfin span	1.30m	4ft 3¼in

Weights		
Weight, without rocket unit	1,180kg	2,601 lb
Explosive weight	710kg	1,565 lb

Messerschmitt Me 163B inverted wind-tunnel model with two BT 700s.

The BT 700 Bomb-Torpedo.

Messerschmitt Me 263

1944

During the last phase of the war, further development of the Me 163B was transferred to Junkers due to lack of capacity at Messerschmitt. The reason for further development was its all-too-short powered duration and lack of an undercarriage. At Junkers, Prof Dr-Ing Heinrich Hertel and his team were able in late 1944 to eliminate the deficiencies by improving the design. By evolving a fuselage of smaller diameter and increased length, coupled with a retractable tricycle undercarriage and increased T-Stoff and C-Stoff fuel tankage of 2,440kg (5,379 lb) capacity, dura-

tion was increased to 15 minutes. The wing remained almost unaltered except that the fixed outboard slots were replaced by automatic leading-edge slots, wing washout was reduced at the tips, and additional landing flaps were added ahead of the inboard trailing-edge elevons. The redesigned cockpit became a pressurised compartment, and power was furnished by a 2,000kg (4,409 lb) thrust Walter HWK 509C rocket motor. Known at Junkers as the Ju 248, the project was handed back to Messerschmitt on 22nd January 1945, carrying the designation Me 263. Flight trials with the first prototype commenced in February 1945 at Dessau with towed take-offs as a glider and with the undercarriage fixed down. Two further prototypes were completed by the end of the war but were not flown.

Left and opposite page bottom: **Messerschmitt Me 263/Junkers Ju 248 V1 (DV+PA).**

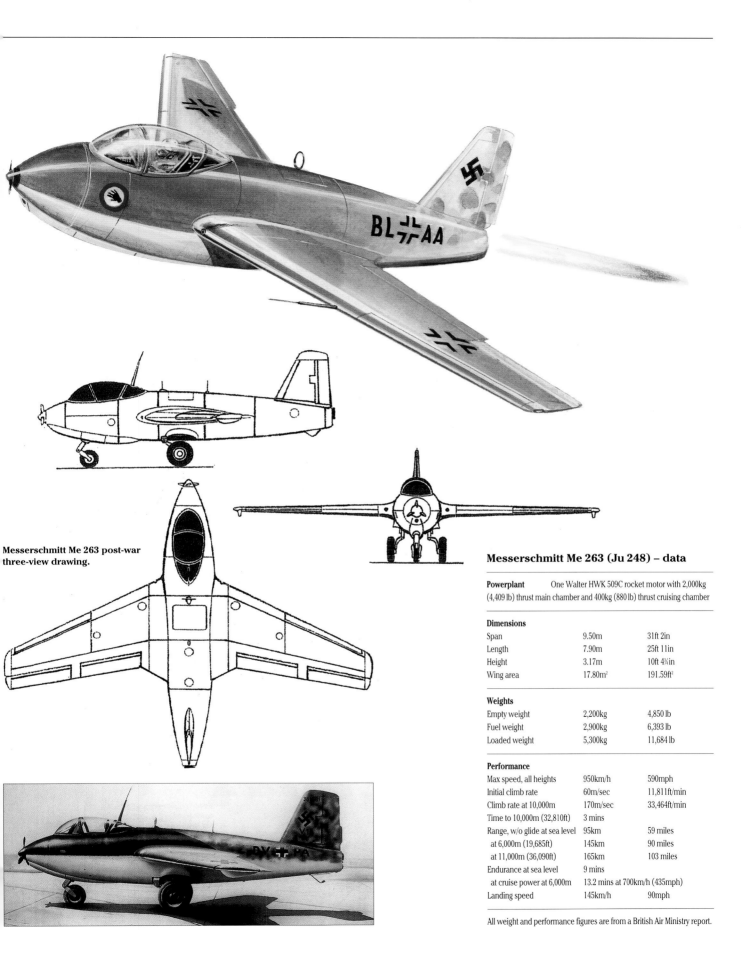

Messerschmitt Me 263 post-war three-view drawing.

Messerschmitt Me 263 (Ju 248) – data

Powerplant   One Walter HWK 509C rocket motor with 2,000kg (4,409 lb) thrust main chamber and 400kg (880 lb) thrust cruising chamber

Dimensions

Span	9.50m	31ft 2in
Length	7.90m	25ft 11in
Height	3.17m	10ft 4¾in
Wing area	17.80m²	191.59ft²

Weights

Empty weight	2,200kg	4,850 lb
Fuel weight	2,900kg	6,393 lb
Loaded weight	5,300kg	11,684 lb

Performance

Max speed, all heights	950km/h	590mph
Initial climb rate	60m/sec	11,811ft/min
Climb rate at 10,000m	170m/sec	33,464ft/min
Time to 10,000m (32,810ft)	3 mins	
Range, w/o glide at sea level	95km	59 miles
at 6,000m (19,685ft)	145km	90 miles
at 11,000m (36,090ft)	165km	103 miles
Endurance at sea level	9 mins	
at cruise power at 6,000m	13.2 mins at 700km/h (435mph)	
Landing speed	145km/h	90mph

All weight and performance figures are from a British Air Ministry report.

Chapter Eight

Air-Launched Fighters and Attack Gliders

Other than 'Mistel' (Mistletoe) composite aircraft dealt with separately, Schleppjäger (towed fighters) and Bombensegler (bomb-carrying gliders) were the last weapon developments that were on the drawing boards in Germany a few months before the end of the Second World War.

By pure chance it has become possible from surviving documents to describe some previously unknown or otherwise little known project studies that were unique in their layout and design. From the numerous documents available, only those developments that were intended for use in the total air war over Germany have been selected for

portrayal in this narrative, among them the Waffenträger (weapons carriers) for Selbst-opfer (self-destruction) missions that had already been forbidden by the Führer, Adolf Hitler.

Besides the DFS 'Eber' (Wild Boar), the Zeppelin 'Rammer' and 'Fliegende Panzer-faust' (Flying Bazooka), there was also the Blohm & Voss BV 40 glider of 1943 which may have served as a precedent for later air-launched jet and rocket-powered bomber interceptors.

Arado E 381 Miniature Fighter

1944

Arado E 381 miniature fighter beneath the Ar 234C-3 parent craft.

Within the framework of the 1944 Jägernot-programm (Fighter Emergency Programme), Arado submitted their E 381 Kleinstjäger (miniature fighter) design to the RLM in November 1944, to be air-lifted in carry-tow to the altitude of enemy bomber formations by the Ar 234C parent craft. At a height of some 1,000m (3,300ft) higher than the bombers, the E 381 was to be released to commence its initial approach in a high-speed shallow dive. At an initial speed of some 820km/h (510km/h),

it was able to glide about 20km (12 miles) towards the target. As compared with the Messerschmitt Bf 109 with a frontal area of 1.8m² (19.375ft²), the E 381 presented a very small target for enemy gunners with its 0.45m² (4.84ft²) frontal area.

In all, three variants of the E 381 were proposed. Common design features were a constant-chord cantilever shoulder wing with a steel tube main spar, twin endplate fins and rudders, prone pilot, and a central fuselage landing skid. Pilot accommodation was in an armour-protected cockpit with a 14cm (5½in) thick vertical plexiglass armoured screen and a 20mm thick armour-plated entrance hatch that could be swivelled aside for exit in an emergency. Protection at the rear for the pilot, fuselage fuel tanks and rocket motor was provided by additional armour plating, the fuselage skin being a 5mm thick armoured steel shell.

In the first two proposals, pilot entry and exit was via the armoured, sideways-hinged dorsal hatch. The low clearance of the aircraft beneath the Ar 234C fuselage meant that the pilot had to lie in the E 381 before it was attached, allowing no means of escape in an emergency. The third proposal with lengthened fuselage permitted a side entry and exit. For propulsive power, only the cruising chamber of the HWK 109-509B of 350kg (772 lb) thrust was selected for the E 381, sufficient to provide the necessary acceleration and duration for his attack. To shorten the landing run on the extensible fuselage landing skid, a brake chute was housed in a container above the rocket motor compartment.

Various armament loads were proposed, ranging from a single MK 108 with 45 rounds in the fuselage spine, to four MK 108s or two MG 131s in external bulges beside the pilot, plus 21cm RB Spr. Gr. rocket projectiles, or six wing-mounted RZ 73 projectiles.

For the estimated 600 man-hours to manufacture a series-produced aircraft in 13 major components, a total of 830kg (1,830 lb) of material was required, of which 42% was steel, 38.6% steel sheet, 14.5% wood and the remainder dural. Construction of a mock-up and a small number of unpowered wooden airframes for the purpose of providing prone pilot training was started, and although it is reported that one unmanned prototype reached the stage of a towed take-off, no examples of the planned powered variant were ever flown, as apart from the unavailability of the rocket motor and an insufficient number of Ar 234C parent craft, added to the war situation, the RLM lost interest as it was viewed in the same light as other similar developments such as the 'Natter' (Viper), 'Eber' (Wild Boar) and 'Fliegende Panzerfaust' (Flying Bazooka).

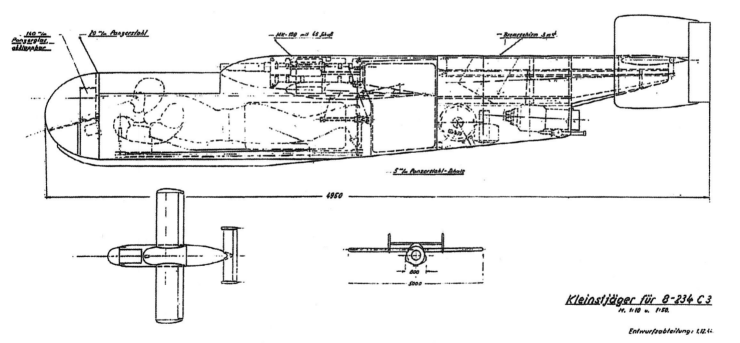

Arado E 381 three-view works drawing of 1st December 1944. Key (left to right): 140mm armoured glass screen (removable); 20mm armoured steel; MK 108 with 45 rounds; 5mm thick armoured steel shell, and 3m (9ft 10in) diameter brake chute.

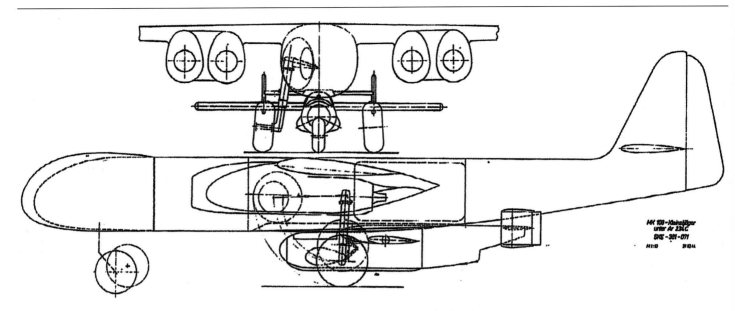

Arado E 381 beneath the Ar 234C-3. Works drawing of 31st October 1944.

Arado E 381 – data as of 31.10.1944

Powerplant	1 x 350kg (772 lb) thrust Walter 109-509B rocket motor	
Dimensions		
Span	5.00m	16ft 5in
Length	4.95m	16ft 3in
Height, fuselage	0.92m	3ft 0¼in
overall	1.22m	4ft 0in
Wing area	5.00m²	53.82ft²
Tailspan	2.00m	6ft 6¾in
Weights		
Empty weight	890kg	1,962 lb
T-Stoff weight	150kg	331 lb
C-Stoff weight	52kg	115 lb
Loaded weight	1,200kg	2,646 lb
Performance		
Max speed	900km/h at 8,000m	559mph at 26,250ft
Armament	1 x MK108 (45 rounds)	

Arado E 381 – data as of 1.12.1944

Powerplant	1 x 400kg (882 lb) thrust Walter 109-509B rocket motor	
Dimensions		
Span	5.00m	16ft 5in
Length	4.95m	16ft 3in
Height, fuselage	0.92m	3ft 0¼in
overall	1.22m	4ft 0in
Wing area	5.00m²	53.82ft²
Tailspan	2.00m	6ft 6¾in
Weights		
Empty weight	890kg	1,962 lb
T-Stoff weight	200kg	441 lb
C-Stoff weight	67kg	148 lb
Loaded weight	1,265kg	2,789 lb
Performance		
Max speed	885km/h at 8,000m	550mph at 26,250ft
Range, with glide	100km at 7,000m	62 miles at 22,965ft
Landing speed	160km/h	99mph

Arado E 381 – data (third variant)

Powerplant	1 x 400kg (882 lb) thrust Walter 109-509B rocket motor	
Dimensions		
Length	5.70m	18ft 8½in
Height, fuselage	0.70m	2ft 3½in
overall	1.22m	4ft 0in
Wing area	5.50m²	59.20ft²
Weights		
Loaded weight	1,500kg	3,307 lb
Performance		
Max speed	895km/h at 8,000m	556mph at 26,250ft

Blohm & Voss BV 40

1943 to 1944

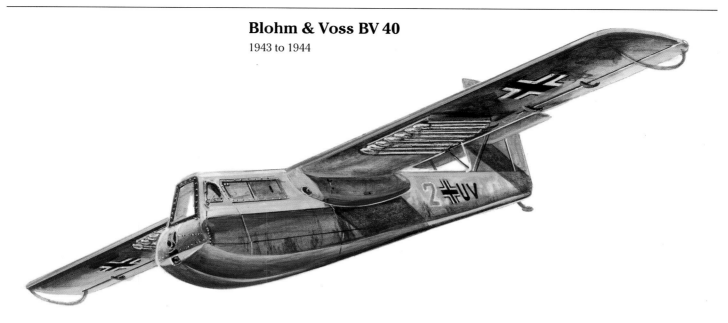

Headed by Dr-Ing Richard Vogt, the Blohm & Voss Project Office in the summer of 1943 proposed a strongly-armoured small glider that could be used as a Gleitjäger (glide-fighter) and later, as a Bombensegler (bomb-carrying glider) to combat Allied bomber formations. For this purpose, the P 186 glider had to be towed to altitude by a powered aircraft such as the Fw 190 and after release in diving flight, make its attack. In the project description Richard Vogt wrote:

'The glide-fighter is a weapon for the rational and undaunted pilot, who, with a clear and cold-blooded daredevilry, is offered the opportunity of bringing his glide-fighter into a position of attack, and protected by armour plating, view the enemy at close range and destroy him. If this prerequisite is practised and upheld, this new weapon will gain an appropriate place beside the established ones.'

Following submission of the P 186 design details to the RLM on 19th August 1943, approval for the manufacture of 6 prototypes was given on 30th October 1943 under the designation BV 40, to be completed the following year. This number was later increased to 12 on 15th December and to 20 on 9th February 1944. Noteworthy in its design and method of construction was that all aerodynamic refinements had been deliberately dispensed with. As a glider, the BV 40 only needed to operate in the 220-250km/h (137-155mph) speed range. Materials selected were wood, beech ply and coarse sheet-metal planking – materials that were still available in sufficient quantities in 1944. The cockpit with its semi-prone position couch, was made of strong, welded armour plating

and thus protected against frontal attack, the 12cm (4¾in) thick armoured windscreen attached to the upper armoured component. The cockpit was attached to the centre fuselage by quick-release bolts to enable pilot escape. The fuselage centre section was covered with 0.8mm thick steel sheet, whilst the wooden rear fuselage and wing surfaces were covered with 4mm thick plywood. For take-off and landing, the flaps were lowered to 50°, and for controlling the glide angle, to

Blohm & Voss BV 40 during flight testing.

80°. Large elevators ensured sufficient stability during the glide towards the target. The undercarriage was as simple as the entire aircraft: a welded tubular axle supported the narrow-track wheels on either side of the central landing skid, the undercarriage being jettisoned on take-off. Cockpit instrumentation was kept to the very minimum.

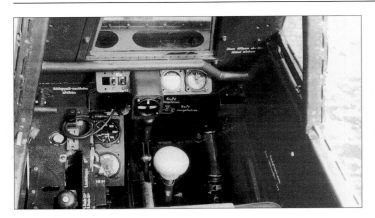

A view of the BV 40 cockpit instruments and controls.

Blohm & Voss BV 40 V1 (PN+UA) port side view.

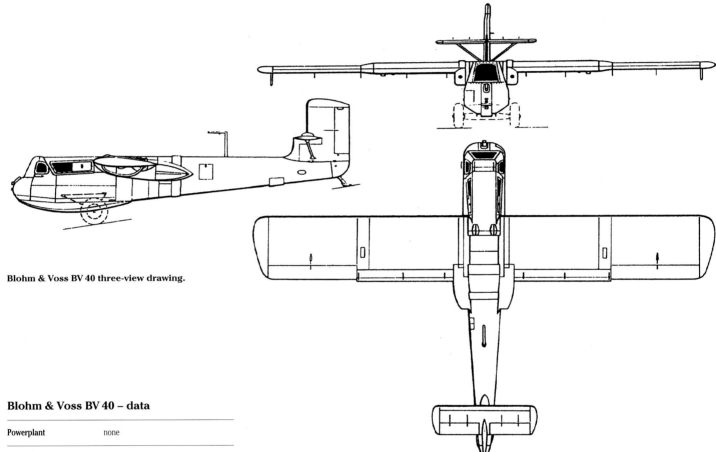

Blohm & Voss BV 40 three-view drawing.

Blohm & Voss BV 40 – data

Powerplant	none	
Dimensions		
Span	7.90m	25ft 11in
Length	5.70m	18ft 8½in
Fuselage width	0.70m	2ft 3½in
Height	1.57m	5ft 1¾in
Wing area	8.70m²	93.64ft²
Tailspan	1.75m	5ft 9in
Weights		
Loaded weight	950kg	2,094 lb
Performance		
Max diving speed	900km/h at 5,000m	559mph at 16,400ft
Landing speed, 80° flaps	140km/h	87mph
Armament	2 x MK108 (35rpg)	

Following completion of the prototype BV 40 V1 (PN+UA) in April 1944 and after an unsuccessful take-off attempt at Hamburg-Finkenwerder, the first successful towed take-off with an Me 110G-0 took place in Wenzendorf on 20th May 1944. Flight trials had been largely completed by mid-July 1944, enabling the RLM Technisches Amt to commence with operational planning. According to available flight-test reports, the BV 40 reached 470km/h (213mph) in diving flight. Flutter tests conducted in April and May 1944 indicated that the aircraft could be flown at speeds up to 650km/h (404mph), reaching diving speeds of 850km/h (528mph) above 4,000m (13,120ft) and 700km/h (435mph) at sea-level, with a maximum of 900km/h (559mph) at 5,000m (16,400ft). After making its attack the glider, flying in pendulum fashion, was to avoid enemy defensive fire and then land. This, however, did not agree with the concept of the RLM which wanted the BV 40 to have maximum effect in total-war operations. Beneath the wings, it was intended to carry two high-explosive bombs equipped with time fuses that would be released from a superior approach altitude over a tightly-knit bomber formation. As a final solution, Blohm & Voss proposed the suspension of two BT 700 bomb-torpedoes beneath the wings, in which case two BV 40s could be suspended beneath the wings of a Heinkel He 177A-5 or He 277B-5 up to release altitude.

It was planned to manufacture two BV 40s per month of which the last was to leave the assembly lines in March 1945. The last BV 40 prototype to have flown was the V6 (PN+UF) on 27th July 1944. Actually, on 18th August 1944, the OKL stopped all further work on the aircraft, and in October 1944 the remaining 14

BV 40 prototypes (to V20) in various stages of completion were destroyed in an air raid. The suggestion to use the Argus As 014 pulsejet was neither favoured nor pursued.[1]

[1] Various armament and auxiliary propulsion systems had been proposed for the BV 40. As well as its two MK 108 cannon (35rpg), the Gerät 'Schlinge' (noose or sling) was proposed, consisting of an explosive-filled cable-suspended 30kg (66 lb) jettisonable sheet-metal container that could be remotely detonated among the bombers. Another was the release of steel cables that would wind themselves around the propellers of the enemy aircraft. The use of an HWK 507 T-Stoff and Z-Stoff 'cold' rocket motor as used in the Henschel Hs 293

missile was also suggested by Dr-Ing Vogt on 15th March 1944. With a burning time of 2 minutes, which could be increased to 4 minutes with further development, an attack from a distance of 15-20km (9.3-18.6 miles) after release could be achieved, but flight tests with the motor were forbidden by the RLM on 10th May 1944. The BV 40 was also intended at one stage as a 'Rammjäger' (ram-attack fighter) to ram the vertical tails of bombers and protect itself from retaliatory fire with its MK 108.

As originally designed, the P 186 had a span of 7m (22ft 11½in), wing area 7m² (75.3ft²), loaded weight 750kg (1,653 lb) and armament one MK 108 cannon. In a later P 186, span was 7.5m (24ft 7¼in), wing area 8.2m² (88.26ft²) and loaded weight 930kg (2,050 lb), increased later to 1,150kg (2,535 lb). Armament was two MK 108 cannon.

Blohm & Voss BV 40 cutaway drawing and parts table.

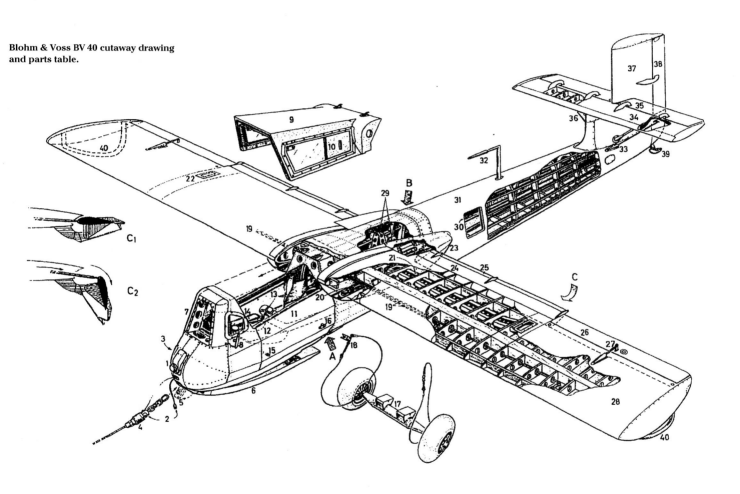

1 Einbauverkleidung für Kompaß
2 Brechkupplung für EiV (Verständigung Bord-Bord), Steckdose in der Schleppkupplung
3 Einbauort des Sammlers
4 Einklinkknopf mit Schleppseil und Öse
5 Schnellverschluß zum Aufbocken des Flugzeuges
6 Gleitkufe (ausgefahren)
7 Stirnpanzerverkleidung mit Panzerglasscheibe (120 mm)
8 Betätigungshebel für Einstieghaube (offen)
9 Einstieghaube
10 Panzerblende (geöffnet)

11 Liegepolster für Pilot
12 Wanne für Brustfallschirm
13 sechs Anschnallgurte
14 Armstütze
15 Verriegelungswelle für Abwurffahrwerk
16 Konterbeschlag für Haltekabel
17 Abwurffahrwerk in Fallstellung
18 Verriegelungsbeschlag mit Spannschloß
19 Steuergestänge für Querruder (oben) und Landeklappe (unten)
20 linke Bordwaffe (MK 108, 30 mm)
21 linker Munitionskanal mit Stirnpanzerung und 35 Schuß Zerfallgurt

22 rechte Munitionskanalabdeckung
23 Waffenwanne
24 Landeklappe
25 Hilfsruder für Landeklappenbetätigung
26 Querruder
27 Stoßstange für Querruderbetätigung
28 Flügel, dreiholmige Holzkonstruktion
29 Seitensteuerpedalen zu den Sauerstoffflaschen
30 Zugang zum Sammler (7,5 Ah) und
31 Rumpfheck in Gonzholz-Bauweise, Steuerseile sind sichtbar
32 Staurohr

33 Steuerseilaustritt für Seitensteuer
34 Höhensteuerflosse
35 Höhenruder
36 Abstrebung Rumpf/Höhensteuer
37 Seitensteuerflosse
38 Seitenruder
39 gefederter Sporn
40 Schutzbügel für Flügelspitzen
A abwerfbare Kanzel aus Panzerstahlblechen (geschweißt), Trennstelle
B Trennspant zur Verbindung von Rumpfmittelstück (Stahlblech, geschweißt und genietet) und Holz-Rumpfheck
C₁ Landeklappe in Normalflug-Stellung
C₂ Landeklappe auf 80° ausgefahren

DFS Bombensegler
(Bomb-carrying Glider)

DFS Bombensegler (bomb-carrying glider) with an SC1000 bomb on target approach.

This bomb-carrying glider, developed by Dr Alexander Lippisch in co-operation with the DFS,* consisted of a small glider designed to transport a 1,000kg (2,205 lb) external bomb towed to altitude in Deichselschlepp (pole-tow) by a Bf109 to the vicinity of the target area and released to continue in a dive towards its target. In order to hit and destroy the target, a simple target-sighting mechanism was installed for the pilot. The ogival-shaped fuselage tapered in cross-section to a small circular tube at the rear, on whose extremity were the delta-shaped cruciform tail surfaces. The likewise delta-shaped wooden wings of 4.28m (14ft 0½in) span and wooden cruciform empennage had a leading-edge sweep of 48°, the SC1000 bomb supported on an ETC suspension beneath the cockpit. Take-off was with the aid of a wheeled dolly. After releasing the bomb, the 7.25m (23ft 9¼in) long glider was to descend with the aid of a parachute. In another variation, the glider was foreseen for use in the suicide role. The RLM's subordinate Entwicklungs-Hauptkommission (Development Main Committee) rejected the Lippisch project as unrealisable.[2]

[2] In the book by J Miranda & P Merkado: *Secret Wonder Weapons of the Third Reich: German Missiles 1934-1945*, Schiffer Publishing Ltd, Atglen, Pa, 1996, p.56, the authors mention that this delta winged and tailed bomb-carrying glider of Lippisch configuration but of *unknown* origin, was first published in the *World War II Investigator* magazine in 1988, stating that the aim was for it to be used against seaborne targets. A Ju 88 was to tow the wooden aircraft covered with metal sheeting, to an altitude of 8,000m (26,250ft) up to 10km (6.2 miles) away from the target which it could reach with a theoretical attack speed of nearly 1,300km/h (808mph) in the target's defensive zone. After the bomb was released at a distance of about 700m (2,300ft) from the target, the aircraft then climbed away in a rising curve. A folded balloon in the fuselage, presumably serving as a dive-brake, could be ejected and stagewise inflated by a compressed air bottle. The method thought of for saving the pilot was not known. No dimensions were given for the glider, carrying a standard SC1000 bomb. In his books *Ein Dreieck Fliegt (A Delta Flies)* and *Erinnerungen (Recollections)*, Lippisch made no mention of the aircraft.

Four colour photographs of a model of this aircraft carrying a suspended BT bomb-torpedo are illustrated but without comment in the book by David Myhra: *Secret Aircraft Designs of the Third Reich*, Schiffer Publishing Ltd, Atglen, Pa., 1998, pp.214-215.

*For a detailed description of DFS-developed projects, see Horst Lommel: *Geheimprojekte der DFS. Vom Höhenaufklärer bis zum Raumgleiter (DFS Secret Projects – From the High-altitude Reconnaissance Aircraft to the Space Glider)*, Motorbuch Verlag, Stuttgart, 2000.

DFS 'Eber'

1944

In its design of the 'Eber' (Wild Boar) ram-attack aircraft, the DFS responded to a TLR requirement of 1944 for an expendable aircraft able to carry out an attack against enemy aircraft in two approaches. The single-seat 'Eber' was to be towed aloft behind an Fw 190 or Me 262 to an altitude some 300m (1,000ft) above an enemy bomber formation and released for its first diving attack using its forward-firing armament. By using the solid-propellant rocket motor installed in the rear fuselage, the second approach and ramming attack was to be accomplished. The impulse provided by the rocket unit was to be so utilised that an attack from an altitude of 700-1,000m (2,300-3,300ft) above the enemy could be carried out. The calculated approach speed of 150-200m/sec (336-447mph) resulted in an acceleration of 100 'g' (one hundred times the force of gravity). Since the pilot was able to withstand a maximum force of 16 g, a special sprung seat was to have been installed to reduce the force to this figure. Of aircraft configuration, the fuselage, wings, and empennage were of simple wooden construction whereby the DFS discarded the use of swept wings and tail surfaces, high speeds being achieved by the use of suitable aerodynamic profiles. The choice of twin fins and rudders served to avoid rolling moments caused by rudder deflection.

Following the initial attack with air-to-air rockets or underwing MK 108 cannon, the prone pilot exited during the ramming approach by jettisoning the cockpit canopy, whereupon a small 'working' parachute would pull the pilot and his couch clear of the aircraft. For the ramming attack, the pilot was afforded protection in an armoured enclosure. For propulsion, two possibilities were foreseen:

- Two Rheinmetall-Borsig RI-502 solid-propellant rocket motors each of 1,500kg (3,307 lb) thrust for 6 seconds duration. Each rocket unit measured 17.8cm (7in) in diameter, 1.27m (4ft 2in) in length, and weighed 48kg (106 lb).
- A DFS-developed pulsejet of 200kg (441 lb) thrust installed in the rear fuselage. Due to the high thermal loads imposed on the aircraft and pilot, it was decided to use solid-propellant rockets.

The 'Eber' project is often confused in the aeronautical press with the DFS 'Jagdsegler' (fighter-glider). Both projects – the 'Eber' and the 'Jagdsegler' resembled each other in design and layout, except that the 'Jagdsegler' was to have carried two 250kg (551 lb) bombs in place of the weapons containers.[3]

[3] The author himself here appears to confuse the rocket-powered DFS 'Eber' which had a conventionally-seated pilot, with the pulsejet-powered

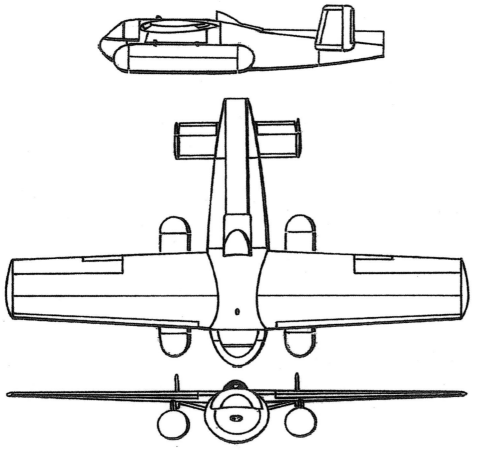

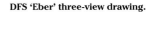

DVL (not DFS) 'Jagdsegler' with prone-pilot arrangement as shown in his drawings, as both were of similar concept.

In a brief account in his book, author Fritz Hahn: *Deutsche Geheimwaffen (German Secret Weapons) 1939-1945*, Erich Hoffmann Verlag, Heidenheim, 1963, p.418, stated that the 'Eber', which leaned heavily on the Bachem 'Natter' development, was to have been towed to an altitude of c.1,000m (3,300ft) above and released some 6-8km (3.7-5.0 miles) behind an enemy bomber formation for the attack, accelerated by its HWK 509 (sic!) rocket motor. Armament considered was the SG119 multiple-shot 30mm battery or three 21cm RB.-Sprenggranate (explosive shells). The project was not realised due to limitations imposed on the manoeuvrability of the towing craft and lack of availability of the rocket motor. Other than various models tested in the wind-tunnel, nothing was manufactured nor were final design and construction details settled. More information concerning the project could not be established by Fritz Hahn due to the paucity of available documents.

Believed to be *published here for the first time*, a more informative account was given in CIOS Report XXXII-66 'Deutsche Forschungsanstalt für Segelflug, Ainring', HMSO, London (June 1945), p.28. The 'Eber' project, undertaken upon the aegis of the Fofü or Forschungsführung (Research Leadership) of the RdL and ObdL, was led by the DFS Department of Aerodynamics director Prof Dr-Ing Paul Ruden and his assistant Dipl-Ing Schieferdecker. Begun in August 1944, it envisaged a short-

range interceptor similar in concept to the Bachem 'Natter', but instead of employing a near-vertical rocket-powered self take-off, was to be towed to altitude by an Me 262 or Fw190 fighter using the Starrschlepp (rigid pole-tow) method of attachment, on which the DFS had accumulated considerable in-flight experience with the Ju 52+DFS 230, Do17+DFS 230, He111+DFS 230 and He177+Go 242 as well as other aircraft combinations. Experience showed that use of a coupling with only two degrees of freedom of movement for the towed aircraft should be used, and that if the natural frequency of oscillation of the rigid pole-tow was increased, the incidence of oscillation at high towing speeds was reduced, enabling heavy towing loads such as a 950kg (2,094 lb) attachment to reach high critical speeds. Upon attaining a height of some 300-400m (1,000-1,300ft) above the bomber formation, the 'Eber' was to be released from 5-10km (3.1-5.0 miles) behind them, the approach being accelerated by its RI-502 solid-propellant rocket motors with which it was calculated that burning time would be sufficient for the 'Eber' to reach the formation with a closing speed of not less than 100m/sec (224mph). Initial considerations for a single ramming attack to be carried out were dropped on closer examination in favour of two separate attacks with its armament of a 30mm salvo-gun or R4M air-to-air projectiles. The pilot was protected by an armoured windscreen and by armour-plating against frontal fire from 0.5 in (12.7mm) machine-gun fire, the remaining rocket fuel sufficient to enable acceleration from 700-1,000m (2,300-3,300ft) below the bombers to execute the second attack. The realisation that the towing fighter with 'Eber' attached could lose as much as 100km/h (62mph) in speed if attacked by enemy fighters from above, meant that even at light loading, the combination was endangered before release.

The question of ram attacks had been studied time and time again, with the simultaneous need of how to provide protection and safe ejection for the pilot. The mechanical process of ramming had been studied by Prof Dr Ruden and Dr Schapitz in a model experiment where a falling weight impacted on a cylinder-shaped fuselage representing the structure, the aim having been to determine the magnitude of the thrust impulse and the effective duration and energy requirement. To protect the pilot from unbearable accelerations of 100 g and more, a spring-dampened catapult-type of *sliding seat* was required that reduced the thrust over a damped spring or suitable energy-destroying apparatus, but construction difficulties were encountered to accommodate a sufficient number of spring tracks in the aircraft. Further consideration on pilot exit led to the concept of a parachute-equipped seat which appeared very promising but due to lack of time, had not been built and tested.

In *Luftfahrt Documente LD 11: Project 'Bewaffnung Eber'*, Verlag Karl R Pawlas, Nürnberg, regarding the proposed armament, Dipl-Ing Schieferdecker, head of the DFS Institute A, Dept. A3, requested Stabsing. Bühler of the RLM GL/C-II VI on 10th October 1944 to arrange for supply and installation drawings of the 30mm MK108 cannon and the 28-round R4M rocket pack. On 9th November 1944, Prof Dr-Ing

Karl Leist of the Fofü Göttingen mentioned in a communication that he would be visiting the DFS Ainring for one week, bringing with him documentation on a similar 'Project Bembo' on which he was working. In a Fofü conference held in Berlin on 7th December 1944, various investigation and clarification tasks were assigned to members of the DFS, DVL, LFA and FGZ by the Fofü on matters of the Eber's armament, reflex sights, dispersion, materials, pilot protection and exit via ejection seat or parachute and the RI-502 propulsion unit. A DFS table of costs dated 8th January 1945 for technical assistance and material storage suggests that although design details had probably neared conclusion in December 1944, the project must have been terminated around that date.

A DFS table of 1944 for a towed aircraft weighing 950kg (2,094 lb) towed by an Fw190, showed the following performance losses (here extracts only):

Fw190 / Fw190+Attachment

Weights

Fuel weight	410kg	410kg
	904 lb	904 lb
Loaded weight	3,900kg	4,850kg
	8,598 lb	10,692 lb

Performance

Max speed, at sea-level	533km/h	490km/h
	331mph	304mph
at 6,000m (19,685ft)	637km/h	597km/h
	396mph	371mph
at 9,000m (29,530ft)	623km/h	583km/h
	387mph	362mph
Climb rate, at sea-level	15.2m/sec	9.8m/sec
	2,992ft/min	1,924ft/min
at 6,000m (19,685ft)	9.5m/sec	4.5m/sec
	1,870ft/min	886ft/min
at 9,000m (29,530ft)	3.6m/sec	–
	708ft/min	–
Time to climb		
to 6,000m (19,685ft)	8.1 mins	14.3 mins
to 9,000m (29,530ft)	19.9 mins	–
Service ceiling	10,300m	8,000m
	33,792ft	26,250ft
Range, at 6,000m (19,685ft)	770km	710km
	478 miles	441 miles
at 9,000m (29,530ft)	1,122km	1,035km
	697 miles	643 miles

Last but not least, it should be mentioned that the DFS had been actively involved with both ground-based and flight-testing of Argus-designed pulsejets ever since the end of 1941, when two early 150kg (330 lb) thrust pulsejets were flown with the DFS 230 glider. By the end of 1944, considerable testing experience had been amassed, and preparations were in hand to commence a comprehensive programme of investigations early the following year on all aspects dealing with pulsejet power, headed by Dr Felix Kracht and Dr-Ing Karl Eisele. For more details see: Anthony L Kay: *German Jet Engine & Gas Turbine Development*, Airlife Publishing Ltd, Shrewsbury, 2002, pp.253-257. With all their accumulated knowledge of existing Argus pulsejets, it is unlikely that the DFS would have striven to develop their own '200kg thrust' unit mentioned by the author for the 'Eber'.

The DVL 'Jagdsegler' (glide-fighter), submitted for consideration in June 1944 and depicted in the three-view drawing which *matches in scale*, had span 5m (16ft 4¾in), length 3m (9ft 10in), fuselage diameter 62.5cm (24½in), wing area 3.5m² (37.67ft²) and loaded weight 640kg (1,411 lb), of which fuel weight was 160kg (353 lb) for its 300kg (661 lb) thrust pulsejet which had a maximum useful life of 90 minutes. Estimated maximum speed was 900km/h (559mph). With two underwing SC 250 bombs, loaded weight rose to over 1,100kg (2,425 lb). Among parent craft considered to bring the glide-fighter to altitude were the Focke-Wulf Ta152 and Ta154, He 219, Ju188, and Me 210 and Me 410.

DFS 'Eber' – data

Dimensions

Span	5.16m	16ft 11in
Length	3.36m	11ft 0¼in
Fuselage diameter	0.87m	2ft 10¼in
Wing area	3.70m²	39.83ft²

Weights

Flying weight	640kg	1,411 lb

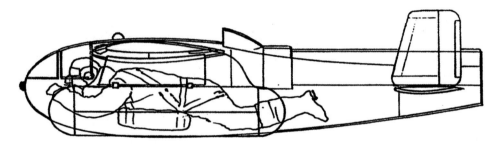

Pilot accommodation in the DFS 'Eber'.

Messerschmitt P1104 – A Rocket-powered Target-Defence Interceptor

1944

In response to an RLM specification calling for a rocket-powered 'Objektschutzjäger' (target-defence interceptor), the Messerschmitt Project Office in mid-1944 drew up the P1104 proposal which could be viewed as a competitor to parallel designs submitted by Heinkel, Junkers and Bachem.

A conventional shoulder-wing wooden monoplane of the simplest construction, the P1104 was designed to fulfil all the requirements contained in the specification. In appearance it seemed rather primitive in terms of aerodynamic finesse, having rectangular wing and tail surfaces and a rotund fuselage of circular cross-section. The pilot was housed beneath a raised cockpit canopy at whose rear was a ribbon parachute, the aircraft landing on a single extensible fuselage skid. Powerplant was an HWK 109-509A-2 rocket motor of 1,700kg (3,748 lb) thrust, provision having been made for an auxiliary 300kg (661 lb) thrust cruising chamber.

Altogether Messerschmitt produced seven different variants of the P1104, ranging from a Sprengstoffträger (high-explosives carrier) housing a 300kg (661 lb) high-explosive load for the SV or SO-Einsatz (suicide) role, to an unpowered, unmanned auxiliary fuel tank to be towed aloft in the Deichselschlepp (pole-tow) mode behind an Me 262 and then

detached as an expendable device. One of the most interesting proposals was as a 'Rammjäger' (ram-attack fighter) equipped with an armoured steel nosecap with which it would ram enemy bombers. Following the RLM decision at the end of 1944 in favour of the Bachem Ba 349 'Natter' (Viper), work on the project was terminated.[4]

[4] The project was preceded by two very similar P1103 design studies that dated from July 1944. In one, of 6th July 1944, the prone pilot was protected against frontal fire by six flat and curved thick armoured-vision panels, the aircraft featuring V-1 type rectangular low wings of constant chord. Powerplants were four Rheinmetall-Borsig RI-502 solid-fuel rockets mounted abreast in a bank beneath the fuselage behind the two fixed MK108 cannon (300rpg) located beneath the pilot's couch. Designed as a 1,100kg (2,425 lb) weight Rammjäger (ram-attack fighter), both pilot and couch were to be extracted by a large parachute in the upper rear fuselage, the remainder of the airframe descending by a second parachute. A second P1103 scheme (drawing XII-283) of 12th September 1944 intended as a Bordjäger (air-launched fighter) had a conventionally-seated pilot beneath a raised cockpit canopy not protected by an armoured nose cone. The shoulder wing and tailplane were likewise of constant chord and thickness but the tail surfaces were of narrower chord. Armament was one MK108 beneath the cockpit floor. Take-off was on an Me163B-type dolly, the aircraft towed to altitude

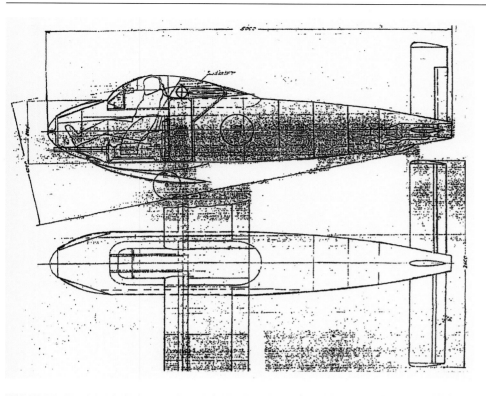

Top: **Messerschmitt P1104 Bordjäger works drawing (XII-283) of 12th September 1944.**

Centre: **Messerschmitt P1104 Bordjäger works drawing, in pole-tow behind an Me 262.**

Bottom: **Messerschmitt P1104 Rammjäger works drawing (XVIII-118) of 22nd September 1944.**

Messerschmitt P1104 – data

Powerplant	1 x 1,700kg (3,748 lb) thrust HWK 109-509A-2 rocket motor	
Dimensions		
Span	5.30m	17ft 4¾in
Length	5.00m	16ft 4¾in
Height	1.56m	5ft 1½in
Performance		
Max speed	840-930km/h	522-578mph

by an Me 262, landing on its extended skid and tail-wheel. Powerplant was a Walter 109-509A-2 and dimensions were as cited above by the author for the P1104.

Other P1104 variants differed in wing and tailplane positions, wing areas and loaded weights, depending on the quantity of fuel carried. For the P1104 of 22nd September 1944, a British Air Ministry report lists the following data: Powerplant: one HWK 509 rocket motor, wing area 6.5m² (69.96ft²), T-Stoff weight 900kg (1,984 lb), C-Stoff weight 300kg (661 lb), take-off weight 2,570kg (5,666 lb), maximum speed 800km/h (497mph), initial climb rate 200m/sec (39,370ft/min), range after climb 87km at 6km (54 miles at 19,685ft) and take-off run 155m (509ft). Armament was one MK108 cannon. Take-off was horizontal, and landing on the extended fuselage skid.

For additional coverage of German rocket-propelled aircraft, see Joachim Dressel & Manfred Griehl: *Die deutschen Raketenflugzeuge 1935-1945*, Motorbuch Verlag, Stuttgart, 1989.

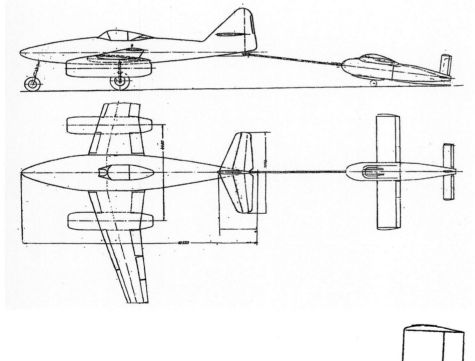

Zeppelin 'Fliegende Panzerfaust'

1944

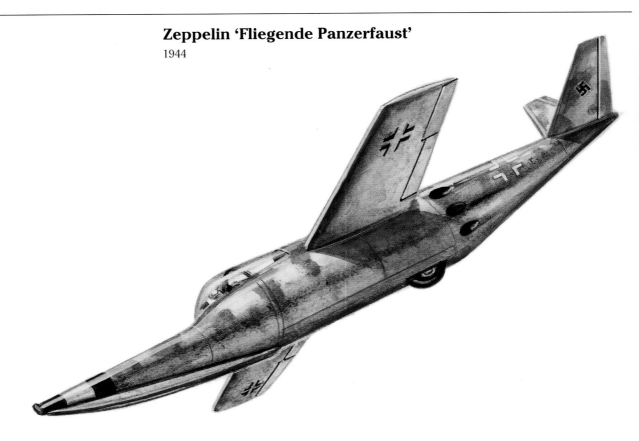

At the outbreak of the Second World War the Luftschiffbau Zeppelin GmbH airship construction company of Friedrichshafen felt compelled to terminate dirigible manufacture in favour of aircraft. In addition to the design and manufacture of Großraum (large capacity) transport aircraft such as the Messerschmitt Me 323 'Gigant' (Giant) and the ZSO 523 transport (six Gnôme-Rhone GR 18R motors), span 70m (229ft 8in), useful load 24,000kg (52,910 lb) and maximum loaded weight 95,000kg (209,437 lb) – a joint undertaking together with the French Société Nationale des Constructions Aéronautiques du Sud-Ouest (SNCASO) concern, the design office headed by Dipl-Ing Arthur Förster in 1944 drew up proposals for miniature aircraft to combat air targets. Code-named the 'Fliegende Panzerfaust' (Flying Bazooka or literally Flying Tank-Fist), details of it were submitted to the RLM Technisches Amt at the end of the year.

The miniature aircraft, with shoulder wing and butterfly tail, was to have been towed to altitude by a Messerschmitt Bf 109 fighter by the Starrschlepp (rigid pole-tow) method. Upon release from its towcraft, the single-seat prone-piloted aircraft was to accelerate towards its target with the aid of its six Schmidding SG 34 rocket propulsion units, and after firing off its rocket armament, was to proceed with a ramming attack. The pilot

would then save himself and land by parachute, the remaining portion of the machine also descending by parachute for re-use.

A special variant was also proposed as a Sprengstoffräger (high-explosives carrier) for use in the self-sacrifice role against bomber formations where the pilot detonated the explosive load. A full-scale mock-up of the Flying Bazooka was completed at Friedrichshafen in January 1945 and inspected by a Sonderkommission (Special Committee) of the SS-Hauptamt (SS Main Office) that was responsible for development of such weapons.

LZ 'Fliegende Panzerfaust' – data

Powerplant   6 x 1,200kg (2,646 lb) thrust Schmidding SG 34 diglycol solid-propellant rockets of total weight 150kg (331 lb)

Dimensions

Span	4.50m	14ft 9¼in
Length	6.00m	19ft 8¼in
Height	1.50m	4ft 11in
Wing area	3.80m²	40.90ft²

Weights

Loaded weight	1,200kg	2,646 lb

Performance

Max speed	850km/h	528mph

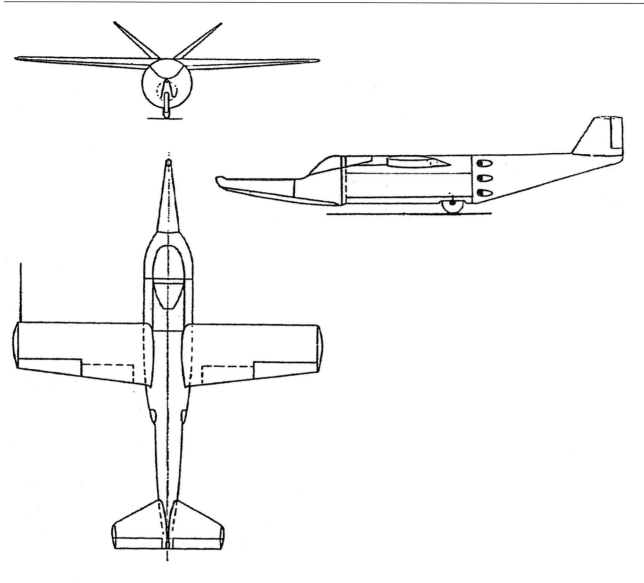

Top: **Luftschiffbau Zeppelin 'Fliegende Panzerfaust' three-view drawing.**

Above: Artist's impression of a **'Fliegende Panzerfaust' in tow by a Bf 109 and flanked protectively by two companion fighters to release altitude.**

Zeppelin 'Rammer'
1944

LZ 'Rammer' – data

Powerplant   1 x 1,000kg (2,205 lb) thrust Schmidding 109-533 diglycol solid-propellant rocket motor of 12 seconds duration. Fuelled weight was 135kg (298 lb), including diglycol weight of 66kg (146 lb).*

Dimensions

Span	4.95m	16ft 2¼in
Length	5.10m	16ft 8¾in
Height	1.75m	5ft 9in
Wing area	6.00m²	64.58ft²
Tailspan	1.80m	5ft 11in

Weights

Loaded weight	860kg	1,896 lb

Performance

Max speed	780km/h	485mph
Landing speed	110km/h	68mph

Armament   14 x R4M or 1 x SG118 (14 x MK108 rounds)

* As an alternative, a 1,200kg (2,646 lb) thrust Schmidding SG 34 of 10 seconds duration and 150kg (331 lb) fuelled weight was also proposed.

Another miniature towed fighter project by the Luftschiffbau Zeppelin GmbH was the 'Rammer.' For ramming purposes, its constant-chord rectangular wing was strengthened by three large spanwise tubular spars along the forward section up to the 40% chord line, the single solid-propellant Schmidding 109-533 rocket motor welded at its forward end of the wing rear spar and serving as a support for the rear fuselage. The entire fuselage and flying surfaces, designed to withstand high stresses, were of all-metal construction. The balloon cable-cutting type of wing leading edge had been tested successfully in combat over England.

As had been planned with the DFS 'Eber' and the 'Fliegender Panzerfaust', the 'Rammer' was to carry out an initial gliding attack from a distance of about 550m (600yds) with its 14 R4M rockets or 30mm-equipped SG118 battery, fired off singly or in salvoes from the protruding nose container before making the ram attack in the second approach.

The conventionally seated pilot was completely protected by armour plating varying from 28mm thickness at the front to 20mm at the rear, the bullet-proof windscreen panel being of 80mm thickness, with the top and side panels each of 40mm thickness. The towed Starrschlepp (rigid pole-tow) take-off was to have been with the aid of a jettisonable dolly, the aircraft landing on its underfuselage skid and brake-chute.

Flight trials carried out at the DFS in January 1945 with an unpowered 'Rammer' were conducted successfully. Production of an initial batch of 16 aircraft, however, did not materialise as the Zeppelin factory facilities were destroyed in bombing raids.

Chapter Nine

'Mistel' Composites

As well as the 'Schräge Musik' (Jazz Music) obliquely inclined upward-firing weapons, the German aircraft industry also developed unmanned explosive-carrying aircraft that were to have been brought by a parent craft to the neighbourhood of enemy bomber formations and released. Guided into the bombers from the parent craft by remote control, an enormous destructive effect was expected from detonation of the thin-walled explosive head. Release and guidance from the controlling aircraft was by means of a periscope or reflex sight (Revi) and transmitted commands. In operations against ship or surface targets, the simple 'target covering' or dog-leg curve method via the remote control system was to have been employed.

In an RLM discussion on 4th and 5th September 1944 between Oberstleutnant Siegfried Knemeyer and the RLM officials Scheibe and Haspel of GL/C-E2/II, the subject of 'manned execution' with a cockpit and control system was also mentioned, and although forbidden by Adolf Hitler, was to have been employed in the course of SO = Selbstopfer (self-sacrifice) or SV = Selbstvernichtung (self-destruction) missions to destroy enemy bomber formations. The possibilities of the pilot leaving the explosive-laden craft were also discussed, but without any clear solution to the problem.

Tests with unmanned explosives carriers had been conducted at the beginning of 1945 in Ötztal. Since the enormous pressure wave caused by a detonation at 3,000m (9,840ft) altitude led to civilian casualties amongst the community beneath, tests were transferred to the Luftwaffe E-Stelle Peenemünde-Karlshagen, in which the RLM explosives specialist Dipl-Ing Demann of GL/C-E7/III played an active part.*

* For a detailed account of the history, development and operational use of the 'Mistel' and other aircraft composites see:
Robert Forsyth: *Mistel – German Composite Aircraft & Operations 1942-1945*, Classic Publications, Crowborough, 2001.
Hans-Peter Dabrowski: *Mistel – Die Huckepack-Flugzeuge der Luftwaffe bis 1945, Waffen-Arsenal Sonderband S-37*, Podzun-PallasVerlag, Friedberg/Hessen, 1993.
Arno Rose: *Mistel – Die Geschichte der Huckepack Flugzeuge*, Motorbuch Verlag, Stuttgart, 1981.
Ernst Peter: *Der Flugschlepp von den Anfängen bis Heute*, Motorbuch Verlag, Stuttgart, 1981 and
P W Stahl: *Geheimgeschwader KG 200*, Motorbuch Verlag, Stuttgart,1977.

The Ar E 377 Explosives Carrier

The remote-controlled Arado E 377 glide-bomb was to have been suitable for combating enemy bomber formations as well as ground targets. This glide-bomb, already described in depth in *Luftwaffe Secret Projects: Strategic Bombers 1939-45*, pp.139-141, was drawn up in three variants:
- Scheme A: Motorless remote-controlled glide missile for use against ground and sea targets.
- Scheme B: Equipped with two Porsche 109-005 (Works drawings show two BMW 003 turbojets – Translator) expendable turbojets for operations against surface area-targets and for combating bombers outside Reich territory.
- Scheme C: As a glide-bomb both with and without propulsion units for so-called SO or SV missions. In this scheme, two auxiliary solid-propellant rocket units mounted on the fuselage sides were to provide acceleration during target approach. The cockpit for the prone pilot took the form of a streamlined teardrop canopy aerodynamically blended into the missile body. A 1,200kg (2,646 lb) warhead of liquid high-explosive was to have been housed in shock-proof containers in the fuselage.

Arado E 377A remote-controlled powered glide-bomb with Heinkel He 162A upper component.

Extensive wind-tunnel measurements in connection with the upper component produced good results but did not lead to actual production.[1]

[1] The Arado E 377, to have been built largely of wood and first proposed in August 1944, was to have been equipped with a standard SC 1800 bomb (less its tailfins) in the fuselage nose. With a special 2,000kg (4,409 lb) hollow-charge warhead, overall length was 10.8m (35ft 5¼in) and fuselage diameter 1.25m (4ft 1¼in). Aft of the 3,500kg (7,716 lb) fuselage and inboard 1,000kg (2,205 lb) wing fuel tanks, a ballast fuel weight of 500kg (1,102 lb) which could eventually be replaced by a liquid explosive charge was located at the rear, ahead of the tail surfaces. As the 'Mistel 6' the Ar E 377 + Ar 234C took off from the 8.4m (27ft 6¾in) long and 6.2m (20ft 4in) track Rheinmetall-Borsig trolley equipped with four 1,500kg (3.307 lb) thrust HWK 501 rocket units burning for 30 seconds, the whole weighing 4,000kg (8,818 lb).

For use with the He 162A or Me 262A as upper component, Junkers also designed the essentially similar but dimensionally slightly smaller Ju 268 lower component (qv).

Arado E 377a unpowered explosives carrier of November 1944 on its five-wheel Rheinmetall-Borsig take-off trolley with superimposed Arado Ar 234C-3.

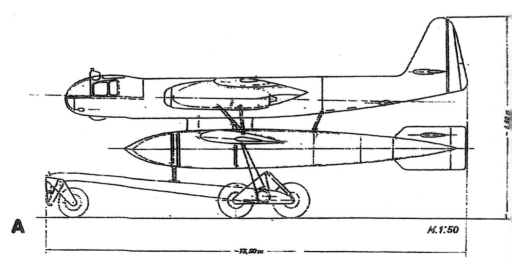

Arado E 377 (lower component) – data

Powerplant	2 x 800kg (1,764 lb) thrust BMW 003A turbojets	

Dimensions

Span	12.20m	40ft 0⅛in
Length, with SC1800	10.90m	35ft 9in
Height	1.40m	4ft 7in
Wing area	25.00m²	269.09ft²
Equipped weight		
with 2 tonne high-explosive	4,000kg	8,818 lb

Weights

Fuel weight	5,000kg	11,023 lb
Loaded weight	9,000kg	19,841 lb

Arado Ar 234C (upper component) – data

Powerplant	4 x 800kg (1,764 lb) thrust BMW 003A turbojets	

Dimensions

Span	14.40m	47ft 2¾in
Length	12.84m	42ft 1¼in
Height	4.15m	13ft 7¼in
Wing area	27.00m²	290.62ft²

Weights

Equipped weight	6,900kg	15,212 lb
Fuel weight	3,100kg	6,834 lb
Loaded weight	10,000kg	22,046 lb
Total without take-off trolley	19,000kg	41,887 lb

Performance

Max speed	720km/h at 6,000m	447mph at 19,685ft
Range, normal	1,300km	808 miles
with extra fuel in E 377	2,000km	1,242 miles

Arado E 377A powered explosives carrier on its Rheinmetall-Borsig five-wheel take-off trolley with superimposed Heinkel He 162A.

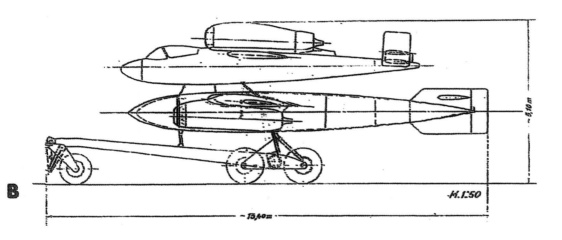

The BV P 214 'Pulkzerstörer' (Bomber-formation Destroyer)

At the end of 1944, Dr-Ing Richard Vogt submitted to the RLM the documentation on a manually-guided Sprengstoffträger (explosives carrier) intended to combat and destroy enemy bomber formations. Onto an 8m (26ft 3in) long bomb body of 1m (3ft 3¼in) diameter was mounted a small, 6m (19ft 8¼in) span aircraft whose prone pilot guided the explosives carrier following their towed ascent to an appropriate altitude by a Dornier Do 217E. After separation from the guidance aircraft, the propulsion unit of the explosive carrier was to be ignited to enable it to fly at high speed into the bomber formation. Detonation of the 1,000kg (2,205 lb) high-explosive charge was expected to disintegrate the formation, resulting in a number of destroyed aircraft.

Both components – the guidance aircraft and the explosives carrier, were to have been powered by ramjets which attained their best performance upon attaining high speeds. The powerplants, however, were still in the development stage and would not have been available before 1947. Additionally, these propulsion units in their design, layout and performance, were intended for high subsonic speeds that could not be reached due to the low speed of the towcraft. Even in diving flight, the necessary air density in the ramjet combustion chamber could not be achieved for ramjet efficiency to enable high speeds to be attained. These were the decisive reasons for the RLM's rejection of the project.[2]

Blohm & Voss P 214 'Pulkzerstörer' (Bomber-formation destroyer).

[2] The 'P 214 manually-guided projectile' description applied to this project dating from the summer of 1944, was first seen by this translator in the book by Karlheinz Kens & Heinz Nowarra: *Die deutschen Flugzeuge 1933-1945*, J F Lehmann's Verlag, Munich (Supplement of 1964, p.849). The BV P 214, however, has been described in earlier published articles as a single-seat tailless 40° swept-wing fighter which featured the Blohm & Voss 'arrow wing' – a constant-chord swept wing planform that had first been used in the P 208 piston-engined fighter (three variants) and featured in most of their subsequent tailless jet aircraft designs up to and including the two-seat P 215 night & bad-weather fighter of 1945.

The P 214, powered by a 1,300kg (2,866 lb) thrust HeS 011 turbojet, was the Blohm & Voss submission in late 1944 in the RLM fighter competition to which the Focke-Wulf P.VI/I and P.VI/II, Heinkel P.1078C, Henschel P.135 and Messerschmitt P.1110/I and P.1116 proposals had been tendered. The task of evaluating the projects had been assigned in OKL conferences from 19th to 21st December 1944 to the DVL Berlin-Adlershof. (see articles by Dipl-Ing Dietrich Friecke, 'Flugwelt' June & July 1953). Salient data for the P 214 were: span 9.5m (31ft 2in), length 7.4m (24ft 3¼in), height 3.1m (10ft 2in), wing area 16.9m² (181.91ft²) gross or 14.54m² (156.5ft²) net, fuel weight 1,200kg (2,646 lb), and loaded weight 4,200kg (9,259 lb). Estimated maximum speed was 966km/h at 7km (600mph at 22,965ft), initial climb rate 21m/sec (4,134ft/min) at loaded weight and 26m/sec (5,118ft/min) at mean (50% fuel) weight. Service ceiling at mean weight was 12,400m (40,680ft) where rate of climb was 2m/sec (394ft/min) and endurance 1 hour. Take-off run was 830m (2,663ft) and landing speed 186km/h (116mph). Armament was three MK 108 cannon and bombload 500kg (1,102 lb).

The combination of guidance aircraft and wing-less explosive-carrying bomb body described and illustrated here as the BV P 214 was one of five

known ramjet-powered studies drawn up by Dipl-Ing Heinz Stöckel from August 1944 onwards, this scheme having been called a 'Torpedoträger mit Lorin-Antrieb' (Ramjet-driven torpedo carrier). The prone-piloted guidance aircraft of span 6m (19ft 8¼in), length 4.8m (15ft 9in), height 1.35m (4ft 5¼in), wing area 6m² (64.58ft²) and loaded weight 500kg (1,102 lb), was mounted above the 1m (3ft 3¼in) diameter and 8m (26ft 3in) long bomb body. Of the total combined weight of 4,000kg (8,818 lb), ramjet fuel comprised 2,300kg (5,071 lb), the bomb-torpedo 1,200kg (2,646 lb) and the guidance aircraft the remainder. Cruising speed was 720km/h (447mph) and range 1,000km (621 miles). The combination was intended to have been air-launched from a Do 217 using a catapult or towed take-off.

Another similar Stöckel-designed torpedo-shaped manually guided rocket projectile, in this instance of 1.8m (5ft 11in) diameter and 11m (36ft 1in) length, was to have been guided by a prone pilot in a miniature aircraft mounted directly above the rear of the torpedo body, the piloted aircraft of 5m (16ft 4¾in) span being ramjet powered. Total weight of the composite was 10,000kg (22,046 lb), of which the fuel comprised 4,500kg (9,921 lb), guidance aircraft 500kg (1,102 lb), and impact weight of the torpedo body 5,000kg (11,023 lb). Performance estimates included a maximum ceiling of 50km (31 miles), range 300km (186 miles), flying duration 5 minutes, and impact velocity 800m/sec (1,790mph). A further Stöckel project is described by the author later on in this chapter.

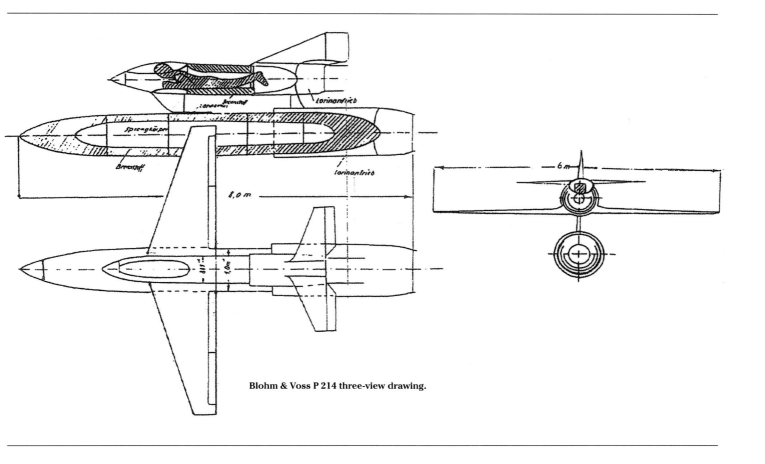

Blohm & Voss P 214 three-view drawing.

The Fi 103 as a Tactical Weapon

In order to be able to deploy the Fieseler Fi 103 – better known as the Vergeltungswaffe 1 (Vengeance Weapon No.1 or V-1) – as a tactical weapon against land and sea targets, the DFS together with the Luftwaffe tested it in carry-tow trials where it was suspended between the port or starboard engine of the Heinkel He 111H-20 on special bomb attachment lugs. At the same time, around the summer of 1944, a further take-off possibility was examined according to plans by Focke-Wulf and the DFS where the Fi 103, cradled beneath an Fw 190A, would be mounted on a three-wheeled jettisonable take-off trolley.

According to documents on the subject, the flying-bomb was to be used in Mistel-type operations against enemy bomber formations whereby the Fi 103 would be steered into the bomber stream after release from the Fw 190 by radio commands. It was planned to use it in a similar manner against ground targets. At the instigation of the SS-Hauptamt (SS Main Office), the RLM put pressure on an early solution to the matter of operational and transport problems, as these were also of rel-

evance for the operational use of the manned version of the V-1 – the 'Reichenberg'. Whereas take-off and launch trials with the He 111 were conducted successfully, the Mistelschlepp (carry-tow) problems and missile separation from the Fw 190 had still not been satisfactorily solved by the end of the war.*

The Fi 103 was also proposed to be towed or carried to altitude by an Arado Ar 234B (1st November 1944) or Ar 234C (25th October 1944).

* For detailed Fi 103 design and developments see:
Botho Stüwe: *Peenemünde-West*, Bechtle Verlag, Esslingen/München, 1995, pp.473-684.
Dieter Holsken: *V-Missiles of the Third Reich: The V-1 and V-2*, Monogram Aviation Publications, Sturbridge, Mass.,1994.
Wilhelm Helmold: *Die V-1 – Eine Dokumentation*, Bechtermünz Verlag, Esslingen, 1999.

Fieseler Fi 103A-1 – data

Powerplant 1 x Argus 109-014 pulsejet of 360kg (792 lb) thrust at S/L static and 254kg (560 lb) at 650km/h at 3,000m (404mph at 9,840ft)

Dimensions		
Span	5.37m	17ft 7½in
Length, overall	8.325m	27ft 3¾in
Fuselage diameter	0.84m	2ft 9in
Height	1.423m	4ft 8in
Wing area	5.40m²	58.12ft²

Weights		
Explosive load, Amatol	830kg	1,830 lb
Loaded weight	2,152kg	4,744 lb

Performance		
Max speed, average	650km/h at 3,000m	404mph at 9,840ft
Range, normal	240km	150 miles

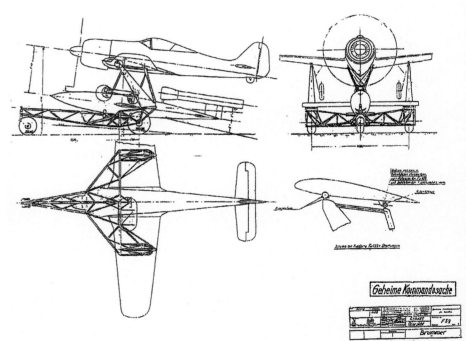

Top: **The Fieseler Fi103 (V-1) as a tactical weapon beneath a Focke-Wulf Fw190 fighter.**

Above: **Works drawing of the Fieseler Fi103 on take-off trolley beneath the Fw190.**

Above left: **The successful sports aircraft pilot and designer Dipl-Ing Robert Lusser (1899-1969) headed the Messerschmitt Project Office from 1933. He later became Technical Director at Heinkel and from 1941 was at the Fieseler-Werke where he was responsible during the last years of the war for series-production of the V-1 flying-bomb.**

Bottom left: **The world-renowned Flugkapitän Hanna Reitsch (1912-1979) as a test pilot at the E-Stelle Rechlin flew numerous types of aircraft, among them the Me163 and the piloted Fi103 'Reichenberg'. As a dedicated patriot, she even volunteered herself as a pilot for a self-sacrifice (suicide) mission.**

Above: **Three views in and outside the cockpit of the Fi103Re-4 'Reichenberg'.**

Top left: **Fieseler Fi103 mounted beneath the starboard wing of a Heinkel He111H-20.**

Centre and bottom left: **Two views of the Fi103 'Reichenberg' captured by the Americans.**

The Henschel Hs 293 as an Anti-bomber Weapon

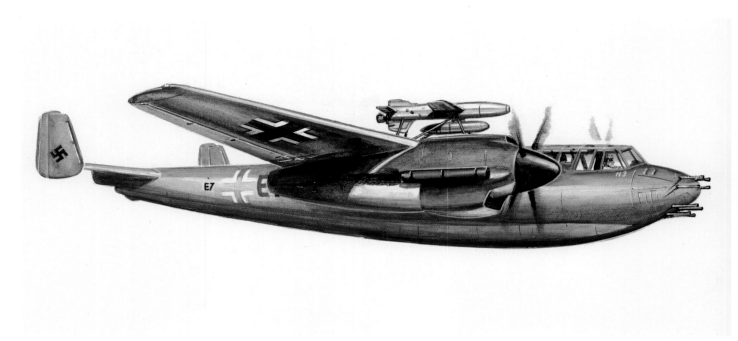

The Henschel Hs 293 in 'Mistel' tow by a Dornier Do 217.

In addition to its previous use for attacking naval vessels, the Henschel Hs 293 missile was intended to attack airborne targets. Flight trials conducted at Anklam in 1944 with the missile proved successful. A Dornier Do 217 modified as a carrier craft for the Hs 293 supported above the fuselage steered the missile after release to its target by means of the FuG 203 'Kehl' transmitter control system. The FuG 230 'Straßburg' receiver in the missile converted the commands transmitted by the FuG 203 'Kehl' missile controller to the elevators. Guidance into the enemy bomber stream was to take place from a height above the bombers as the missile had no rudder of its own. Detonation of the 510kg (1,124 lb) weight Trialen 106 warhead was to be accomplished via radio commands or by means of the built-in proximity fuse. Propulsion was provided principally by the Walter 109-507 liquid-propellant 'cold' rocket motor producing 600kg (1,323 lb) thrust for 10 seconds, accelerating the missile to an impressive 600km/h (373mph).[3]

[3] The Hs 293A was normally air-launched from the He 111, He 177, Fw 200 and Do 217 and could reach a maximum speed of over 860km/h (534mph) during descent, reducing to 560km/h (348mph) when pulling up towards its naval target. The model intended as a Pulkzerstörer (bomber formation destroyer) was the Hs 293H, similar in dimensions and construction to the Hs 293A-2. Equipped with the E-230H/I command receiver, it was to have been powered by two Schmidding 109-543 solid-propellant rockets or the Schmidding 109-513 (SG 9) liquid-propellant M-Stoff (65% methanol + 35% water) and A-Stoff (liquid oxygen) rocket motor which produced 420kg (926 lb) thrust for 9 secs. By increasing methanol strength to 88% (intended for the production engine), thrust was increased to 610kg (1,345 lb) for 11 secs, and with 98% methanol, rose to 990kg (2,183 lb) for 10 secs. Weight of the 109-513 was 96kg (212 lb) empty, and with fuel 133kg (293 lb). Length was 2.355m (7ft 8¾in) and diameter 82cm (32¼in). Only a small number of Hs 293H prototypes were built from March 1944 and tested, among them the Hs 293H V1, the improved V2, the V3 with the 'Marder' (marten) proximity fuse, the V4 with the 'Kakadu' (cockatoo) proximity fuse, the V5 with television guidance, the V6 with barometric fuse and the V7 with an infra-red fuse. Production missiles were to have been the Hs 293H-1a. The Hs 293A-2 (1,200 built but only 500 used), also intended as a 'Pulkzerstörer', was to have been air-launched from the Ar 234C-5, Me 262 and He 343. On a typical operational mission, the Hs 293H would have been detached at a distance of some 1,000-3,500m (1,100-3,750yds) away and 600-1,000m (2,000-3,300ft) above the bombers and guided towards them by the 'target-cover' or dog-leg curve method. To detonate the explosive charge, the 'Marder' proximity fuse had to be less than 50m (165ft) away. Acoustic fuses and other guidance systems had also been considered, but only the radio command system had been tested in the prototypes.

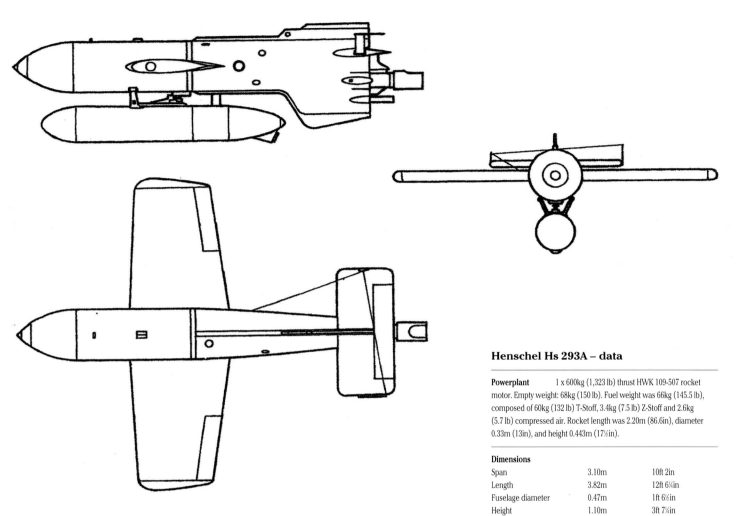

Henschel Hs 293 three-view drawing.

Henschel Hs 293 missile with its Walter 109-507
rocket motor partially uncovered.

Henschel Hs 293A – data

Powerplant   1 x 600kg (1,323 lb) thrust HWK 109-507 rocket
motor. Empty weight: 68kg (150 lb). Fuel weight was 66kg (145.5 lb),
composed of 60kg (132 lb) T-Stoff, 3.4kg (7.5 lb) Z-Stoff and 2.6kg
(5.7 lb) compressed air. Rocket length was 2.20m (86.6in), diameter
0.33m (13in), and height 0.443m (17½in).

Dimensions
Span	3.10m	10ft 2in
Length	3.82m	12ft 6¼in
Fuselage diameter	0.47m	1ft 6½in
Height	1.10m	3ft 7¼in
Wing area	1.92m²	20.67ft²

Weights
Explosive weight	295kg	650 lb
Loaded weight	1,045kg	2,304 lb

Performance
Max speed	120-250m/sec	270-559mph
Range, maximum	15km	9.32 miles

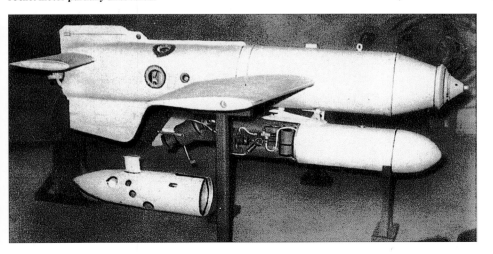

The Ju 268 Explosives Carrier

The Junkers Ju 268 explosive-laden missile with its Heinkel He 162A guidance aircraft.

The Junkers Project office proposal for an explosives carrier, similar to those of Arado and Messerschmitt and guided to its land or air target, was the Ju 268. Of the simplest wooden construction, the mid-wing aircraft of 13.05m (42ft 9¾in) and length 12.95m (42ft 5¾in) had a cruciform empennage and a fixed tricycle undercarriage jettisoned upon take-off. Powered by two 600kg (1,323 lb) thrust Porsche 109-005 turbojets, it was to have been guided towards its target by the pilot of the He 162A-1. To serve the purpose of its mission, it carried an appropriate amount of explosives, distributed in various fuselage compartments and detonated from the control aircraft when used against air targets. The Ju 268 was also proposed by its designers as a manned SO (suicide) aircraft. The end of the war prevented the project proceeding to the construction stage.

The He 162A (one BMW 003E) and Ju 268 (two BMW 003A) 'Mistel 5' of November 1944 was proposed with three types of explosive load. The first consisted of an SC1800 bomb or a special 2,000kg (4,409 lb) charge cradled in the Ju 268 lower forward fuselage. The second, a steel-cased 3,500kg (7,716 lb) hollow-charge warhead was installed near the

aircraft's centre of gravity, and the third, a steel-cased solid-nose warhead of the same weight that was likewise positioned. Take-off from the jettisonable tricycle undercarriage was assisted by six solid-fuel RATO units As visible in Drawing 2, the Ju 268 piloted version was operated with conventional controls.

The project was active between January and April 1945, but other than wind-tunnel tests, no hardware was built.

Junkers Ju 268 (lower component) – data

Powerplant	2 x 800kg (1,764 lb) thrust BMW 003A turbojets	
Dimensions		
Span	11.50m	37ft 8¾in
Length	11.60m	38ft 0¾in
Fin height	2.50m	8ft 2½in
Fuselage diameter	1.40m	4ft 7in
Wing area	22.00m²	236.86ft²
Tailspan	3.50m	11ft 5¾in
Weights		
Equipped weight	4,300kg	9,480 lb
Fuel capacity	5,050 litres	1,111 gals
Warhead weight	2,000kg	4,409 lb
Loaded weight without RATO and wheels	10,500kg	23,148 lb

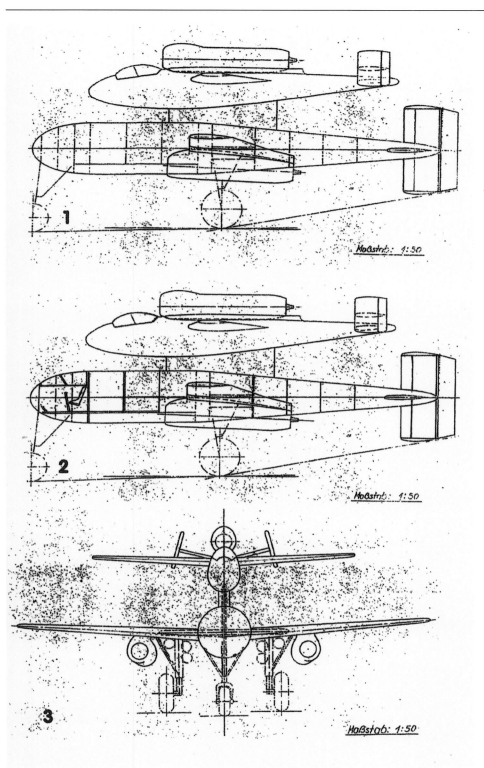

Heinkel He 162A (upper component) – data

Powerplant	1 x 800kg (1,764 lb) thrust BMW 003E turbojet	
Dimensions		
Span	7.20m	23ft 7½in
Length	9.25m	30ft 4¼in
Wing area	11.15m²	120.01ft²
Weights		
Equipped weight	1,725kg	3,803 lb
Fuel load	1,530 litres	337 gals
Loaded weight	3,100kg	6,834 lb
Performance		
Max speed at sea level	790km/h	491mph
at 6,000m (19,685ft)	840km/h	522mph

Ju 268 and He 162A 'Mistel 5' – data

Dimensions			
Wing area	Both	33.15m²	356,85ft²
Weights			
Equipped weight		6,025kg	13,283 lb
Bombload	Ju 268	2,000kg	4,409 lb
Flying weight		13,600kg	29,982 lb
Performance			
Max speed at sea level		780km/h	485mph
at 6,000m (19,685ft)		830km/h	516mph
Cruising speed		800km/h at 11,000m	497mph at 36,090ft
Rate of climb at sea level		16m/sec	3,150ft/min
at 6,000m (19,685ft)		8.5m/sec	1,673ft/min
Range		1,600km	994 miles
Take-off run with RATO		1,400m	4,593ft

Works drawings of the Ju 268 and He 162 composite.
Drawing 1: The unmanned Ju 268 and He 162.
Drawing 2: The manned Ju 268 variant with He 162
Drawing 3: Head-on view of the Ju 268 and He 162.

The Ju 287 Explosives Carrier

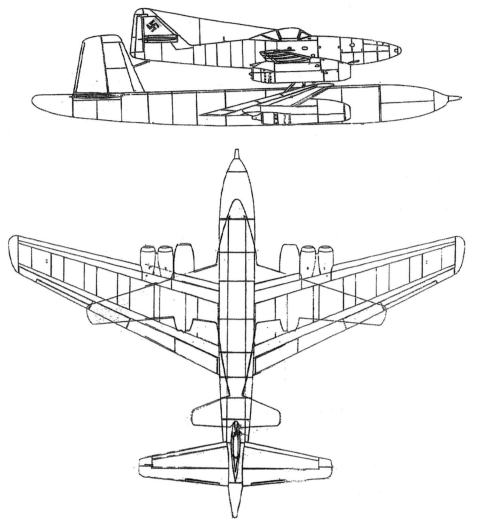

[4] A top plan-view drawing of an Me 262A above a Ju 287, apparently with a 3,500kg (7,716 lb) globular hollow-charge high-explosive warhead detonated by a long 'elephant's trunk' nose impact fuse – of the type experimentally fitted to the Ju 88A-4 and other early 'Mistel' lower components and powered by four (presumably) Jumo 004B turbojets in paired wing nacelles but projecting *behind* the trailing edges of the 20° forward swept wings, appeared in an anonymous article entitled 'Composite Aircraft for Combat' in the Swiss magazine *Interavia* Vol.4, No.7 (July 1949) pp.434-435, but with no details in the text.

A side-view drawing of the Ju 287 + Me 262 composite on its five-wheel take-off trolley and labelled as the 'Mistel 4' was illustrated by author Thomas Hitchcock in *Monogram Close-Up 1: Ju 287* (1974), p.27. Stated to have been conceived in late 1944 as a means of hitting well-fortified prime targets with its hollow-charge warhead, the take-off trolley was equipped with four HWK 501 RATO units to reduce take-off distance by one-third. The side-view drawing in *Close-Up 1* and that illustrated here differs from *Interavia* in the shape of the forward fuselage and warhead nose probe. Presumably due to considerable variances published since the end of the Second World War regarding Ju 287 dimensions, weights and performances, none have been given for this 'Mistel' composite.

Since inception of the Ju 287 bomber project in late 1943, a variety of turbojets had been proposed for it, featuring two-, four-, and six-jet layouts. These included four Jumo 004B, two Jumo 004B plus four BMW 003A, six BMW 003A, six Jumo 004, four HeS 011, two Jumo 012 or two BMW 018 turbojets. Only two prototypes were completed – the Ju 287 V1 (RS+RA) powered by four 900kg (1,984 lb) thrust Jumo 004Bs and first flown on 8th August 1944, and the Ju 287 V2 (RS+RB) powered by two fuselage-mounted Jumo 004Bs and four BMW 003As in paired nacelles beneath the wings. Prior to the arrival of the Russians at Dessau, the nose and cockpit of the fully completed Ju 287 was blown up to prevent it being flown. Photographs of this aircraft in its damaged state have already been published.

Another Junkers explosive-laden missile proposed for use against land and air targets was a reduced-size variant of the Ju 287 bomber, where the similarity lay principally in the wing planform and empennage surfaces. The overall configuration of the missile was the result of the good flying characteristics of the Ju 287. The fuselage, designed as a low-drag body of circular cross-section, besides its large explosive load and incendiary fluid, also housed the fuel to be carried. Power was to have been provided by two or four turbojets in underslung nacelles projecting ahead of the wing leading edges. The control aircraft mounted on pylons above the Ju 287 was an Me 262A. As with the Ju 268, the Ju 287 was to have been detached in a diving attitude and reach its target under its own power via remote control. For take-off, the undercarriageless lower component was to have been mounted on a five-wheel Rheinmetall-Borsig jettisonable trolley.[4]

Purely as a guide, the little-known planned initial production version powered by four 1,300kg (2,866 lb) thrust HeS 011A turbojets had the following data: span 19.4m (63ft 7¾in), length 19.5m (63ft 11¾in), and wing area 58.5m² (629.67ft²). With 10,000kg (22,046 lb) fuel and bombload 4,000kg (8,818 lb), maximum take-off weight was 31,150kg (68,673 lb). Estimated maximum speed was 885km/h at 7km (550mph at 22,965ft) and cruising speed 800km/h (497mph). With bombload, service ceiling was 10,800m (35,435ft) and optimum range 4,400km (2,734 miles). Take-off run with six 1,200kg (2,646 lb) thrust RATO units each of 10 seconds duration was 2,300m (2,515yds) and with 10 such RATO units, was reduced by 37% to 1,450m (1,516yds). Landing speed was 170km/h (106mph).

Messerschmitt Me 262
Twin Composite

Messerschmitt Me 262A-2/U2 + Me 262A-1 'Mistel' composite.

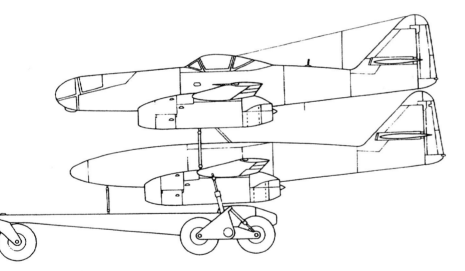

Side-view of the twin Me 262 'Mistel' composite proposal.

One of the most unusual composite projects was the Me 262A-2/U2 mounted above an unmanned Me 262A-1 explosive filled airframe for use against ground and air targets. The upper two-seater component housed the missile controller lying prone in the slightly-lengthened fuselage nose, from where the missile was steered towards its target via the 'Tonne-Seedorf' (Barrel-Seaside Village) television transmitter-receiver system. Project data submitted to the RLM at the end of 1944, however, remained unanswered.[5]

[5] By 28th November 1944, Messerschmitt had drawn up three proposals for the 'Mistel 4' composite. The upper component, armed with two 30mm cannon and carrying 2,133kg (4,702 lb) of fuel, had a loaded weight of 6,985kg (15,399 lb). For the lower component, three alternative explosive loads had been planned.

In one, with 1,494kg (3,294 lb) fuel and explosive load 4,460kg (9,833 lb), loaded weight was 9,917kg (11,650 lb). A second, with the same fuel load but with 6,030kg (13,294 lb) of explosives, loaded weight was 11,650kg (25,684 lb). In the third, carrying the same quantity of fuel but with explosive load reduced to 5,210kg (11,486 lb), loaded weight was 10,125kg (22,322 lb). Together with the manned upper component, total fuel weight was 3,627kg (7,996 lb) of J2, total take-off weight in each case being 16,902kg (37,262 lb), 18,635kg (41,083 lb) and 17,110kg (37,721 lb) respectively. The five-wheel Rheinmetall-Borsig trolley weighed an additional 2 tonnes (4,409 lb). For a more detailed coverage of 'Mistel' developments, see Robert Forsyth and others mentioned in the footnote at the beginning of this chapter.

Works drawing (XII-362) of
the twin Me 262 Mistel
composite on its five-wheel
Rheinmetall-Borsig take-off
trolley accelerated by four
HWK 501 rocket motors.

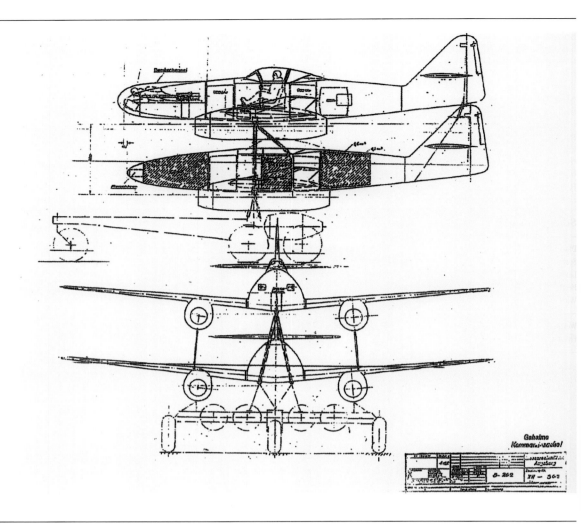

Sombold So 344 'Rammschussjäger'

At the beginning of January 1944, the Inge-
nieurbüro R Bley* of Naumburg/Saale submitted a project study drawn up by Dipl-Ing
Heinz G Sombold for a so-called Rammschussjäger (ram-attack interceptor) to the
RLM. Designated So 344, this piloted concept
was intended to combat approaching bomber
formations with an explosive-laden warhead
fired into them.

The So 344, a mid-wing aircraft of largely
wooden construction had a simple one-piece
wing, conventional tail surfaces and fixed
landing skids. A jettisonable 400kg (882 lb)
explosive load formed the fuselage nose portion, the cockpit located at the rear ahead of
the empennage surfaces, the bomb body
having four stabilising fins. The aircraft was to
have been brought, 'Mistel' fashion, to a suitable altitude above the bombers and
released, whereupon the pilot accelerated
using his solid-fuel rocket propulsion units to

approach the bombers in a dive. After firing
off the nose charge, the explosive load was
detonated by an automatic fuse on reaching
the target.

Using his two forward firing machine-guns
or a heavier-calibre weapon, the pilot could
select further targets before landing as a
glider on its skids. The So 344 was capable of
being dismantled into two sections for trans-
portation on a truck for re-use.[6]

[6] Author Manfred Griehl in: *Jet Planes of the Third
Reich*, pp.149-151, states that the So 344 project,
approved by the RLM in early 1943, ended with the
general mission description of 22nd January 1944.
Powered by a bi-fuel rocket motor, it was to have
been towed to 6,000m (19,685ft) and released, the
aircraft climbing under rocket power to about
1,000m (3,300ft) above the bombers. The SC 500
fragmentation bomb filled with 400kg (882 lb) of
Amatol explosive, released beyond the range of
enemy defensive fire, was propelled by its solid-fuel

booster rocket into the bomber formation where the warhead was deemed sufficient to destroy three or four bombers. Estimated flying time of the piloted aircraft was about 25 minutes.

*The Bley Segelflugzeugbau sailplane construction firm of Naumburg became well known in the 1930s for its design and manufacture of powered gliders such as the Motor-Mücke (one 18hp Koeller M3), its export version the Kormoran, and the Motor-Condor (one 18hp Koeller M3) of 1935. The designation 8-344 was later re-applied to the X4 air-to-air missile.

Sombold So 344 – data

Dimensions

Span	5.70m	18ft 8½in
Length	7.00m	22ft 11½in
Height	2.18m	7ft 1¾in
Wing area	6.00m²	64.58ft²

Weights

Flying weight	1,350kg	2,976 lb

The Sombold So 344 complete and after discharge of the explosive-laden nose portion.

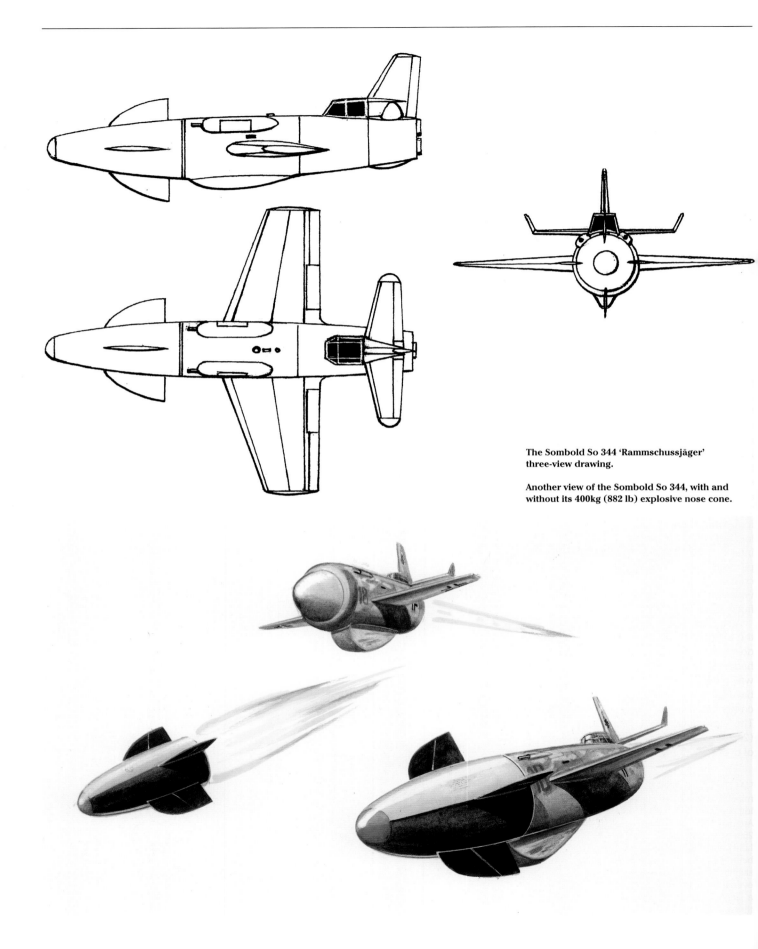

The Sombold So 344 'Rammschussjäger'
three-view drawing.

Another view of the Sombold So 344, with and
without its 400kg (882 lb) explosive nose cone.

Stöckel 'Rammschussjäger'

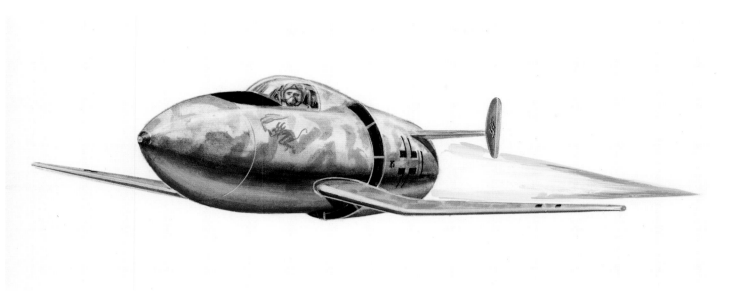

In mid-August 1944, Dipl-Ing J Stöckel, development engineer with Dr-Ing Richard Vogt at Blohm & Voss, proposed his single-seat project for a 'ram-shot interceptor' powered by a mixed ramjet-rocket propulsion unit to the RLM. Designed to be of simple construction, this cranked-wing aircraft of 7m (22ft 11½in) span and 7.2m (23ft 7½in) length had twin fins and rudders and a bomb-shaped fuselage. The mixed propulsion system consisted of four small liquid-propellant rocket motors located within the larger ramjet duct whose intake formed an annular ring aft of the cockpit. For the ramming attack, the fuselage nose contained 200kg (441 lb) of explosive, whereby both pilot and aircraft were sacrificed.

A second variant envisaged the rescue of the pilot, who was to have been catapulted downwards with his seat after the steel nose had been blown off shortly before the target was reached. The 'Ramschussjäger' was to have been brought to altitude as a 'Mistel' combination prior to the start of the attack. The estimated total thrust of 10,000kg (22,046 lb) was calculated to provide a speed of over 1,000km/h (621mph).[7]

[7] As already mentioned, this project was one of five mixed ramjet-rocket proposals drawn up by Dipl-Ing Stöckel, the first of which dated from 20th August 1944. This cranked-wing project dating from 25th August 1944 and termed a 'Raketenjäger mit Luftstrahlantrieb' (rocket fighter with jet drive) had been preceded by an essentially similar aircraft of slightly smaller dimensions. Both were of the same overall layout except that the earlier design featured a single large liquid-propellant 1,500kg (3,307 lb) thrust rocket motor concentric with the ramjet,

instead of the smaller liquid-propellant rockets inside the ramjet diffuser shown in the three-view drawing. In both project variants, the reclining pilot was in an armoured jettisonable cockpit behind the armoured nose cone housing the 200kg (441 lb) of explosive, the self-sealing fuel tank compartment protected by 20mm amour plating. Both were capable of a vertical take-off or else brought to altitude by a parent aircraft.

A three-view drawing of these three Stöckel-designed aircraft is to be found in *'Der Flieger' Luft und Raumfahrt International*, Vol.47, No.1 (January 1968). To a lesser extent, three of the five projects with three-views of two are in Manfred Griehl: *Jet Planes of the Third Reich – The Secret Projects, Vol.l*, Monogram, 1998, pp.120-121 & 123.

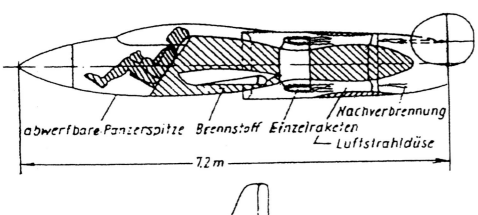

Nachverbrennung

abwerfbare Panzerspitze   Brennstoff   Einzelraketen
└ Luftstrahldüse

7.2 m

1m

7 m

Stöckel 'Rammschussjäger' – data

	Small Rockets/Ramjet	Large Rocket/Ramjet
Dimensions		
Span	7.00m	6.60m
	22ft 11⅛in	21ft 7¾in
Length	7.20m	6.80m
	23ft 7½in	22ft 3¾in
Ramjet diameter	1.00m	1.00m
	3ft 3½in	3ft 3½in
Wing area	10.00m²	9.00m²
	107.64ft²	96.87ft²
Armoured nose	200kg	200kg
	441 lb	441 lb
Weights		
Explosive weight	200kg	–
	441 lb	–
Fuel weight	1,500kg	1,500kg
	3,307 lb	3,307 lb
Bombload	–	200kg
	–	441 lb*
Loaded weight	3,000kg	6,618 lb
	3,000kg	6,618 lb
Landing weight	1,300kg	1,300kg
	2,866 lb	2,866 lb
Performance		
Max speed	1,000km/h	1,000km/h
	621mph	621mph
Initial climb rate,		
VTO, to 10,000m†	200m/sec	200m/sec
	39,370ft/min	39,370ft/min
Ceiling, with VTO	20,000m	20,000m
	65,620ft	65,620ft
Range at 10km		
with ramjet	600km at 720km/h	500km at 720km/h
	373 miles at 447mph	310 miles at 447mph

*1 x 100kg (220 lb) bomb beneath each wing.
† consuming 1,100kg (2,425 lb) of fuel.

Chapter Ten

Airborne Weapons and Special Equipment

The items in this final chapter comprise a variety of aircraft weapons, including oblique upward-firing shells and heavy projectiles, large calibre Bordkanone (aircraft cannon), Bordraketen (rockets), Bordwaffen (weapons), and an experimental airborne flame-thrower, interesting jet propulsion schemes, mobile missile launchers, television guidance systems and night-fighter radars, some of which were used operationally whilst others were still in the testing stage or otherwise proved unsuitable for their intended purpose before the war ended.

Many other weapons and devices existed or were planned that are not mentioned in this book but which by no means detract from the diversity of ideas – some certainly scurrilous – that were under development. Had those that were successful been introduced in numbers at an earlier stage of the conflict, very many surprises could have been in store for Germany's opponents, but as had happened with the advent of their superior jet and rocket-propelled aircraft, were 'too little and too late' to turn the tide of the war. Among the most important innovations of considerable post-war significance for the major powers were the introduction of the cruise missile, liquid-propellant rocket motors, the design and construction of high-speed high-altitude wind tunnels, and swept- and delta-winged technology.

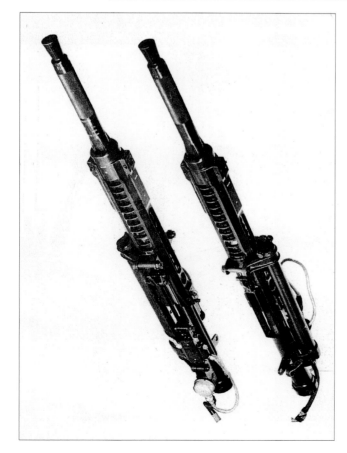

The 20mm MG FF/M with electro-pneumatic triggering. Its rate of fire was 540 rounds per minute.

The Automatic Upward-firing Twin MG FF/M

Commonly known as the 'Schräge Musik' (Jazz Music), this oblique upward-firing weapon developed in 1942 is reported to have been based upon a suggestion by night-fighter pilot Major Rudolf Schoenert and approved by General Kammhuber. Reflected light beams from an enemy aircraft, captured by a caesium photocell in the infra-red 'Zossen A', activated the twin 20mm MG FF/M weapons when flying beneath the enemy aircraft. Jointly developed by the AEG and Carl Zeiss firms, only a few examples of the 'Zossen A2' were used operationally. This infra-red sensor was also considered at one stage for installation in the Me 163B.

Twin MG FF/M – data

Calibre	20mm	0.79 in
Weapon length	1.37m	4ft 6in
Weapon weight	28.0kg	61.73 lb
45-round magazine	16.5kg	36.38 lb
60-round magazine	20.3kg	44.75 lb
100-round magazine	33.1kg	72.97 lb
Rate of fire (average)	540 rds/min	
Muzzle velocity	585-718m/sec	1,919-2,356ft/sec

The oblique (70°) upward-firing weapon system was activated by the AEG/Zeiss-developed 'Zossen A' infra-red sensor (arrowed), seen here on a Junkers Ju 88G.

Perspective view of a twin upward-firing MG FF/M mount in the Messerschmitt Me 110G-4/R8. Key: 1. Twin MG FF/M; 2. main ammunition drum magazines; 3. reserve ammunition drums; 4. compressed-air bottle with pressure reducer and stop valve; 5. spent cartridge containers; 6. electro-pneumatic loading valve; 7. weapon anchorage; 8. weapon inclination adjuster (up to 70° from the horizontal).

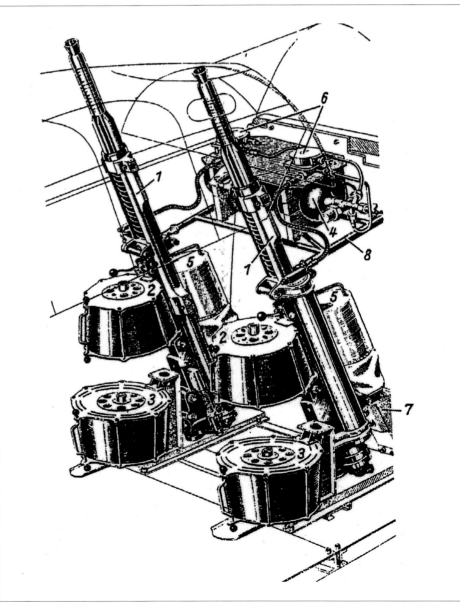

The 3.7cm BK 3.7 Bordkanone

Developed by the Rheinmetall-Borsig AG, this fixed forward-firing cannon, also known as the Flak 18 when used by ground anti-aircraft batteries, was a recoil-loading automatic weapon with a sliding barrel and a central locking mechanism. When installed in an aircraft, the 6-round clip was turned through 90° to the right in a weapons bay. In the Me 110G-2/R1, it was installed in a gondola beneath the fuselage.

Loading of the weapon in flight was performed by the radio operator. Firing followed electro-pneumatically after the pilot activated the safety switch on the SKK safety switch console via the B-button on the KG 13E control column. Target sighting was by the Revi C/12D. Although the BK 3.7 was a very effective weapon against four-engined bombers, it achieved wide success when used against Russian tanks. As mentioned earlier in this volume, its initiator, the Stuka-Ace and 'tank-buster' Hans-Ulrich Rudel, equipped with two of these weapons beneath the wings of his Ju 87D-3 and G-2, was able to destroy over 500 Russian tanks. The BK 3.7 was also fitted to the Hs 129B-2/R3 (12-round clip), the Me 110G-2/R1, R3 and R4 (two 6-round clips), and the Ju 88P-2 and P-3.

BK 3.7 Bordkanone – data

Calibre	37mm	1.46in
Weapon length	3.75m	12ft 3½in
Barrel length	2.106m	6ft 11in
Weapon weight	275kg	606 lb
6-round magazine	12.5kg	27.56 lb

Top right: **The BK 3.7 Bordkanone cannon.**

Centre right: **The BK 3.7 Bordkanone fitted to a Messerschmitt Me 110G-2/R1.**

Bottom left and right: **The BK 3.7 Bordkanone as a Waffenbehälter (weapons container) on the Junkers Ju 87D-3, shown here without the usual aerodynamic streamlined pod surrounding the weapon.**

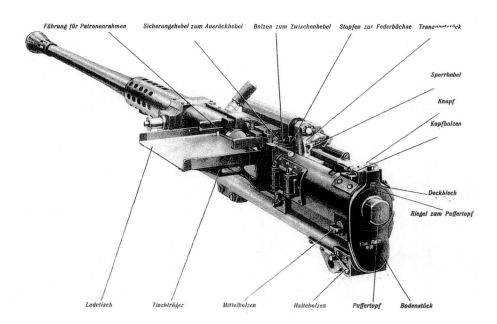

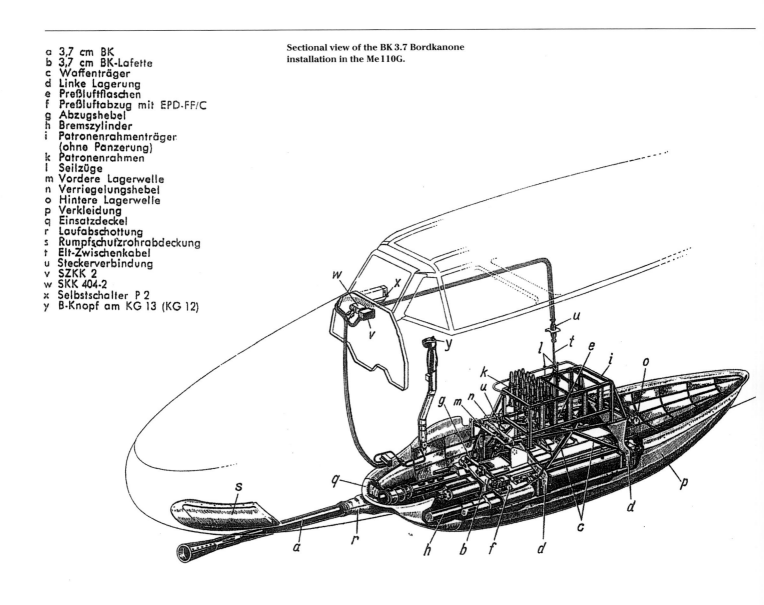

a 3,7 cm BK
b 3,7 cm BK-Lafette
c Waffenträger
d Linke Lagerung
e Preßluftflaschen
f Preßluftabzug mit EPD-FF/C
g Abzugshebel
h Bremszylinder
i Patronenrahmenträger
   (ohne Panzerung)
k Patronenrahmen
l Seilzüge
m Vordere Lagerwelle
n Verriegelungshebel
o Hintere Lagerwelle
p Verkleidung
q Einsatzdeckel
r Laufabschottung
s Rumpfschutzrohrabdeckung
t Elt-Zwischenkabel
u Steckerverbindung
v SZKK 2
w SKK 404-2
x Selbstschalter P 2
y B-Knopf am KG 13 (KG 12)

Sectional view of the BK 3.7 Bordkanone installation in the Me110G.

The 5cm BK 5 Bordkanone

Also developed by Rheinmetall-Borsig in response to the demand for a heavier-calibre weapon for anti-tank use on the Eastern Front in 1943, the BK 5 stemmed from the KWK 39 Kampfwagenkanone (armoured-car or tank cannon). For effective use against enemy bombers, the BK 5 was installed on the Me 410A-2/U4 operated by Zerstörer squadrons.

In order to modify the KWK 39 into an aircraft weapon, the most significant alteration was the provision of an automatic cartridge feed which formed a circular belt at the rear. Since the BK 5 did not have a reciprocating belt to feed the cartridges, a compressed air-

driven rammer was built onto the rear of the gun, the rammer driving a 50mm cartridge out of the feed belt into the gun breech and ramming it into the chamber for firing. Weapon reload was possible during flight by the crew when installed in the Ju 88. On other types of aircraft, a pre-loaded quantity of ammunition was carried: on the Me 410A-2/U4 (with a 22-round magazine) it weighed 650kg (1,433 lb). The BK 5 was also installed in the Ju 88P-4 and for the Me 262, was to have been overcome by use of a 22-round magazine.

Air-firing trials against a ground target in 1944 were highly successful, but plans to equip two other Me 262s were not carried out. For attacking heavily armoured targets, the BK 5 was also planned for installation in the He177A-3, Ju188S and Ju 288 bombers. A single hit with a c.1.6kg (3.5 lb) weight projectile was sufficient to bring down a four-engined heavy bomber. At a greater range, the trajectory of the heavy cannon shell fell off appreciably. After the war, a captured BK 5 was put on display in the USAF Museum in Dayton, Ohio.

BK 5 Bordkanone – data

Calibre	50mm	1.97in
Weapon length	4.342m	14ft 3in
Barrel length	3.04m	9ft 11¾in
Weapon weight	540kg	1,190 lb
Rate of fire	45 rds/min	
Muzzle velocity	920m/sec	3,018ft/sec

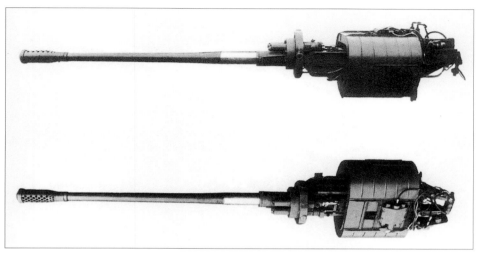

The BK 5 Bordkanone seen from two perspectives.

The BK 5 Bordkanone installed in a Me 410A-2/U4.

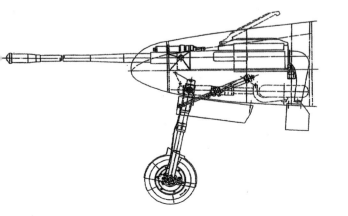
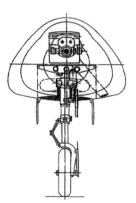

Cross-sectional arrangement of the BK 5 in the Me 410.

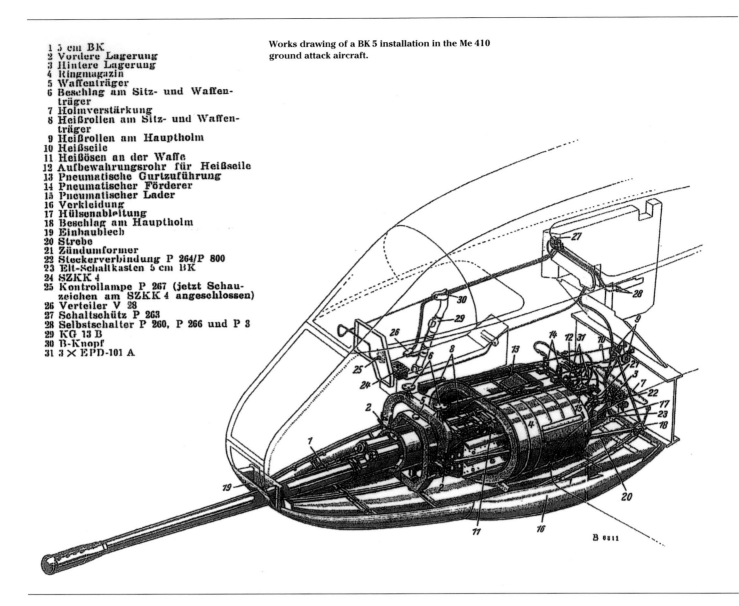

Works drawing of a BK 5 installation in the Me 410 ground attack aircraft.

The 5cm Mauser MK 214A Bordkanone

The Rheinmetall-Borsig BK 5 and the Mauser MK 214 were being developed in parallel. Because of the excessively high 990kg (2,183 lb) installation weight of the 55mm Rheinmetall-Borsig MK114, the two firms were given contracts to produce an improved and lighter weapon. The BK 214A weapon demonstrated by Mauser at the end of 1944 represented largely a further development of the BK 5, but due to its 390 individual parts and various alterations to the locking and loading procedures, became too complex for aircraft installation. The 3,500kg (7,716 lb) recoil force was absorbed by a hydraulic braking system.

The second prototype of the MK 214A was installed in the Me 262A (Werk Nr. 111899), first flown by Karl Baur in late February 1945, and both ground and air firing trials were conducted during March 1945. Another Me 262A (Werk Nr. 170083) was equipped with the third prototype MK 214A, but although both machines were captured by US Forces, it is not known if the latter aircraft was flown with the weapon.

Prior to installation in the Me 262As, the MK 214A had only been test-flown in a Ju 88. Proposals also existed for its installation in the Dornier P 252/3, Heinkel He162A, Messerschmitt Me 262A-1a/U4 and Me 262E-1. As I indicated in the data table, work on an improved fully-automatic MK 214B of 55mm calibre of lower weight and higher rate of fire was under way in April 1944, but had not been tested up to the end of the war.

Mauser MK 214 Bordkanone – data

Calibre	MK 214A	50mm	1.97in
	MK 214B	55mm	2.17in
Weapon length		4.16m	13ft 7¾in
Barrel length		2.825m	9ft 3¼in
Weapon weight	MK 214A	718kg	1,583 lb
	MK 214B	630kg	1,389 lb
Rate of fire	MK 214A	150 rds/min	
	MK 214B	180 rds/min	
Muzzle velocity		930m/sec	3,051ft/sec
Cartridge weight		3.80kg	8.38 lb
Projectile weight		1.54kg	3.40 lb
Belt with 100 rounds		494kg	1,089 lb

Below: **Two views of the 5cm Mauser MK 214A Bordkanone, a further development of the BK 5.**

Below centre: **The MK 214A in the modified Me 262A-1a, where the nosewheel when retracted rotated to lie flat beneath the weapon.**

Bottom left: **The nose-mounted MK 214A in the Me 262A-1a (Werk Nr. 111899).**

Bottom right: **Loading a single round of the MK 214A in the Me 262A-1a.**

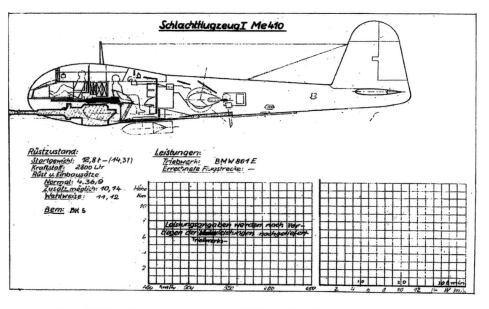

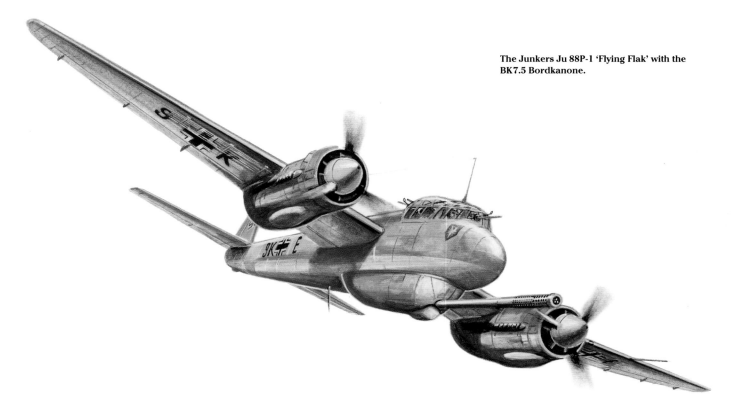

The 7.5cm BK7.5 Bordkanone

From 1942 onwards, German fighter pilots increasingly established that the Allied bombers sported stronger armour protection and increased armament positions. The 13mm MG131 and 20mm MG151/20 weapons were bordering on their limits insofar as the enemy forced the fighter pilots to open fire from long range and remain 'on the ball' until the enemy was destroyed. Fast-firing Bordkanone of heavier calibre and armour-piercing shells were already under development in 1942, but were not yet ready for service use. To overcome this desperate situation, the German weapons experts sought after solutions.

In the summer of 1942, the first trials were undertaken by Junkers with a 7.5cm (2.95in) anti-tank projectile – the Pak 40L equipped with a 10-round magazine, and especially suited for the anti-tank role. The weapon was housed in a ventral gondola beneath a Ju 88A-4 fuselage, and in an emergency, could be completely jettisoned. Production aircraft so equipped bore the RLM designation Ju 88P-1. From its 3.625m (11ft 10¾in) long container, the Pak 40 gun barrel projected 85cm (33⅓in) ahead of the aircraft nose, ending in the over-dimensioned V13 muzzle brake which deflected the gas pressure and together with the sliding barrel, captured the recoil force. The projectile could be fired singly or in series. After each firing, the barrel travelled

back 90cm (35½in), opening the seal and ejecting the spent cartridges, manual loading being undertaken by the crew mechanic. The weapon was centred for a target distance of 650m (711yds) ahead, with sighting by the Revi C12D. Besides Sprenggranate (high-explosive shells), the Pak 40L and the improved BK7.5 fired the Panzergranate 39 (armour-piercing shell) that could pierce 13cm (5in) thick armoured steel at a 90° angle at 1,000m (1,093yds).

Only four Ju 88P-1s were equipped with this weapon as its rate of fire was too low, and in practice was limited to only a few rounds capable of being fired during a single target approach. For day-fighting purposes, the Ju 88P-1 was too slow, especially when it had to escape from enemy fire and was therefore only used in the anti-tank role.

In addition to the Ju 88P-1, the Pak 40L was fitted beneath the nose of the He177A-1/R1. The BK7.5 was installed in some 20 Hs129B-2/R4 aircraft and the Me 410A-2. After only five He177A-3/R5s had been so fitted, the plan to equip further examples was abandoned on account of strong vibrations. The Me 262 'Schnellbomber II' and Ju 388J were also planned to be fitted with this weapon. Because of its limited possibilities of installation and use by the end of 1944, less than 30 examples of the BK7.5 were completed.

BK7.5 Bordkanone – data

Calibre	75mm	2.95in
Weapon length	6.105m	20ft 0¼in
Barrel length	3.92m	12ft 10¼in
Weapon weight	705kg	1,554 lb
Rate of fire	30 rds/min	
Muzzle velocity	790m/sec	2,591ft/sec
Pz Gr 39 cartridge weight	11.9kg	26.23 lb
Pz Gr 39 shell weight	6.8kg	14.99 lb

Pak 40L Bordkanone – data

Calibre	75mm	2.95in
Weapon length	5.78m	18ft 11½in
Weapon width	63.5cm	25in
Weapon height	1.605m	5ft 3¼in
Weight, as A/A gun	1,500kg	3,307 lb

The Ju 88P-1. The BK7.5 multi-orifice muzzle brake is clearly visible.

A Ju 88P-1 damaged on landing due to its nose-heavy BK7.5 cannon.

The Boeing B-17G – a Flying Fortress with nine defensive gun positions which often made it impossible for German fighter pilots to shoot down the bomber with their light ammunition.

A Boeing B-17G Flying Fortress showing heavy damage inflicted by a German fighter's weapons.

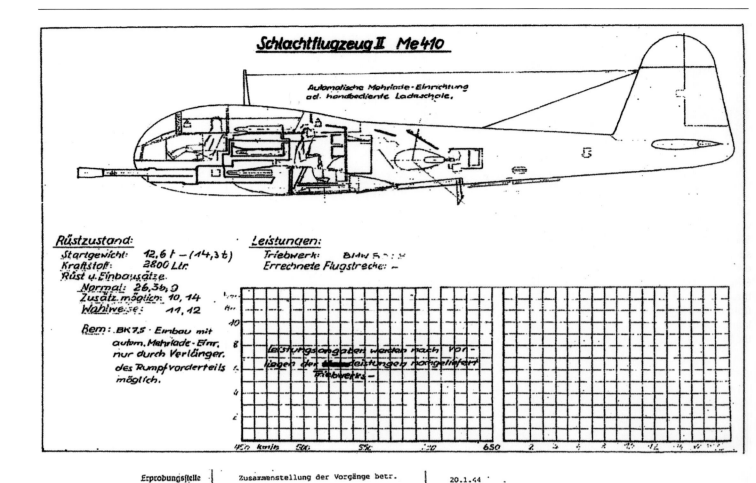

Schlachtflugzeug II Me 410

Automatische Mehrlade-Einrichtung od. handbediente Ladeschale.

Rüstzustand:
Startgewicht: 12,6 t – (14,3 t)
Kraftstoff: 2800 Ltr.
Rüst u. Einbausätze
Normal: 26,36,9
Zusätz. möglich: 10, 14
Wahlweise: 11, 12

Bem: BK 7,5 · Einbau mit
autom. Mehrlade-Einr.
nur durch Verlänger.
des Rumpfvorderteils
möglich.

Leistungen:
Triebwerk: BMW ...
Errechnete Flugstrecke: –

Leistungsangaben werden nach vor-
liegen der ...leistungen nachgeliefert
Triebwerks –

450 km/h   500   550   ...   650

| Erprobungsstelle der Luftwaffe Tarnewitz | Zusammenstellung der Vorgänge betr. Erprobung Ju 88 P-1 mit 7,5 cm BK. | 20.1.44 Blatt |

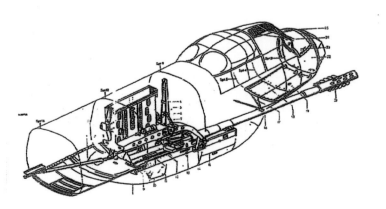

1. Rohrstrebe
2. Beschlag an Rumpfspant 12
3. Stangenmagazin
4. Beschlag an Rumpfspant 9
5. Pressluftflasche
6. Pressluftanschluss
7. Vordere Aufhängung
8. Wartungsklappe und Hülsenauswurf
9. Hintere Aufhängung der Waffe
10. Lafette
11. Lafettenträger
12. Ladeschale
13. Verschluss
14. Wiegenträger
15. Wartungsklappe
16. Rohr der BK 7,5
17. Rohrstütze
18. und 19. Panzerplatten
    nicht belegt
21. Mündungsbremse
22. Panzerplatten an Spant 3
23. Panzerhaube
24. Panzerscheiben
25. Panzerscheiben

Top: **Works illustration of the BK 7.5 installation in the Messerschmitt Me 410A-1.**

Above left and right: **E-Stelle Tarnewitz document dated 20th January 1944 of the BK 7.5 in the Ju 88P-1.**

The R4M 'Orkan' Bordrakete

Development of this 55mm calibre rocket-propelled unguided folding-fin air-to-air rocket projectile was begun by the Deutsche Waffen- und Munitionsfabriken (German Ordnance & Ammunition Factories) in Lübeck-Schlutup upon a development proposal of the Kurt Heber firm in Osterode/Harz under the designation R4M, where R = Rakete (rocket), 4 = its weight in kilograms, and M = Minenkopf (high-explosive warhead). Designed as simply as possible for mass-production, it consisted of a nose impact fuse, an HA 41 warhead (75% Hexogen, 20% aluminium and 5% Montan wax) enclosed in a Nitropenta shell, and a diglycol-dinitrate solid-fuel rocket cylinder 37.4cm (14⅞in) in length and 44mm (1⅜in) in diameter, exhausting through the thrust nozzle 12mm in length that expanded from 10mm to its final 25mm diameter. For launching from beneath the wings of aircraft, the R4M 'Orkan' (Typhoon or Hurricane) was attached to a wooden rack in multiples of 6 or 12 fired at intervals of 0.07 secs. At a distance of 1,000m (1,093yds), its dispersion covered an area 15m (49ft) high by 30m (98ft) wide, the 8 folding fins opening immediately after launch.

Initial trials were undertaken with a Bf110 and an Me 163A, the R4M and EZ 42 gunsight fitted in 1945 to about 60 examples of the Me 262A-1b. Operational use of this weapon brought about a dramatic increase from 1:1 to a 7:1 ratio of kills in the Luftwaffe's favour. On one occasion, Me 262s equipped with R4Ms reportedly destroyed 25 B-17s out of a total of 425 bombers. Enclosed in various designs of honeycomb containers for fuselage nose installation, the R4M was studied for use by the Ba 349, BV P 212, He 162, Hs P 136 and other projects. Of the initial order for 20,000 rockets, some 2,500 examples of the c.12,000 built had been fired off. One post-war development of the R4M was the US 'Mighty Mouse' rocket projectile.

R4M 'Orkan' – data

Calibre	55mm	2.17in
Length	81.2cm	32in
Tailfin span	24.2cm	9½in
Projectile weight	3.85kg	8.48 lb
Explosive weight	520gm	1.15 lb
Diglycol weight	815gm	1.80 lb
Rocket thrust	245kg	540 lb
Duration	0.75 secs	
Max speed	525m/sec	1,722ft/sec
	1,889km/h	1,174mph
Impact velocity at 1,000m	125m/sec	410ft/sec
(3,280ft)	450km/h	280mph
Range	1,500m	1,640yds

Twelve R4M rockets on their launching rack beneath the starboard wing of the Me 262A-1b fighter.

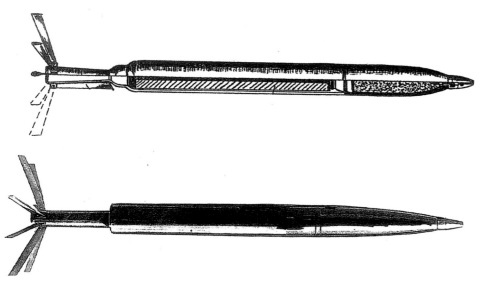

Two views of the R4M with its 8 fins in the folded and extended positions.

The RZ 65 'Föhn' Bordrakete

This solid-propellant rocket-propelled projectile had been initiated and tested as early as 1937 by Rheinmetall-Borsig for the HWA (Army Ordnance Office). With an improved propellant charge, dispersion was reduced from 7m (23ft) at 100m (109yds) distance to a rectangle of 2.6m (8ft 6½in) height and 3.6m (12ft) width at this distance by 1939.

Developed under the leadership of Dr-Ing Heinrich Klein from November 1941, the improved RZ 65A, which stood for Rauchzylinder (smoke cylinder) and 65mm (2½in) body diameter, was succeeded soon afterward by the RZ 65B. Maximum diameter was 73mm (2⅞in) at the warhead and cartridge

attack aircraft) were equipped with this and the essentially similar RZ 73 weapon.

Flight trials were undertaken with individual examples of the Me109F-2, Me110G-4, He111, Ha137, Me 210 V4 and Ju 88A-2 in various wing-mounted and ventral fuselage multi-round arrangements. It was also planned for installation on the Ju 88P-4, Me 210, Ar 240 and the Me163B but were never fitted.

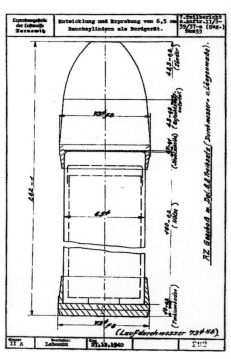

Above: **RZ 65 launchers beneath a Messerschmitt Bf 109F-2.**

Right: **Overall and perspective views of the RZ 65, of 27th December 1940.**

base. Like the R4M, it also had an HA 41 warhead and the WASAG R 61 diglycol-dinitrate rocket charge, the projectile being stabilised in flight through its high rotational speed. By increasing the thrust nozzles from 15 to 20, dispersion was reduced from a 2 x 1.65m (6ft 6¾ in x 5ft 5in) rectangle to only 1.65 x 1.25m (5ft 5in x 4ft 1¼in). Several alterations were necessary during the course of development to turn it into a viable aircraft weapon. Upon detonation, the 130gm (0.29 lb) explosive warhead generated a gas cloud that set alight everything within a radius of 6m (19ft 8¼in). The fragmentation effect can be likened to that of a circular saw which shattered everything into minute particles.

Despite favourable judgements passed on it, the RZ 65 Rüstsatz (equipment set) did not enter series production and only a small number of Zerstörer (heavy fighters), Jäger (fighters) and Schlachtflugzeuge (ground

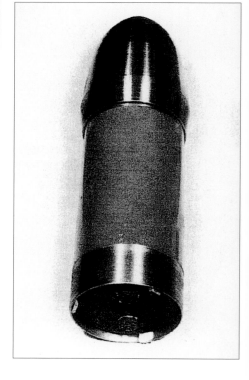

RZ 65 'Föhn' (Warm Wind) – data

Model	RZ 65	RZ 65A	RZ 65B
Calibre	73mm	73mm	73mm
	2.87in	2.87in	2.87in
Length	26.2cm	26.2cm	26.2cm
	10¼in	10¼in	10¼in
Total weight	2.78kg	2.46kg	2.38kg
	6.13 lb	5.42 lb	5.25 lb
Explosive weight	130gm	130gm	130gm
	0.29 lb	0.29 lb	0.29 lb
Propellant weight	685gm	445gm	390gm
	1.51 lb	0.98 lb	0.86 lb
Maximum thrust	300kg	360kg	340kg
	661 lb	794 lb	750 lb
Burning time	0.8 secs	0.4 secs	0.2 secs
Max speed	260m/sec	280m/sec	275m/sec
	853ft/sec	919ft/sec	902ft/sec
Effective range	250-300m	250-300m	250-300m
	273-328yds	273-328yds	273-328yds

Above: **Twelve RZ 65s in two rows beneath an Me 110F-2 fuselage.**

Below left: **The RZ 65 on a Bf 109 with lower panel removed.**

Bottom: **The RZ 65 with 8-round magazine.**

Below: **An RZ 65 pair with 8-round magazines.**

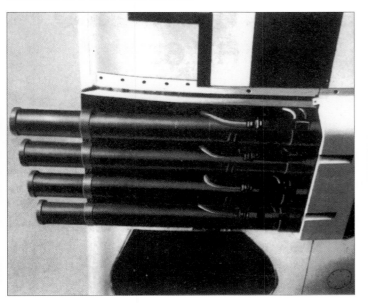

The RZ 100 Bordrakete

One of the oft-cited German artillery weapons of First World War vintage was the 'Dicke Berta' (Big Bertha) mortar shell built by the Krupp firm. With a calibre of 42cm (16.54in) and total weight 922kg (2,035 lb) of which 106.25kg (234 lb) comprised the warhead, it had an initial velocity of 452m/sec (1,483ft/sec) and attained a range of 14,250m (15,580yds). In 1914 it played a special role in the destruction of Belgian fortifications.

The aim behind development of this rocket-propelled shell of the same calibre by Rheinmetall-Borsig in 1941 was to achieve maximum destructive effect from fragmentation of a powerful warhead within a bomber

The airborne 'Dicke Berta' (Big Bertha). The heaviest aircraft weapon ever developed, the RZ 100 is seen here on an experimental launching ramp beneath the dummy Messerschmitt Me 210 fuselage nose.

formation even when accuracy to be imparted was not high. Nicknamed the 'Dicke Berta zur Luft' (Aerial Big Bertha) in association with its forbear, two variants of the RZ100 were developed. The first, 1.474m (4ft 10in) in length, was propelled by a solid-propellant diglycol rocket charge weighing 82kg (181 lb) having 18 discharge nozzles distributed over two concentric rings each of 9 nozzles, inclined at an angle to provide spin stabilisation. In the second, of slightly increased length and diglycol weight, the propelling nozzles were divided for spin stabilisation into a larger ring of 12 angled nozzles and 45 smaller ones inset within the outer ring. This 730kg (1,609 lb) shell – the heaviest Bordwaffe (airborne weapon) ever developed, was planned to be carried by a Messerschmitt Me 210 in a specially-designed launching trough having guide rails. Ground firing tests were conducted by Erprobungskommando

(Test Detachment) EK 25 at the E-Stellen (Test Centres) Rechlin and Tarnewitz. As a result of a decision taken by the RLM Department GL/C-E6, trials were stopped at the end of 1944. Among the reasons given was the inability to steer the weapon via its thrust nozzles.

It was probably also due to the extensive damage caused to the Me 210 ground-test fuselage that would have adversely affected the aircraft's flying capabilities after an air firing. Two other projectiles under development by Rheinmetall-Borsig were the RZ73 (an improved RZ 65) with a propellant thrust of 680kg (1,499 lb) for 0.18 seconds which gave the 3.2kg (7 lb) weight shell a maximum velocity of 360m/sec (1,181ft/sec) and an effective range of 400m (437yds), and the RZ15/8 based on experience with the RZ73. The 25 angled diglycol rocket nozzles provided spin stabilisation for this weapon of 158mm (6.22in) calibre, length 85cm (33½in) and weight 50kg (110 lb), of which 5.8kg (12.8 lb) was the propellant charge. Some examples were tested on a Bf110 but did not pass the experimental stage.

An interesting account of weapons development by the firm is given in the book by Dr-Ing Heinrich Klein: *Vom Geschoß zum Feuerpfeil (From Projectile to Fire-Arrow)*, Kurt Vowinckel Verlag, Neckargemund, 1977.

RZ 100 Bordrakete – data

Calibre	420mm	16.54in
Length	1.65m	5ft 5in
Weight	730kg	1,609 lb
Explosive weight	245kg	540 lb
Propellant weight	85kg	187 lb

The canted rocket exhaust nozzles provided spin stabilisation for the RZ100.

Ground test-firing ramp for the RZ100 in Tarnewitz.

The X7 'Rotkäppchen' Bordrakete

As early as 1941, BMW proposed a guided rocket-powered anti-tank missile. As there was no real need for such a weapon at the time, the suggestion was not pursued. In 1943, Dr Max Kramer at the DVL Berlin-Adlershof commenced work on development of a small winged fighter rocket, the X7, which with a 2.5kg (5.5lb) warhead, was more suited as an anti-tank artillery weapon fired from a dissectable take-off ramp similar to a machine-gun mount.

In 1944, the Luftwaffe became interested in the X7 'Rotkäppchen' (Little Red Riding Hood) as an anti-tank weapon for ground attack aircraft. Launching trials took place with the X7 mounted beneath the wings of a Henschel Hs129 in Tarnewitz at the beginning of 1945. Guidance was by means of the 'target-cover' or dog-leg curve method, commands being transmitted by electrical impulses along the cables spooled out from bobbins at the wingtips.

The X7 'Rotkäppchen' missile.

As can be seen from the three-view drawing, the main body of the X7 had an impact fuse projecting from the missile nose ahead of the hollow-charge warhead, the WASAG rocket motor consisting of two concentric cylinders. The initial high-thrust stage was separated from the second by a Polygan layer – a mixture of asbestos, graphite and calcium silicate. Upon developing a sufficient thrust, the missile was released from its mounting and rotated twice per second around its longitudinal axis to compensate for production inaccuracies. Directional corrections for the rear spoiler to keep it on a collision course by the missile controller with the aid of the tail flare were transmitted over the wire spools to the 0.9kg (1.98lb) control system in the missile, for which various proposals had been made by the FGZ, the RPF and the Askania firms. The hollow-charge warhead was claimed to be able to penetrate all types of tank armour plating, including 20cm (8in) at a 30° angle of impact.

Another variant of the X7 built had a body diameter of 14cm (5½in), overall length 75.8cm (29¾in) and height 30cm (11¾in), the 60cm (23½in) span wings having outward-curved elliptical leading and trailing edges. The main data table represents the status as of September 1944.

In all, some 300 examples were manufactured by the Ruhrstahl AG in Brackwede and the Mechanischewerke in Neubrandenburg. A few firing trials were conducted with the Fw190, but due to stability problems, did not see operational use. For anti-tank use, effective range lay between 1,000-1,500m (1,090-1,640yds) and for air-to-air use, was 2,500-3,000m (2,735-3,280yds). It was still planned in 1945 to use the X7 as a surface-to-air weapon against low-flying aircraft, but the generally chaotic conditions in the last few months prevented any such proposals being realised. To date, no photographs of the X7 appear to have been discovered.

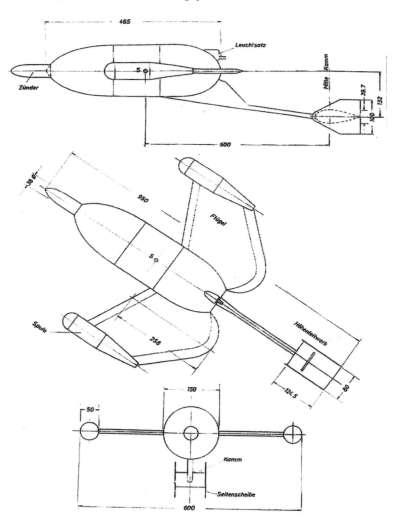

X7 'Rotkäppchen' – data

Powerplant   A two-stage WASAG 109-506 diglycol rocket motor. 1st stage thrust: 68kg (150lb) for 2.5 secs; fuel weight 3.0kg (6.6lb). 2nd stage thrust: 5.5kg (12.1lb) for 8 secs; fuel weight 3.5kg (7.72lb).

Calibre	150mm	5.90in
Wingspan	60cm	23½in
Length, overall	95cm	37½in
Missile weight	9.0kg	19.84lb
Warhead weight	2.5kg	5.51lb
Max speed	100m/sec	328ft/sec
Range	1,200m	1,312yds

The 21cm 'Wurfgranate 42' Bordrakete

The 21cm (8.27in) Wurfgranate 42/Spreng (Mortar Shell 42/high-explosive) counts as one of the most successful rocket projectiles used against air targets from 1943 onwards. One need only recall here the air battle over Schweinfurt on 14th October 1943 when, of 228 Boeing B-17 bombers, no less than 62 were shot down, 17 others crash-landed in England and a further 121 others were so heavily damaged that they also had to be considered as losses.* Exactly how many could be attributed to this 21cm weapon in retrospect, is a matter of conjecture.

The fact is that the 21cm 'Wurfgranate' (usually abbreviated to Wfr.Gr. 42) initially acquired a name for itself as the ammunition for the 21cm Nebelwerfer 42 (smoke-thrower) rocket launcher used by the Army. The five-tube Raketenwerfer (barrage rocket), an example of which is currently on display at the Militärhistorisches Museum in Dresden, fired high-explosive and Flammöl (napalm) projectiles at distances ranging from 500m (547yds) to 7,850m (8,585yds). Altogether, the Maschinenfabrik Donauwörth is reputed to have produced 1,487 examples of the Nebelwerfer 42 as well as 17,000 individual

firing tubes for the Luftwaffe, which mounted the dreaded single-round Wfr.Gr. tubes as auxiliary armament beneath the wings of the Bf109 and Fw190 fighters. Because of its high drag, it was later replaced by a salvo-type drum magazine or rectangular box similar to the Nebelwerfer 42 used by the Army, enabling several rounds to be fired in succession. Designated as the BR-Gerät Drehring (BR rotatable chamber apparatus), this 6-round nose-mounted weapon was experimentally fitted to an Me 410A-2. Ground trials with the weapon took place at the E-Stelle Tarnewitz but due to damage suffered by the airframe, were not concluded at the end of the war.

In April 1944, the weapon was also fitted to the Ju 88 for use against naval targets. Due to the trajectory drop with increasing distance, the firing tubes were aligned at 7° to the fuselage longitudinal axis and centred for a range of 1,400m (1,530yds). Installed as Rüstsätze on the Me109G-6/R2, Me110F-2/R2, Me110G-2/R2 and Fw190A-4/R6.

*According to Allied figures, 186 German fighters were also lost.

Scene from the air battle over Schweinfurt on 14th October 1943: a Boeing B-17 bomber is hit by a Wfr.Gr. 42. The FAG and VKF firms in Schweinfurt produced over 45% of the total ball-bearing production in Germany.

A Heinkel He 177 firing a battery of Wfr.Gr. 42 projectiles from its oblique upward-firing fuselage array.

A Wfr.Gr. 42 single-round tube beneath a Focke-Wulf Fw 190.

Experimental installation of a Wfr.Gr. 42 multi-round magazine as a fuselage nose weapon on a Messerschmitt Me 410A-2.

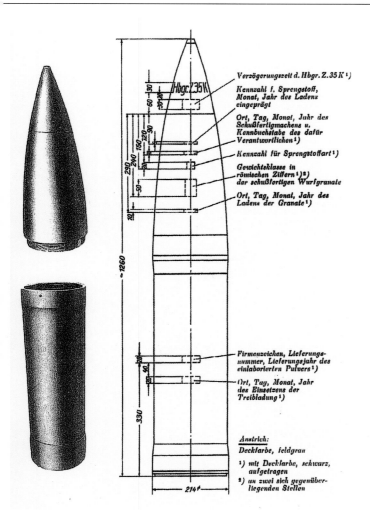

Above: **Descriptive drawing of the 21cm Wfr.Gr. 42.**

21cm Wurfgranate 42/Spreng – data

Powerplant   7-hole diglycol solid-fuel rocket charge with 21 nozzles inclined at 14° producing 1,720kg (3,792 lb) thrust x 2 secs. Max diameter was 14.5cm (5¾in), length 55cm (21¾in) and weight 18.4kg (40.56 lb).

Calibre	214mm	8.43in
Projectile length	1.26m	4ft 1½in
Projectile weight	111kg	245 lb
Rocket motor weight	49kg	108 lb
Explosive weight	38.6kg	85 lb
Max speed	315m/sec	1,033ft/sec
Range	2,200m	2,406yds

Below: **Two views of a 6-round Wfr.Gr. 42 installation shown covered and uncovered, in a Messerschmitt Me 410A-2.**

The Wfr.Gr. 42 as a Special Air Combat Weapon

The success of the Wfr.Gr. 42/Spreng in the air battle over Schweinfurt on 14th October 1943 motivated the OKL (Luftwaffe Supreme Command) to employ the weapon in a further tactical variation, namely, in multi-tube ejectors as oblique upward-firing weapons when flying beneath bomber formations. Accordingly, a Ju 88A-5 was experimentally fitted with a multi-round 'Rohrbatterie' on 19th October 1943. Following the successful conclusion of initial trials, further testing was conducted by the newly-established Erprobungskommando 25 (Test Detachment 25) with multi-round containers in which trajectory measurements and fuse timings were carried out at the E-Stelle Tarnewitz with three He 177s assigned by the General der Jagdflieger. The weapon containers, aligned at 60° forward and upward, were located roughly at the centre of the fuselage. In order to provide space for the 33 tubes, each of 2.3m (7ft 6½in) length and 24cm (9½in) diameter, the forward fuel tanks and the three bomb bays had to be removed. The 15mm (0½in) space between the projectile and firing tube wall served to house the guide rails. The projectiles protruded about 10cm (4in) above and below the fuselage contours to protect the skinning from the hot exhaust gases.

Measurement results showed that the barrage arrangement was eminently suitable for combating large-area targets when fired both singly or in salvoes. On receipt of an order from Luftgaukommando IX (Air Zone Detachment IX), the He 177 and Ju 88 contingent moved on 2nd April 1944 to the E-Stelle Udetfeld, where firing trials in conjunction with the FuG 217 sighting equipment for height determination took place. Available documents do not indicate how the test firings progressed and whether operational use of the obliquely-mounted weapon with live ammunition actually occurred.

The 28/32cm Wurfkörper as a Bordwaffe

Among the most powerful close-combat weapons fitted to ground attack aircraft for use against surface area-targets and fortified bunkers were the 28cm (11in) and 32cm (12.6in) calibre Wurfkörper (high-explosive mortar shells). Stemming from the Army's 21cm (8.27in) Wfr.Gr. 42/Spreng described earlier, they were fired from the 28/32cm Nebelwerfer 41. On the southern sector of the Eastern Front, especially at Kertsch in May 1942 and in the capture of Sevastopol – the world's largest naval fortress – in June 1942, the 28/32cm Nebelwerfer 41 was employed on a massive scale with considerable success. By reason of its enormous effect, it was referred to by the troops as the 'Stuka on Foot'. The Luftwaffe mounted them as single-round units, one beneath each wing, on the Fw 190A-5/U4 ground attack aircraft, firing them in low-level flight.

The 28cm Wurfkörper was activated via a press-button fuse primer and attained a speed of 145m/sec (476ft/sec). Propulsion was from a diglycol solid-fuel charge exhausting from the propelling nozzles at the base of the projectile. Its range lay between 750m (820yds) and 1,225m (1,340yds), the bomb body holding 50kg (110 lb) of TNT. On detonation at impact, it generated pressure waves that destroyed everything within a circle of 2,000m (2,187yds). Overall length was 1.26m (4ft 1½in) and external base diameter 15.6cm (6in).

The 32cm (12.6in) Wurfkörper Flamm (combustible) was an incendiary projectile of equally enormous destructive effect. Its 50 litre (11 gallon) warhead, consisting of an oil mixture and other additives, set fire on impact to an area of over 200m (656ft) radius. An auxiliary explosive load of 1kg (2.2 lb) of TNT in conjunction with the oil mixture had an enormous destructive effect over a wide area, and at the centre of the conflagration, there was a tremendous air suction. The 32cm projectile used the same diglycol rocket propellant and fuel quantity as the 28cm projectile. Approximately 160m (175yds) after launch, the 79kg (174 lb) weight projectile reached a speed of 145m/sec (476ft/sec) and had a range of up to 2,500m (2,734yds). Overall length was 1.3m (4ft 3¼in), maximum diameter (at the surrounding bands) of 33.7cm (13¼in) and base diameter 15.8cm (6¼in).

The 28/32cm Wurfkörper as a Bordwaffe (airborne weapon) beneath the port wing of a Focke-Wulf Fw 190A-5/U4.

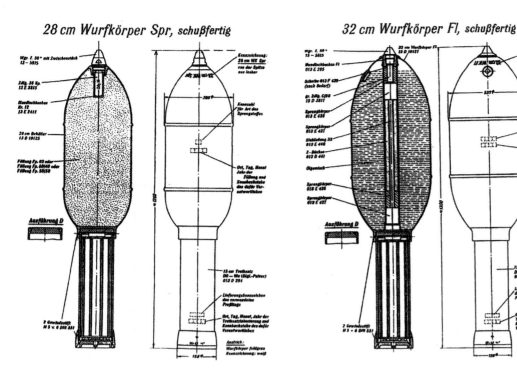

28 cm Wurfkörper Spr, schußfertig

32 cm Wurfkörper Fl, schußfertig

Sectional drawings of the 28cm Wurfkörper/Spreng (left) and the 32cm Wurfkörper/Flamm (right) in their ready-to-fire mode.

The Gerät 104 'Münchhausen' Bordwaffe

Christened 'Münchhausen' after the fabled Baron of that name – usually pictured riding a cannonball, the Rheinmetall-Borsig firm had begun design of the Gerät 104 as early as 1939, intended by the RLM to be carried by an aircraft for use against fortified bunkers and naval targets. The RLM had specified that a shell of 700kg (1,543 lb) weight be capable of penetrating deck armour up to an angle of 60°. The recoil problem of firing a large-calibre heavy shell was overcome by propelling a counterweight of equal mass in the opposite direction simultaneously along the barrel. As shown in the illustration, this recoilless weapon was to have been mounted beneath the Dornier Do 217. Due to the weight of the shell and the counterweight, re-loading in flight was not possible. Initial velocity to the shell was imparted by a black powder charge contained in the counterweight, which was propelled out into the airstream. A ground static test-firing with a Do 217 mounted on a movable trolley was carried out in 1941, but although the weapon functioned as planned, the aircraft suffered heavy damage to the rear fuselage and elevators from the exhaust gases. Improvements in the form of additional protection for the skinning and side openings for the gas exhaust showed no acceptable results.

The RLM had ordered three examples of the Gerät 104 in 1939, intended to be retracted hydraulically into the ventral fuselage of the Do 217 and Ju 288G. In all, 14 rounds were fired, but the installation of an additional 110kg (242 lb) of extra strengthening to the Do 217 airframe was never carried out before the weapon was cancelled in 1941. The recoilless Gerät 104, however, did serve as a useful forerunner for smaller and lighter weapons working on the same principle – the Sondergeräte (special devices) SG113 'Förstersonde', SG116 'Zellendusche', SG117 'Rohrblock', SG118 'Rohrblock' and the SG119 'Rohrbattserie'.

Gerät 104 'Münchhausen' – data

Calibre	355.6mm	14.00in
Length, overall	11.25m	36ft 11in
Ready-to-fire weight	4,837kg	10,663 lb
Projectile weight	700kg	1,543 lb
Counterweight	700kg	1,543 lb
Projectile length	1.2m	3ft 11¼in
with c/w and charge	3m	9ft 10in
Propellant weight	70kg	154 lb
Explosive weight	35kg	77 lb
Still-air range	4,000m	4,374yds
Initial velocity	280m/sec	919ft/sec
In-flight velocity	125m/sec	410ft/sec
Gas pressure	2,200 atmospheres	

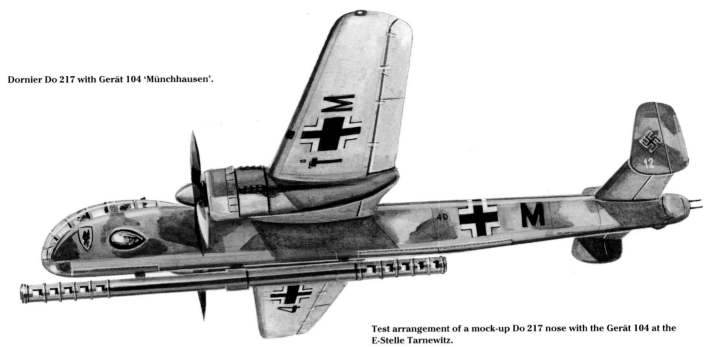

Dornier Do 217 with Gerät 104 'Münchhausen'.

Test arrangement of a mock-up Do 217 nose with the Gerät 104 at the E-Stelle Tarnewitz.

Dornier Do 217 with Gerät 104 'Münchhausen'.

Cross-section of the Gerät 104 with its projectile and counterweight.

Düsenkanone Düka 88
Bordwaffe

Preceded by the Düsenkanone (jet cannon) Düka 75 used by mountain troops and paratroopers, this recoilless weapon lent itself to the development of heavier calibre Bordwaffen (aircraft weapons). The Düka 75, developed by Rheinmetall-Borsig in 1936, had a calibre of 75mm (2.95in) and barrel length 3.067m (10ft 0¾in). Installed beneath a Bf110 fuselage, in static tests it fired a 2.4kg (5.29 lb) warhead in the 6.5kg (14.33 lb) projectile at an initial velocity of 540m/sec (1,771ft/sec) from the 4.85m (15ft 11in) long weapon. Despite 3mm thick additional pro-

tective plating, the Bf110 fuselage suffered considerable damage from the exhaust gases of this 650kg (1,433 lb) weight article. As it was not of the highest priority, development was passed to the Skoda firm in Prague where it was modified to have two 30° angled exhaust gas channels to exhaust above and below the fuselage, and with its 6-round magazine, paved the way for design of the Düka 88 and Düka 280.

The Düka 88, of 88mm (3.46in) calibre, was likewise developed at Rheinmetall-Borsig by a team headed by Dipl-Ing A Kleinschmidt in

1944. The non-automatic 10-round magazine was reloaded in flight by the air gunner, ignition taking place electrically. For test purposes, an experimental installation was made in a Junkers Ju 88A-5 (Werk Nr. 2079) where static firings with 20 rounds were conducted at Unterlüss. In this instance, the exhaust tubes were inclined at 51° to the weapon longitudinal axis, the upper tube being relatively long as it had to pass through the aircraft's fuselage structure. Although firing trials were regarded as satisfactory, the Düka 88 was removed from the weapons development programme. The Düka 88 had also been proposed for use in the Me 262 'Schnellbomber II' project.

Düsenkanone Düka 88 – data

Calibre	88mm	3.46in
Length, overall	4.705m	15ft 5¼in
Barrel length	3.403m	11ft 2in
Ready-to-fire weight	1,000kg	2,205 lb
Rate of fire	10 rounds/min	
Initial velocity	605m/sec	1,985ft/sec
Range (theoretical)	12,000m	13,123yds
Cartridge length	1.245m	4ft 1in
Cartridge weight	18.55kg	40.90 lb
Projectile length	39.6cm	15½in
Propellant charge	4.35kg	9.59 lb
Explosive weight	0.70kg	1.54 lb
Gas pressure	2,000 atmospheres	

Junkers Ju 88 with the Düka 88 Bordwaffe.

Above: **Junkers Ju 88A-4 with Düka 88 in a ventral container.**

Right: **Düka 88 layout and installation works drawings.**

Below: **The Düka 88 10-round magazine as installed in the Ju 88.**

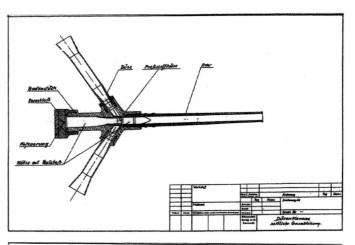

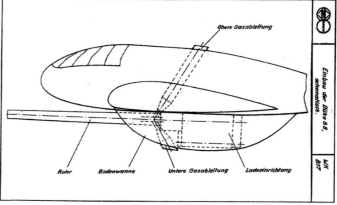

Düsenkanone Düka 280 Bordwaffe

Düka 280-equipped Junkers Ju 288 with its projecting gun barrel and lateral exhaust ports.

Experience gained with the Düka 75, Düka 88 and the Düsenkanone-Marine (Navy) DKM 280 for naval use, led to the development under Dipl-Ing A Kleinschmidt of a heavy, large calibre weapon whose barrel could be extended and retracted telescopically in its housing in the lower fuselage of a Junkers Ju 288 provided with lateral openings in the fuselage sides to lead away the exhaust gases. This 400kg (882 lb) weight 'one-time shot' was designed to penetrate up to 20cm (8in) armoured steel, and when fired in a dive from an altitude of 2,000m (6,560ft) with an initial velocity of 560m/sec (1,857ft/sec), was calculated to have an impact velocity on the target of 530m/sec (1,739ft/sec = 1,186mph) after a 3.7 second trajectory. Corresponding figures for its release in horizontal flight and from 4,000m (13,120ft) altitude, and initial and impact velocities are shown in the accompanying table. The resulting recoil was to have been considerably reduced by the

diverted gases and a pneumatic brake. A prototype Düka 280 installation, however, did not take place due to the calculated high stresses imposed on the airframe, not to mention the need for a fuselage length of 8.5m (27ft 10½in) required for its installation. Theoretical studies by the Rheinmetall-Borsig WKB-Gruppe (Ballistics, Ammunition & Gun Design) headed by Dipl-Ing Kleinschmidt had calculated the physical requirements for weapons with calibres ranging from 210mm (8.27in) to 305mm (12in) and corresponding projectile weights from 105kg (231 lb) to 700kg (1,543 lb). For a calibre of 300mm (11.81in) and projectile weight 425kg (937 lb), the weapon required a barrel length of 6.1m (20ft), a tube weight of 4,600kg (10,414 lb) and an impossibly high length of 14.7m (48ft 2¾in) to absorb and deflect the recoil in an aircraft fuselage, so that such weapons were understandably not pursued by the RLM for aircraft installation.

Düka 280 installation in the Ju 288 and its penetration depth on a ship, compared with that of a normal bomb (Fliegerbombe vs Granate). Whilst a normal bomb only penetrated the first 30mm armoured deck, the Düka 280 penetrated the next two levels of 60mm and 150mm armour-plated decks to explode beneath the latter.

Sonderausführung Ju 288 (mit Kanone)

Vergleich zwischen Fliegerbombe u. Granate von gleichem Gewicht

		Schuß aus Sturzfl.		Abwurf aus Sturzfl.	Abwurf aus Horizontalfl.
Geschoßgewicht	kg	400	400	400	400
Geschoßdurchmesser	mm	280	280	430	430
Abwurfhöhe	m	2000	4000	2000	4000
Anfangsgeschw.	m/sec	560	560	240	0
Auftreffgeschw.	m/sec	530	500	240	230
Fallzeit	sec	3,7	7,5	8,4	31
durchschlagene Panzerstärke	mm	200	200	32	30

GERO FmW-51 Flame-thrower

Between 1940 and 1941 the Luftwaffe conducted trials with a Flammölwerfer (flame-thrower) for use as a defensive Bordwaffe (aircraft weapon) that had previously been successfully employed by the Army and parachute troops against fortifications. Flight trials undertaken at the E-Stelle Tarnewitz in co-operation with the firm of DWM, Berlin-Borsigwalde, were stopped by mid-1941 as the Flammenwerfer (flame-thrower) or Flammölwerfer (inflammable oil thrower) turned out to be unsuitable for its intended purpose.

The twin flame tubes were primarily intended as a defensive weapon to combat enemy aircraft attacking from the rear. The tubes, positioned beneath the fin and rudder of the Junkers Ju 88A-4, led to supply tanks inside the fuselage. Each tank held 120 litres (26.4 gals) of 'Flammöl 19' – a mixture of light and heavy Teeröl (coal-tar oil) that was led out of the tanks under 21 atmospheres nitrogen pressure. Whether ignition of the Flammöl took place by means of gaseous hydrogen or an electromagnetic igniter, is not known. The length of the jet flame was only 40m (131ft).

Top: **The GERO FmW-51 flame-thrower at the rear of a Junkers Ju 88.**

Below: **The GERO FmW-51 flame-thrower in action. A Heinkel He 111H-16 ejecting the burning oil over the E-Stelle Rechlin.**

V-1 Mobile Launching Ramp
for Aircraft and Missiles

The Rheinmetall-Borsig firm in 1942 produced an experimental mobile accelerator for launching the pulsejet-powered V-1 flying-bomb. The missile was mounted on a rocket sled equipped with four 1,200kg (2,646 lb) thrust Schmidding 109-533 solid-propellant rocket motors that launched it at a 6° angle of inclination. The Startwagen (take-off trolley), running along three slits, was braked at the top of the ramp by an automatic brake, releasing the missile whose speed was 400km/h (242mph) at that point.

The mobile Startrampe (take-off ramp) was in experimental use by various Army and Luftwaffe E-Stellen in mid-1943, after which time further launching trials with it were broken off. In its stead, the Walter-developed rocket piston accelerator became the standard take-off launching apparatus.

Besides the Rheinmetall-Borsig take-off ramp, other similar ramps were manufactured by the Maschinenfabrik in Esslingen,

A mobile firing ramp for launching jet-propelled aircraft and missiles, in this instance a US-built copy of the V-1 flying-bomb, accelerated by a battery of four solid-fuel rocket motors.

Fries und Sohn in Frankfurt am Main, and the Meiller Fahrtzeugwerke (Vehicle Works) in Munich which also developed the transport trailer and take-off gantry for the V-2 rocket. The mobile take-off ramps proposed by the Heinkel and Junkers firms received little recognition from the RLM. For the Heinkel P 1077 'Romeo' interceptor projects and the Junkers EF 126 'Elli' low-level attack aircraft described earlier on, take-off ramps of the above type were required as the aircraft were each powered by the Argus 109-014 or 109-044 pulsejets which required considerable ram pressure to be able to take off with these propulsion units. Once started by a supply of compressed air, the pulsejet continued to operate at full thrust. The aircraft or missile, however, needed to be accelerated to an adequate climbing speed.

The Rheinmetall-Borsig take-off ramp and the abovementioned developments became the booty of the Americans and Russians who

recognised the high strategic value of these mobile take-off vehicles and with further development, included them in their strategic concepts.[1] Concerning the hydraulics and other details, no documentation is available.

[1] Sufficient information and actual parts of the V-1 flying-bomb had come into the hands of the British and Americans by June 1944 from specimens that had crashed in Sweden and Poland to be able to reproduce the missile and its pulsejet. Whereas the British were primarily interested to combat the flying-bomb menace and destroy its launching sites, the Americans were keen to build it for eventual use against Japan.

On 12th July 1944, 2,500 salvaged V-1 parts left England, arriving next day at Wright-Patterson AFB in the USA, and within three weeks the first US-built V-1 (known under the USAAF designation JB-2 Thunderbug and US Navy designation KUW-1 Loon) was completed. Before the end of July 1944, orders had been placed for mass-production of the Argus As 014 pulsejet (built by the Ford Motor Company as the PJ-31) and the V-1 airframe (built by the Republic Aviation Company, later sub-contracted to Willys-Overland). By the end of the war in the Pacific, of the tens of thousands ordered and planned, 2,401 pulsejets had been manufactured by Ford and 1,385 JB-2 airframes had been delivered to the US War Department.

Whereas the US-built flying-bomb differed in minor external details from the German original, the guidance and launching procedures were different. Because neither the actual parts, technology, nor the T-Stoff (hydrogen peroxide) and Z-Stoff (calcium permanganate) for the Walter rocket launching piston and ramp were available, alternative means of launching the missile had to be considered. At first, a 400ft long 6° inclined ramp was used, followed by a rocket-propelled level railtrack, and finally, a 50ft ramp mounted above a multi-wheeled trailer. The ramp length was progressively reduced to twice the length of the flying-bomb fuselage, the missile perched on top of a sled driven by four solid-propellant booster rockets. The sled was not braked at the top of the ramp but flew attached to the flying-bomb before being released during the climb.

The US-built launching ramps had already been considered before the end of the Second World War and were thus independently developed from the German-designed ramps.

V-2 Transport Trailer and Launch Platform

Bearing the designation MA-3, the Meiller Fahrzeugwerke (Vehicle Works) in Munich in 1943 developed a transport trailer and launching platform for the HWA Peenemünde A4 ballistic missile – better known as the V-2.

The tri-axle eight-wheeled transporter had an overall length of 14.4m (47ft 3in) with towing axle but without the V-2, maximum width 2.8m (9ft 2¼in), forward wheel track 1.6m (5ft 3in), rear wheeltrack (rotatable pairs) 2m (6ft 6¾in), wheelbase (to axis of rotation) 9m (29ft 6¼in) and vertical height (to V-2 clasps) 2.8m (9ft 2¼in). With the V-2 rocket in place, overall height was 4m (13ft 1½in). Turning radius varied from an outermost 16.7m (54ft 9½in) to the innermost of 5.8m (19ft 0¼in), the forward wheel fork being able to rotate 40° on either side of the trailer axis. Ground clearance was 35cm (13¾in). Other dimensions are as visible in the drawing.

Left: **The erected Starttisch (take-off table) with the hydraulic components and upper fold-out accessory platform.**

Top: **Launch preparations for the V-2 standing on its take-off rig surrounded by propellant bowsers.**

Above: **The Meiller MA-3 transporter with a partly-covered V-2. Beside the German-language wall instruction that 'Entrance to the assembly hall is only permitted to authorised personnel,' the English-language 'No admittance except on duty,' etc, indicates that this V-2 must have been an Allied captured example.**

Junkers Aircraft Steam Turbine Project

Beginning in 1940, the Junkers Motoren-werke engineers were engaged on the development of steam turbines suitable for installation in large aircraft. The work was led by Dipl-Ing von Schlippe who worked on a 3,000hp steam turbine which ran on the engine test-bed in mid-1941. Besides the Junkers aero-engine division, Professor Lösel and Dipl-Ing Paucker of the Technische Hochschule, Vienna, developed a dimensionally-small 4,000hp steam turbine for aircraft use.

In a document dated 12th April 1941, Dr Adolf Baeumker, head of the RLM Research Dept., proposed that they undertake development of a steam turbine under the highest 'Sonderstufe SS' priority in order to accelerate the work. In this, Dr Baeumker referred to a discussion with Dipl-Ing Schmedemann of the Messerschmitt firm, who envisaged steam turbine installation in the large-capacity Me 321 glider. An important factor for the development of such powerplants was the requirement for high-performance long-range aircraft intended for use in the North Atlantic for submarine supply and reconnaissance duties. The advantage of the steam turbine lay in its high efficiency and higher power to weight ratio over existing reciprocating engines. Calculated operating life of the steam turbine lay between 4,000 and 6,000 hours as against only 500 hours for the airscrew engine. The question of fuel for the steam turbine was also much simpler, since it could operate on heavy inflammable fuels such as tar-oil and coal dust as opposed to high-octane fuels needed for the orthodox internal-combustion engine.

In an RLM decision of 21st August 1942, development work on steam turbines was terminated in favour of the turbojet as its power to weight ratio was far more favourable than the steam turbine.[2]

[2] Not so for the early BMW 003 and Jumo 004 turbojets whose thrust to weight ratio was just over unity. Even the higher thrust BMW 018 and Jumo 012 did not have a thrust/weight ratio greater than 2.5:1.

Not mentioned by the author is that interest in high-power steam turbines for aircraft was revived in 1944. In August 1944, the Osermaschinen GmbH founded by Prof Lösel was commissioned to design and manufacture a steam turbine intended for installation and flight-testing in a Me 264. The design called for 6,000hp at 6,000rpm, with a weight to power ratio of 0.7kg (1.54 lb) per hp and an sfc of 190gm (0.42 lb) per hp/hr. Two types of airscrew drive were proposed: one of 5.3m (17ft 4¼in) diameter rotating at 400-500rpm, and a smaller one of only 2m (6ft 6¾in) diameter revolving at 6,000rpm. One scheme was designed to use 65% pulverised coal and 35% petrol, but it was intended to use all-liquid fuel when sufficient quantities became available. A start had been made with auxiliary and main turbine assembly, and one boiler of the total of four had been completely manufactured. At the end of the war, many other components had been completed and ready for use including the turbine blades, combustion and air-draught fan, and condensate pump. The Me 264 intended for it, however, was destroyed in an air raid on the Messerschmitt works. Prof Lösel was captured by the Russians in 1945 and taken to Moscow for interrogation.

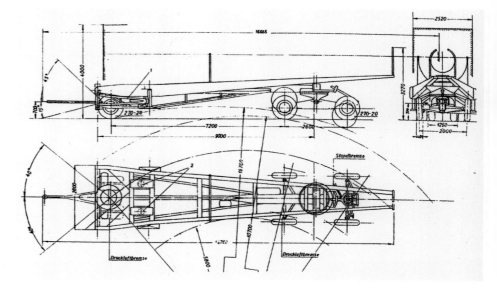

A Junkers steam turbine design proposal for 3,000 shaft hp. Drawing notations: Power: 3,000hp; turbine speed 8,000rpm; propeller speed 950rpm; turbine pressure 100atm; temperature 550°C; exhaust pressure 0.15atm; weight 800kg (1,764 lb). Power to weight ratio was therefore 3.75:1.

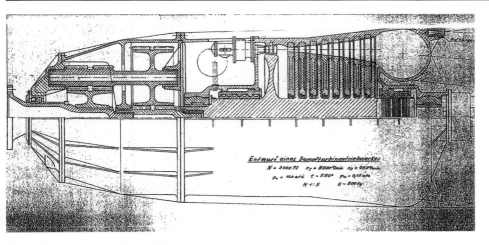

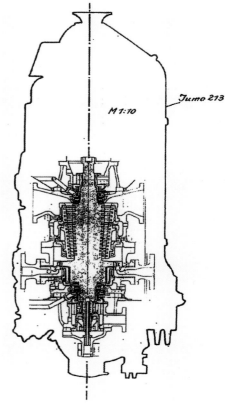

Above: **Size comparison of the 4,000hp steam turbine developed by Prof Lösel and Dipl-Ing Pauker with the Junkers-Jumo 213. Its overall length is only 54% and maximum height only 68% of the Jumo 213.**

Right: **The large-capacity Messerschmitt Me 321 glider for which the steam turbine installation was proposed.**

Below: **Schematic arrangement drawing dated 8th November 1940 of the 3,000hp steam turbine proposal for wing installation.**

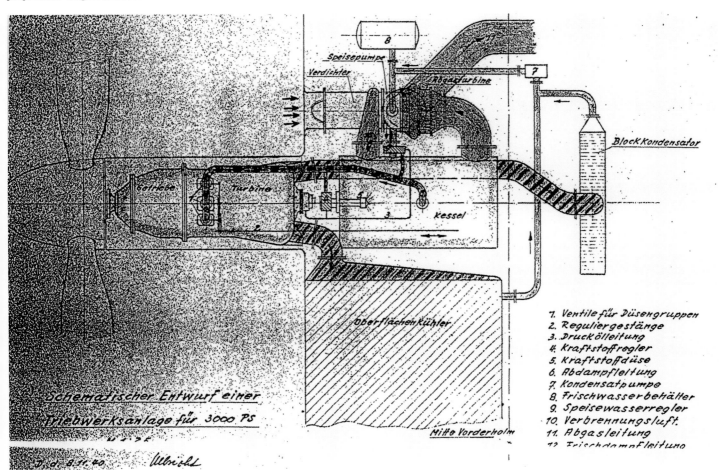

1. Ventile für Düsengruppen
2. Reguliergestänge
3. Druckö/leitung
4. Kraftstoffregler
5. Kraftstoffdüse
6. Abdampfleitung
7. Kondensatpumpe
8. Frischwasserbehälter
9. Speisewasserregler
10. Verbrennungsluft
11. Abgasleitung
12. Frischdampfleitung

The 'Tonne-Seedorf' TV Guidance System

During the last months of the war, the development of remote control systems had made considerable progress, manifested in high performance and small dimensions. Between 1943 and 1945, the Fernseh GmbH (Television Co Ltd) of Darmstadt had developed the 'Tonne' (Cask or Barrel) TV transmitter and 'Seedorf' (Sea- or Lakeside Village) TV receiver with an operating range of 150km (93 miles) for installation in explosive-laden guided missiles. The TV camera and transmitter worked on a frequency of 400 MHz (megacycles), a scan of 441 lines at 25 pictures/second and a power of 20 Watts, transmitting the target image over a Yagi antenna to the controlling aircraft. The 'Tonne' equipment was housed in the nose of the missile with the Yagi antenna attached to the fuselage tail. In addition to the standard FuG 203 'Kehl'/FuG 230 'Straßburg' radio-command equipment, the target image was viewed in the controlling aircraft on the 'Seedorf' TV screen measuring 8 x 9cm (3¼ x 3½in), enabling the appropriate course-correction commands to be given to the missile.

As well as the 'Tonne' and 'Seedorf', the Fernseh GmbH had also developed the 'Adler' (Eagle) and 'Sprotte' (Sprat) TV cameras – to mention but a few, that enabled missiles to be guided to the target with greater precision. The high standard of German technology at that time was confirmed by further development and manufacture of these German devices by the former opponents after the end of the war.[3]

The 'Tonne' television camera and transmitter which captured an image of the target ahead of the guided missile.

The 'Seedorf' television receiver displayed the target picture transmitted from the 'Tonne' to the missile guidance controller who gave the appropriate radio commands to the missile via his joystick.

[3] Not mentioned in the author's narrative is that the 'Tonne-Seedorf' equipment was flight-tested exclusively in the Henschel Hs 293D missile (shown in side profile in *Luftwaffe Secret Projects: Strategic Bombers 1935-1945*, p.137). This consisted of a standard Hs 293A-1 where the 'Tonne' TV equipment was installed in the fuselage nose ahead of the warhead, behind which was an additional fuselage collar, with the horizontal Yagi antenna at the rear of the missile just beneath the tailplane and elevators. TV guidance thus allowed the controller to guide the missile without needing to actually see the target from the aircraft. The pilot was hence able to keep at a longer distance and take evasive action or hide in cloud, the target image

increasing in size on the TV screen the closer the missile approached its target.

Whereas the German *commercial TV* broadcasting system prior to the Second World War was based on a scan of 441 lines and 25 pictures per second, the 'Tonne' TV camera developed by the Fernseh GmbH in co-operation with the German Post office, operated on *224* lines at a frequency of *50* pictures per second because of the fast-moving missile. Since the (ship) target presented its main dimension horizontally, after some tests the lines were posed vertically for better resolution. The optical lens was the Zeiss 'Biogon' of focal length 35mm with an image angle of +/-13°. The 'Tonne' chassis measured 17 x 17 x 40cm (6¾ x 6¾ x 15¾in) and only two types of valve were used. The camera was enclosed in a frame of cast magnesium alloy, replacing the trim weight formerly installed and had an anti-dim glass, battery-heated to avoid condensation and icing. Missile length at the nose was increased by 45cm (17¾in). To accommodate the DEAG battery, Osmig converter and TV transmitter, the centre section was lengthened by 23cm (9in). The Yagi antenna was positioned where the rear guide flare was normally located.

The 'Seedorf' cathode-ray TV receiver tube had a diagonal of 13cm (5in), the chassis measuring 17 x 22 x 40cm (6¾ x 8¾ x 15¾in). The whole conversion added about 130-150kg (287-331 lb) to the missile weight. The Blaupunkt firm was given a contract to build 1,000 sets but this was later cancelled. During the missile guidance phase, only three picture adjustments were necessary: to the screen brightness, contrast, and picture phase. For the 70 flight trials, two He 111s had been converted, followed later by a Do 217. Prior to actual airborne tests, the missile controller received training on a ground simulator developed by the DFS Ainring.

In addition to the above equipment, in co-operation with the Telefunken firm and the Reichspost Forschungsanstalt research institute, the Fernseh GmbH had developed the 'Sprotte' TV camera which contained a miniature iconoscope with a reduced number of lines and smaller dimensions, intended for installation in anti-aircraft rockets, but did not reach the flight-test stage. Another development by the Fernseh GmbH was the FB 50 of even less weight. It used a 50-line scan with a picture frequency of 25 per second, but likewise did not reach the testing stage. One other TV camera was the 'Falke' (Falcon) designed by the Loewe-Opta company of Berlin-Steglitz. Simple in design, it used a spiral scanning system but was given up when the 'Tonne' was adopted instead. The Gollnow & Sohn firm of Stettin had also developed a missile bomb which was to guide itself onto a ship target with its built-in TV camera which utilised spiral scanning. A prototype TV set with 100-200 spiral lines was developed for Gollnow & Sohn by the Fernseh GmbH. The firm itself had built another camera with improved characteristics and tested it with satisfactory results in 1942 in a sea-and ship-model test, but this was also dropped in favour of the 'Tonne'.

For a detailed account of all these developments, see Dipl-Ing Fritz Münster: *A Guidance System Using Television, AGARDograph 20*, Verlag E Appelhans & Co, Brunswick, 1957, pp.135-161.

The British H2S Panoramic Radar

From the wreckage of a downed British bomber, German troops in February 1943 captured the damaged British H2S radar equipment working on the 9cm (3½in) wavelength which displayed an electronic image of the land and sea features beneath it in flight. Named the 'Rotterdam' after the location where it was captured, despite extensive damage it was subsequently brought to functioning order in the Telefunken AG laboratories as a model for the manufacture of German centimetre-wave equipment based on this find. Up to that point, the German military leadership had consistently advocated the necessity to use cm-wave technology, but those responsible in the RLM had rejected these demands since authoritative personalities in the economics and technology sector represented the viewpoint that centimetric-wave systems were of no significance as in their character, bore too great a similarity with light waves and upon impact with the target, produced scattered reflections. Comprehensive laboratory tests with the captured 'Rotterdam', however, showed that the advantages of cm-wave technology were much greater than envisaged. For this reason, the head of radar development at that time, Dipl-Ing Brandt, proposed that all German radar sets be converted to centimetre wavelengths.

The first German unit working on the centimetre wavelength was developed by Telefunken in 1943 as the FuG 240 'Berlin A' and experimentally installed in a Junkers Ju 88G-6 night-fighter. Despite its purely experimental nature, the 3,300 MHz device was smaller and lighter than the captured British H2S and showed a noticeable improvement in performance. The parabolic reflector dish, similar to that used in the H2S, had a diameter of 70cm (27½in) enclosed beneath a wooden casing on the Ju 88G-6. The common dipole for the sender and receiver was rotated by a means of a motor. Connected to the motor drive was the adjustment switch for the antenna dish, which transmitted the programmed values to the receiver exit that channelled the signals to the deflection plate of the indicator tube of the sighting device. The antenna dish reflector thus gave the same picture on the approach of the aircraft to its target as with the FuG 220 'Lichtenstein SN-2'. The successor model of the 'Berlin A' experimentally introduced into operational service bore the designation 'Berlin N 1a' and was successfully employed for the first time in March 1945 with NJG 1 based in Gütersloh. Its target-recognition range was 4-5km (2.5-3.1 miles) with a near-resolution of 350m (1,150ft). Out of the 350 sets ordered, 25 examples had been delivered by March 1945, of which ten came to be installed in night-fighter aircraft. The 'Berlin N 1a' was succeeded by the FuG 244 'Bremen O' night-fighter radar working on the 9cm wavelength at 3,300 megacycles at a power of 10 kilowatts. The search range was 5km (3.1 miles), with near resolution improved down to 200m (656ft). The parabolic reflector, the rotating system and the indicator screens were identical to the 'Berlin N 1a', the armament activated by the EG3 'Elfe 3'. The entire 'Bremen O' equipment was smaller and lighter than the initial 'Bremen' model. Only one 'Bremen O' had been delivered and was undergoing trials at the E-Stelle Diepensee at the end of the war. The codename 'Elfe' (elf) applied to the weapon activator which, on receipt of a target echo, automatically fired the aircraft's armament.

Radar image and plan of the Lower Elbe wharves by Hamburg as seen by the British H2S radar.

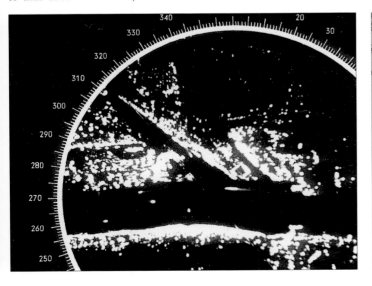
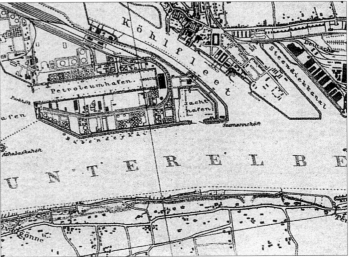

The British H2S centimetre-wave panoramic radar equipment.

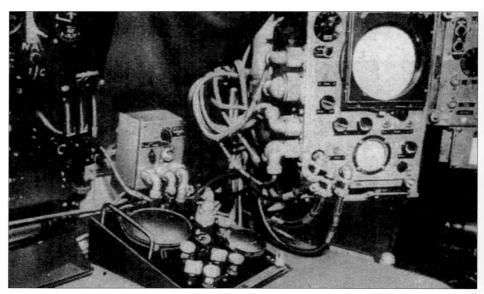

Partial view of the directional, sighting, and control equipment of the German FuG 240 'Berlin 1a' radar.

Control console with twin screens for the FuG 240 'Berlin 1a' radar.

Top: (Parabolic reflector and dipole of the FuG 240 'Berlin 1a' radar.

Above: A captured Junkers Ju 88G-6, Werk Nr.622811 (Air Min 48) with its FuG 240 'Berlin 1a' beneath the wooden nosecone. In the background is a captured Me 262A to the left of the de Havilland Mosquito at the RAE Farnborough.

INDEX

Index of Personalities

We hope you enjoyed this book…

Midland Publishing book titles are carefully edited and designed by an experienced and enthusiastic team of specialists. We always welcome ideas from authors or readers for books they would like to see published.

In addition, our associate company, Midland Counties Publications, offers an exceptionally wide range of aviation, military, naval and transport books and videos for sale by mail-order worldwide.

To order further copies of this book, and any of many other Midland Publishing titles, or to request a copy of the appropriate mail-order catalogue, write, telephone, fax, or e-mail to:

Midland Counties Publications
4 Watling Drive, Hinckley,
Leics, LE10 3EY, England
Tel: 01455 254 450 Fax: 01455 233 737
e-mail: midlandbooks@compuserve.com
www.midlandcountiessuperstore.com

LUFTWAFFE SECRET PROJECTS: Fighters 1939-1945
by Walter Schick & Ingolf Meyer

Germany's incredible fighter projects of 1939-45 are revealed in-depth – showing for the first time the technical dominance that their designers could have achieved – shapes and concepts that do not look out of place in the 21st century.

With access to much previously unpublished information the authors bring to life futuristic shapes that might have terrorised the Allies had the war gone beyond 1945. Full colour action illustrations in contemporary unit markings and performance data tables show vividly what might have been achieved. Careful comparison with later Allied and Soviet aircraft show the legacy handed on, right up to today's stealth aircraft.

Hardback, 282 x 213 mm, 176 pages, 95 full colour artworks, over 160 diagrams and over 30 photographs. 1 85780 052 4 £29.95

LUFTWAFFE SECRET PROJECTS: Strategic Bombers 1935-1945
By Dieter Herwig and Heinz Rode

Reichmarshal Hermann Goering was so convinced in 1940 that his Luftwaffe was invincible he decreed that only refinements of existing types need be considered. The Battle of Britain and the flawed invasion of the Soviet Union changed all that. From mid-1941 some of the finest minds in aerodynamics turned their thoughts to a new generation of aircraft.

In this companion to the enormously popular volume on fighters, Germany's incredible strategic bomber projects 1935-1945 are revealed showing the technical dominance that their famed designers could have achieved if time had allowed, including even transatlantic jets and shapes and concepts that do not look out of place in the 21st century. The authors bring to life futuristic shapes that might have terrorised the Allies had the war gone beyond 1945. Full colour action illustrations in contemporary unit markings show vividly what might have been. Comparison with later Allied and Soviet aircraft show the legacy handed on, right up to today's stealth aircraft.

Hardback, 282 x 213 mm, 144 pages, 100 colour artworks, 132 b/w photographs, plus 122 line drawings. 1 85780 092 3 £24.95